Lecture Notes in Artificial Intelligence   12598

Subseries of Lecture Notes in Computer Science

More information about this subseries at http://www.springer.com/series/1244

Zygmunt Vetulani · Patrick Paroubek ·
Marek Kubis (Eds.)

Human
Language Technology

Challenges for Computer Science and Linguistics

8th Language and Technology Conference, LTC 2017
Poznań, Poland, November 17–19, 2017
Revised Selected Papers

Springer

Editors
Zygmunt Vetulani 🆔
Adam Mickiewicz University
Poznań, Poland

Patrick Paroubek 🆔
Laboratoire d'Informatique pour la Méca
Orsay, France

Marek Kubis 🆔
Adam Mickiewicz University
Poznań, Poland

ISSN 0302-9743   ISSN 1611-3349 (electronic)
Lecture Notes in Artificial Intelligence
ISBN 978-3-030-66526-5   ISBN 978-3-030-66527-2   (eBook)
https://doi.org/10.1007/978-3-030-66527-2

LNCS Sublibrary: SL7 – Artificial Intelligence

This Springer imprint is published by the registered company Springer Nature Switzerland AG
The registered company address is: Gewerbestrasse 11, 6330 Cham, Switzerland

Preface

This book present a selection of the refereed papers of the 8th Language and Technology Conference: Challenges for Computer Science and Linguistics, LTC 2017, held in Poznań, Poland, in November 2017, in memoriam Alain Colmerauer (1941–2017), pioneer of Logic Programming in natural language processing.

People started to use rules to describe language several thousand years ago, going back to Plato in the western tradition, and it was only much later that computer scientists joined them. Among the numerous contributions at the crossing of the two communities, Prolog is one of the most noticeable. For Prolog people 2017 is a special year since it is the year of the demise of Alain Colmerauer, one of the two inventors of Prolog and a member of the LTC Program Committee in 2005. Naturally, the 2017 LTC conference was dedicated to him. It is remarkable that this edition of LTC was also the first one to have the word "deep" in some of the paper titles, not associated with "parsing" as it is usually but with "neural network" or used in reference to deep learning, acknowledging the fact that computational linguistics is entering a new era.

The selection of updated papers from the LTC 2017 proceedings that we present in this book therefore offers a view of our domain where the classical approaches lie next to the most recent developments in computational linguistics.

In this book the reader will find a selection of 25 revised and in most cases substantially extended and updated versions of papers presented at the 8th Language and Technology Conference in 2017. The reviewing process was done by the international jury, composed of the program committee members, or experts nominated by them. The selection was made among 97 contributions presented at the conference and is indicative of the preferences of the reviewers. Totaling 72, the authors of the selected contributions represent research institutions from 14 countries: Canada, Czech Republic, France, India, Ireland, Japan, Nigeria, Norway, Poland, Spain, Switzerland, UK, Ukraine.

What are the presented papers about?

To try to make the presentation of the papers transparent we have organized them into seven parts. These are:

1. Language Resources, Tools, and Evaluation (8)
2. Less-Resourced-Languages (LRL) (2)
3. Speech Processing (4)
4. Morphology (2)
5. Computational Semantics (3)
6. Machine Translation (1)
7. Information Retrieval and Information Extraction (5)

The clustering of the articles is approximate as many papers address more than one thematic area. The ordering of the chapters has no "deep" meaning: it roughly approximates the order in which humans proceed in natural language production and processing: starting with language resources, speech and text processing, to LT applications. Within these parts we ordered contributions in alphabetical order with respect to the family name of the first author.

We start this volume with the **Language Resources, Tools, and Evaluation** part, containing eight contributions. In the paper "Creating Norwegian valence resources from a deep grammar" the authors (Lars Hellan, Dorothee Beermann, Tore Bruland, Tormod Haugland, and Elias Aamot) propose a procedure for generating valence resources for the Norwegian language from a deep grammar, which was intended to make grammatical information encoded in a deep parser more accessible for humans and for further processing. The aim of the paper "How to Improve Optical Character Recognition of Historical Finnish Newspapers Using Open Source Tesseract OCR Engine – Final Notes on Development and Evaluation" (Mika Koistinen, Kimmo Kettunen, and Jukka Kervinen) is to present experiments aiming at improvement of optical character recognition (OCR) technology applied to a 500,000 word sample from the historical Finnish newspaper collection for the period from 1771 to 1910. In the next paper "Fine-tuning Tree-LSTM for phrase-level sentiment classification on a Polish dependency treebank" the authors (Tomasz Korbak and Paulina Żak) describe a variant of Child-Sum Tree-LSTM deep neural networks fine-tuned for working with dependency trees and morphologically rich languages using the example of Polish (presented at the LTC evaluation challenge PolEval). The fourth contribution of this part is "Supervised Transfer Learning for Sequence Tagging of User-Generated-Content in Social Media" (Sara Meftah, Nasredine Semmar, Othmane Zennaki, and Fatiha Sadat). In this work, the authors analyse the impact of supervised sequential transfer learning to overcome the sparse data problem in the *Tweets-domain* by leveraging the huge annotated data available for the *Newswire-domain*. What follows is the paper "Investigating the Lack of Consensus among Sentiment Analysis Tools" (Marco A. Palomino, Aditya Padmanabhan Varma, Gowriprasad Kuruba Bedala, and Aidan Connelly), which contains a survey on the current state of the art in the field of sentiment analysis in which the authors compare the performance of several Sentiment Analysis systems on a Twitter-based corpus. In the sixth contribution "Automated normalization and analysis of historical texts" (Paweł Skórzewski, Krzysztof Jassem, and Filip Graliński) the authors introduce a method for processing historical texts that applies a list of diachronic pairs found with the aid of word distribution vectors in historical corpora. The seventh article is "PADI-web: an event-based surveillance system for detecting, classifying and processing online news" (Sarah Valentin, Elena Arsevska, Alize Mercier, Sylvain Falala, Julien Rabatel, Renaud Lancelot, and Mathieu Roche). Here the authors present a platform for automated extraction of animal disease information from the Web (PADI-web) which is a multilingual text mining tool for automatic detection, classification, and extraction of disease outbreak information from online news articles. In the last paper of the first part "KRNNT: Polish Recurrent Neural Network Tagger Extended", another contribution to the PolEval evaluation challenge, its author (Krzysztof Wróbel) presents the morphosyntactic tagger KRNNT for Polish based on recurrent neural networks, and argues for the superiority of neural bidirectional approaches over other existing tagging methods.

Less-Resourced Languages are considered of special interest for the LTC community and since 2009 the LRL workshop has constituted an integral part of LTC meetings. The Less-Resourced-Languages (LRL) part contains two papers. The first one "Experiments with automatic and semi-automatic detection of sparse word forms in Old Braj" (Rafał Jaworski and Krzysztof Stroński) presents the authors' work on automatic converb detection in Old Braj poetry from the 15th-17th centuries and is a part of research on non-finite verbal forms in early New Indo-Aryan (NIA) language corpora comprising data from Old Rajasthani, Awadhi, Braj, Dakkhini, and Pahari. In the second one, titled "Towards Better Text Processing Tools for the Ainu Language" (Karol Nowakowski, Michał Ptaszyński, and Fumito Masui), the authors present their research devoted to the development of Natural Language Processing technologies for the Ainu language, a critically endangered language isolate spoken by the Ainu people, the native inhabitants of northern parts of the Japanese archipelago.

The **Speech Processing** part contains four papers. The contribution "The Harmonia Corpus – a Dialogue Corpus for Automatic Analysis of Phonetic Convergence" (Jolanta Bachan, Mariusz Owsianny, and Grażyna Demenko) describes the Harmonia spoken dialogue corpus created for analysis and objective evaluation of phonetic convergence in human-human communication with the goal to build convergence models which could be implemented in spoken dialogue systems. The authors of the second article in this part, "Resources and tools for Automated Speech Segmentation of the African Language Naija (Nigerian Pidgin)" (Brigitte Bigi, Oyelere S. Abiola, and Bernard Caron), present the development of HLT resources and tools for the African language Naija (Nigerian Pidgin), spoken in Nigeria, focusing on language resources for a tokenizer, an automatic speech system for predicting the pronunciation of words and their segmentation. In the next paper, "Speaker Variability for Emotions Classification in African Tone Languages" (Moses Ekpenyong, Udoinyang Inyang, Nnamso Umoh, Temitope Fakiyesi, Okokon Akpan, and Nseobong Uto), the authors examine the effect of speaker variability on emotions and languages, and propose a classification system based on the study of speech characteristics such as fundamental frequency and intensity for two languages, Ibibio (New Benue Congo, Nigeria) and Yoruba (Niger Congo, Nigeria), from voice recordings of native speakers of these languages. The last paper of this part "Analysis of Polish nasalized vowels based on spatial energy distribution and formant frequency measurement" (Anita Lorenc, Katarzyna Klessa, Daniel Król, and Łukasz Mik) offers the results of the analysis of F1 and F2 frequency measurements in Polish nasalized vowels represented in writing by the graphemes *e* and *a* (realized before voiceless fricatives).

The **Morphology** part is composed of two papers. The authors of the first one, "RNN Language Model Estimation for Out-of-Vocabulary Words" (Irina Illina and Dominique Fohr), propose new approaches to out-of vocabulary proper noun probability estimation using a Recurrent Neural Network Language Model. The second paper, "Automatic Pairing of Perfective and Imperfective Verbs in Polish" (Zbigniew Kaleta) presents an algorithm that automatically detects morphological dependencies between verbs in Polish and uses them to match corresponding perfective and imperfective verbs.

The **Computational Semantics** part is composed of three papers. This part opens with the text "Transforming Syntactic Relations in Attributive Groups" (Iuliia

Romaniuk, Nina Suszczańska, and Przemysław Szmal). In their paper on the Thetos translation system from Polish into sign language, the authors present recent translation quality improvements resulting from deepened syntactic and semantic analysis of the source text. Then the paper "Syntactic-Semantic Classes of Context-Sensitive Synonyms Based on a Bilingual Corpus" (Zdenka Urešová, Eva Fučíková, Eva Hajičová, and Jan Hajič) summarizes findings of a three-year study on verb synonymy in Czech-English translation based on both syntactic and semantic criteria on the basis of existing CL resources, including the Prague Dependency Treebank-style valency lexicons, FrameNet, VerbNet, PropBank, WordNet, and the parallel Prague Czech-English Dependency Treebank. In the third contribution, "Towards the evaluation of feature embedding models of the fusional languages" (Alina Wróblewska, Katarzyna Krasnowska-Kieraś, and Piotr Rybak), the authors investigate features to be used for estimating Neural-Networks-based NLP models of the fusional languages.

The **Machine Translation** section contains one contribution: "Syntactic and Semantic Impact of Prepositions in Machine Translation: An Empirical Study of French-English Translation of Prepositions 'à', 'de' and 'en'" (Violaine Prince), where the author presents a study about ambiguous French prepositions, stressing their role as dependency-introducers in order to derive some French-English MT translation heuristics, based on a French-English set of parallel texts.

The **Information Retrieval and Information Extraction** part contains five contributions. The first one, "Sentence Answer Selection for Open Domain Question Answering via Deep Word Matching" (Fabrizio Ghigi, Diana Turcsany, Thomas Kaltenbrunner, and Maurizio Cibelli), proposes an unsupervised approach for sentence answer selection (called Deep Word Matching) that uses both the string form and distributed representations of words, thereby capturing their hidden semantic relatedness. In the second paper "On the contribution of specific entity detection in comparative constructions to automatic spin detection in biomedical scientific publications" (Anna Koroleva and Patrick Paroubek), the authors address the problem of providing automated aid for the detection of misrepresentation ("spin") of research results in scientific publications from the biomedical domain. In the next paper "Automatic Taxonomy Generation: A Use-Case in the Legal Domain" (Cécile Robin, James O'Neill, and Paul Buitelaar), the authors describe a methodology for generating a taxonomy of legal concepts based on the analysis of a collection of official legal texts from Great Britain and Northern Ireland. The next paper, "Title Categorization based on Category Granularity" (Kazuya Shimura and Fumiyo Fukumoto), focuses on a problem of short-text categorization (newspaper titles), and presents a method that maximizes the impact of informative words due to the titles' sparseness. This part ends with the article "Identification of Domain-Specific Senses based on Word Embedding Learning" (Attaporn Wangpoonsarp and Fumiyo Fukumoto). This paper is about the domain-specific meaning of a word and proposes a machine-learning approach for detecting the main meaning of a word given the domain.

We wish you all interesting reading.

Zygmunt Vetulani
Patrick Paroubek

Tyson Roberts	Google, Japan
Piotr Rychlik	IPI PAN, Poland
Rafał Rzepka	Hokkaido University, Japan
Kevin Scannell	Saint Louis University, USA
Sanja Seljan	University of Zagreb, Croatia
Elizabeth Sherley	IIITM-Kerala, India
Zhongzhi Shi	Institute of Computing Technology, Chinese Academy of Sciences, China
Marcin Skowron	Johannes Kepler University Linz, Austria
Claudia Soria	CNR-ILC, Italy
Virach Sornlertlamvanich	NECTEC, Thailand
Janusz Taborek	Adam Mickiewicz University, Poland
Ryszard Tadeusiewicz	AGH, Poland
Dan Tufiş	RACAI, Romania
Yuzu Uchida	Hokkai-Gakuen University, Japan
Tamás Váradi	RIL, Hungary
Andrejs Vasiljevs	Tilde, Latvia
Zygmunt Vetulani	Adam Mickiewicz University, Poland
Dusko Vitas	University of Belgrade, Serbia
Aleksander Wawer	IPI PAN, Poland
Katarzyna Węgrzyn-Wolska	Efrei/Esigetel, France
Adam Wierzbicki	Polish-Japanese Academy of Information Technology, Poland
Rodrigo Wilkens	Universidade Federal do Rio Grande do Sul, Brasil
Marcin Woliński	IPI PAN, Poland
Bartosz Ziółko	AGH, Poland
Mariusz Ziółko	AGH, Poland
Andrzej Zydroń	XTM-INTL, UK

EDO Workshop Organizers

Michał Ptaszyński	Kitami Institute of Technology, Japan
Rafał Rzepka	Hokkaido University, Japan
Paweł Dybała	Jagiellonian University, Poland

EDO Workshop Program Committee

Alladin Ayesh	De Montfort University, UK
Karen Fort	Sorbonne University, France
Dai Hasegawa	Aoyama Gakuin University, Japan
Magdalena Igras-Cybulska	AGH, Poland
Yasutomo Kimura	Otaru University of Commerce, Japan
Paweł Lubarski	Poznań University of Technology, Poland
Fumito Masui	Kitami Institute of Technology, Japan
Mikołaj Morzy	Poznań University of Technology, Poland
Koji Murakami	Rakuten, USA

Noriyuki Okumura	National Institute of Technology, Akashi College, Japan
Michał B. Paradowski	University of Warsaw, Poland
Tyson Roberts	Google, Japan
Marcin Skowron	Johannes Kepler University Linz, Austria
Yuzu Uchida	Hokkai-Gakuen University, Japan
Zygmunt Vetulani	Adam Mickiewicz University, Poland
Katarzyna Węgrzyn-Wolska	Efrei/Esigetel, France
Adam Wierzbicki	Polish-Japanese Academy of Information Technology, Poland
Bartosz Ziółko	AGH, Poland

LRL Workshop Organizers

| Girish Nath Jha | JNU, India |
| Claudia Soria | CNR-ILC, Italy |

PolEval Workshop Organizers

Maciej Ogrodniczuk	Polish Academy of Sciences, Poland
Łukasz Kobyliński	Polish Academy of Sciences, Poland
Aleksander Wawer	Polish Academy of Sciences, Poland

Contents

Speech Processing

Morphology

Computational Semantics

Machine Translation

Information Retrieval and Information Extraction

Language Resources, Tools and Evaluation

Creating Norwegian Valence Resources from a Deep Grammar

Lars Hellan$^{(\boxtimes)}$, Dorothee Beermann, Tore Bruland, Tormod Haugland, and Elias Aamot

NTNU, 7491 Trondheim, Norway
{lars.hellan,dorothee.beermann}@ntnu.no, t-brul@online.no, tormod.haugland@gmail.com, elias.aamot@gmail.com

Abstract. We present a procedure for generating a valence resources for Norwegian (Bokmål) from a deep grammar. The corpus is presented in the form of IGT (interlinear glossed text) augmented by valence information. Our deep parser is the HPSG-based computational grammar Norsource (Hellan and Bruland 2015), our online IGT repository is TypeCraft (Beermann and Mihaylov 2014), while the sentences of the corpus are taken from the Leipzig Corpus Collection (Goldhahn et al. 2012). We create a common structure for the resources. Our aim is to make the grammatical information encoded in a deep parser more readily accessible for humans and for further processing.

Keywords: Online valence resources · Norwegian · Deep parsing · Interlinear glossed text

1 Introduction

We describe the Norwegian (Bokmål) valence resource cluster *NorVal*, which consists of various tools and applications. One of them is a *valence corpus*, where valence information is offered in a multilayered format combining sentence level valence information and a morpho-syntactic, functional, structure rendered as IGT (interlinear glossed text).[1] The tool we chose for our corpus resources is TypeCraft[2] (TC) (Beermann and Mihaylov 2014), since it allows for an easy access to the data in a general linguistic format. The query interface allows users to carry out various types of searches in a predefined search interface. This has important advantages as we will see, but the simplicity also comes with a price. The format is shown in Fig. 1. The valence frame instantiated by each occurring verb in a sentence is shown underneath the IGT in a field named 'Comment', using three different types of code for marking a valence frame ('SAS' for 'Syntactic Argument Structure', 'FCT' for 'functional label' and 'CL' for 'ConstructionLabel'). The verb annotated is indicated in the form it has in the sentence, while its base form (infinitive) is retrievable from the IGT. The IGT tiers are, from top, the sequence of words, the words with morphological break-up, base forms of content words, morphological and functional gloss, and parts of speech (POS).

---

[1] See https://typecraft.org/tc2wiki/Norwegian_Valency_Corpus.

[2] https://typecraft.org.

© Springer Nature Switzerland AG 2020
Z. Vetulani et al. (Eds.): LTC 2017, LNAI 12598, pp. 3–16, 2020.
https://doi.org/10.1007/978-3-030-66527-2_1

String: Jeg vet at hun forbauset ordføreren
Free translation: I know that she surprised the mayor

Jeg	|vet	|at	|hun		|forbause	|t	|ordføreren	
	|vite				|forbause		|ordfører	|en
	|know				|surprise		|mayor	
1.SG.NOM	|PRES	|DECL	|3.SG.FEM.NOM	|		|PAST	|	| DEF.SG.MASC
PN	|V	|COMP	|PN		|V		|N	

Comment
vet: SAS: NP+Sdecl
          FCT: transWithSentCompl
          CL: v-tr-obDECL

forbauset: SAS: NP+NP
             FCT: transitive
             CL: v-tr

Fig. 1. Valence and IGT information in the TypeCraft editing interface.

The valence labels represent three different styles of indicating valence frames, which we describe with reference to the annotation for *vet*. The SAS labels 'NP' and 'Sdecl' name the subject and object constituents, respectively, in the order in which the arguments represented occur in a sentence; we may call this an *alignable* specification. The FCT label 'transWithSentCompl', in contrast, being short for 'transitive with sentential complement', names the frame by a single expression, although with letter-size indication of notional decomposition. The CL label 'tr-obDECL' being short for 'transitive whose object is a declarative clause', is likewise non-alignable, but with hyphenated indication of notional decomposition. These three styles of naming valence frames are understood as equivalent in content in each case. It may be noted that the SAS specification, although alignable, is not actually aligned with the linear representation in Fig. 1 – in this they are all used as *en bloc* labels; see Sect. 4.1 for an example of a format where alignable labels are actually aligned.

The valence-IGT corpus consists at present of about 22,000 sentences annotated in this style, of length up to 20 words. It is generated through automated procedures, with critical use of a deep grammatical parser, namely the HPSG-based grammar *Norsource*-cf. Sect. 2. Lexical, syntactic and morphological features of the parse of each sentence are ported via XML and conversion scripts into the format of TypeCraft. The sentences are from the Leipzig Corpus Collection (LCC) (Goldhahn et al. 2012).

A further resource integrated in NorVal is a *valence file repository* which has been derived in a series of steps from the verb lexicon of the grammar. Here each verb is specified with regard to each valence frame in which it can occur, so that there is one entry per frame per verb. The specification includes the lemma, possible items (such as prepositions or adverbs) selected by the verb, a code for the frame in question, and illustrative sentences. For the present, the latter is restricted to a short sentence illustrating the frame, but we envisage the possibility of automatic access to corpora whereby naturally occurring sentences can be used as illustrations, such as the present corpus. The valence code uses the code style 'CL' exemplified in Fig. 1, which is also the code style used in the grammar; in this way each valence annotation in the corpus will, through its CL label and its lemma, have a unique counterpart in the dictionary.[3]

NorVal will be available as an online tool, the overall structure depicted in Fig. 2, of which the resources so far developed constitute the lower half.[4]

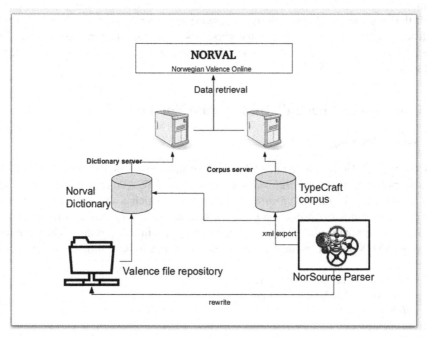

Fig. 2. NorVal overall design, with parser feeding valence file repository and the TypeCraft IGT, in turn feeding the NorVal dictionary, with steps for the online resource in the upper half.

---

[3] The CL specifications in the dictionary and the corpus are thus more or less 'orchestrated' by their common provenance from the grammar. For descriptions of the system, see Hellan and Dakubu (2010), Dakubu and Hellan (2017), Hellan (2019, 2020). A conversion code between the CL, SAS and FCT format is given at https://typecraft.org/tc2wiki/Valency_code_illustrat ions, with illustrating examples. A related resource is an abstract *Valence Profile*, a systematic listing all valence frame types used in a language; for Norwegian, see https://typecraft.org/tc2 wiki/Valence_Profile_Norwegian. The relevant valence specifications using the SAS and FCT codes can be seen at https://typecraft.org/tc2wiki/Valency_label_%27SAS%27 and https://typ ecraft.org/tc2wiki/Valency_label_%27FCT%27.

[4] A further project envisaged to be part of NorVal is a bilingual valence dictionary for Norwegian and German. A resource realizing the upper half is described in Beermann et al. (2020).

Among the main existing valence dictionaries can be mentioned, for English: FrameNet;[5] VerbNet[6]; PropBank;[7] EngVallex;[8] for German: Evalbu;[9]. for Czech: Vallex;[10] for Polish: Walenty.[11] In most of these resources the dictionary is separate from, but explicitly linked to a corpus.[12]

Among differences between these and the present resource is that—to our knowledge—none of them have been produced by means of a computational grammar.[13] [14]

In the following, we describe the main steps in producing the TC corpus from the input sentences from LCC via the Norsource parser (Sect. 2), and then the search facilities offered for the corpus (Sect. 3). Some comparison with other systems, and general assessments, follow in Sect. 4.

2   Generating TypeCraft Tokens from Norsource Parses

2.1   General Setting

Norsource[15] (cf. Hellan and Bruland 2015) is constructed within the theoretical formal framework *Head-Driven Phrase Structure Grammar* (*HPSG*) (Pollard and Sag 1994, Sag et al. 2003), using the *LKB platform* (Copestake 2002), which is a general computational platform for grammar construction using the formal format of *typed feature-structures* (*TFS*). Work within this format is supported by the Delph-In consortium.[16] The grammar has about 85,000 lexical entries, 250 syntactic rules, and 40 lexical rules for derivation

---

[5] https://framenet.icsi.berkeley.edu/fndrupal.

[6] http://verbs.colorado.edu/~mpalmer/projects/verbnet.html.

[7] https://propbank.github.io/.

[8] http://ucnk.ff.cuni.cz.

[9] http://hypermedia2.ids-mannheim.de/evalbu/.

[10] http://ucnk.ff.cuni.cz.

[11] http://clip.ipipan.waw.pl/Walenty; cf. Przepiórkowski et al. (2014)

[12] The main exception is Evalbu, where a dictionary entry consists of a set of annotated sentences, which constitutes a small corpus by itself.

[13] In some cases valence corpora, possibly in conjunction with tree-banks, are used in the construction of computational grammars Cf., Osenova (2011), Patujek and Przepiórkowski (2016). Here we go the opposite way.

[14] A case where a general purpose treebank is built by means of a computational grammar is described in Dyvik et al. 2016. Distinct from the present resource, the LFG resource does not include a valence corpus; moreover, reflecting that its lexicon is a full-form lexicon, inflectional and morphological information is provided exclusively in terms of features on words, not in the form of IGT.

[15] Norsource was started in 2001. The Norsource code files are downloadable from GitHub: https://github.com/Regdili-NTNU/NorSource/tree/master. Online access, for description: http://typecraft.org/tc2wiki/Norwegian_HPSG_grammar_NorSource Web demo: http://regdili.hf.ntnu.no:8081/linguisticAce/parse. NorSource's most actively used application is the online grammar tutor *A Norwegian Grammar Sparrer*—http://regdili.hf.ntnu.no:8081/studentAce/parse; cf. Hellan et al. (2013).

[16] http://moin.delph-in.net.

and inflection (Norsource having a lemma-based lexicon). It moreover, has about 300 valence frame types, covering nearly all morpho-syntactic construction types in the language, with lexical specifications and selection defined for about 12,900 verb entries. A trial version from 2017, resulting from batch parses of the 22,000 sentences, is currently being replaced by a version produced from batch parses of the same sentences[17] with improvements in both the grammar and in the conversion procedures.

Although the grammar as a parser is not robust enough to process free text with more than 60% coverage,[18] whenever a sentence is parsed, this entails – given the HPSG design for full sentence parsing[19] – that each verb in the sentence is assigned one of the valence frames defined in the grammar for that verb, which means that every pairing of verb with a valence is a pairing declared as correct by the grammar. In that sense, the valence information supplied for each successful parse is accurate.

The parser in many cases will provide very many parses for a given sentence, but relative to a statistically based ranking defined for the Delph-In grammars, only the two top ranked parses are accepted for further processing. The full parse result of a sentence within a large scale HPSG grammar is a rather huge feature structure, only a few aspects of which correspond to grammatical information, while the rest is procedural information reflecting the way the parser builds up structure. The task of the conversion and import process is to separate procedural from grammatical information and assign the latter information to the appropriate TC tags.

To exemplify, the following is a schematic display of the analysis tree from a parse for the sentence *Jeg vet at hun forbauset ordføreren*' I know that she surprised the mayor', from which the XML for the sentence is constructed, and exported to TC.

```
(1)
head-subject-rule
 jeg_perspron
  jeg
 head-verb-inf-or-s-comp-rule
  pres-infl_rule
   vite_subord_vlxm
    vet
  head-complementizer-comp-fin-rule
   at_subord
    at
   head-subject-rule
    hun_perspron
     hun
    head-verb-comp-rule
     pret-nonfstr-et_infl_rule
      forbause_tv_vlxm
       forbauset
```

---

[17] See https://typecraft.org/tc2wiki/Norwegian_Valency_Corpus for updates.

[18] For a large part this is because the grammar does not include mechanisms for automatic accommodation of names, noun compounds and numbers and number combinations which are not already represented in the lexicon.

[19] In this project, fragment parses are excluded.

```
sg_def_m_final-full_irule
sg-masc-def-noun-lxm-lrule
ordfører_n_masc_nlxm
ordføreren
```

We describe how information related to valence gets extracted from such a tree, together with morphological information and POS.

2.2 Mapping of Valence Information

We assign a valence value to every verb occurrence in a sentence. Illustrating for the item *vet* ('knows') in (1), its identifier `vite_subord_vlxm` in (1) carries the type *v-tr-obDECL*, and this type, which is also the CL code, is mapped to the SAS and FCT codes as in (2a), yielding the part of Fig. 1 indicated in (2b):

```
(2)    a. v-tr-obDECL =>
          SAS: "NP + Sdecl";
          FCT: transWithSentCompl
       b. vet: SAS: NP + Sdecl
          FCT: transWithSentCompl
          ConstructionLabel: v-tr-obDECL
```

The file providing the mapping between the identifier `vite_subord_vlxm` in (1) and the type *v-tr-obDECL* is derived from the verb lexicon of the grammar, with about 12,000 entries,[20] and the file of conversion rules like (2a) counts about 400 conversions.

2.3 Mapping of GLOSS and POS Information

GLOSS Specification for Inflection
Norsource has a lemma-based lexicon, which means that inflectional processing is done via 'rules', stated in a form exemplified in (3) (for verbs there are 22 such rules, for nouns 28, and for adjectives 38).

```
(3)
pret-nonfstr-et_infl_rule:=
%suffix (e a) (e et) (es es) (es edes)
infl-pret-verb-word &
[ARGS < [ INFLECTION nonfstr-et] >].
```

This rule is named in the tree for *forbauset* in (1), reflected in the lines

---

[20] For convenience provided in the pdf v-type-to-id, at https://typecraft.org/tc2wiki/Norwegian_ Valency_Corpus. Reference may also be made to the online parallel valence lexicon MultiVal, cf. Hellan et al. (2014), and http://regdili.hf.ntnu.no:8081/multilanguage_valence_demo/multiv alence. (The lexicons are based on computational HPSG grammars for Norwegian, Ga, Spanish, Bulgarian, with the same labels for valence frame encoding as are used presently.)

```
pret-nonfstr-et_infl_rule
forbause_tv_vlxm
forbauset
```

stating that the form *forbauset* has been derived from the lemma form *forbause* by the application of this rule in the parse. ARGS in (3) stands for the lexeme to which the inflection applies, and introduces with the line INFLECTION nonfstr the inflection paradigm type which must be carried by the lexeme in order for it to satisfy the inflection rule. (4) being the specification of this lexeme, the requirement is met (*nonfstr-et* being a subtype of *nonfstr*):[21]

```
(4)
forbause_tv_vlxm: = v-tr &
[STEM < "bruke" > ,
INFLECTION nonfstr,
SYNSEM.LKEYS.KEYREL.PRED "forbause_v_rel"].
```

The rule (3) having generated a *preterite* (i.e., past tense) form, the appropriate GLOSS tag in TC will be PRET, and this is assigned through the rule (5) mapping the relevant Norsource specification to the tag in the GLOSS line in TypeCraft:

```
(5) pret-nonfstr_infl_rule = PRET
```

Most of the mapping rules from Norsource inflection rules to TC GLOSS tags apply simply to rule names, like (5), more examples are given in (6) for perfect participle form with '-dd' and for passive formation with – s:

```
(6)
a. ppart-finalstr-dd_infl_rule = PRF
b. s-passive_s_infl_rule = PASS.PRES
```

Similar rules for nouns are shown in (7):

```
(7)
a. pl_def_m-or-f_light-e_irule = PL.DEF
b. sg_def_n_light-e_irule = SG.DEF.NEUT
```

Only in one case is an inflection GLOSS assigned on the basis of more information than just the rule name of an inflection. This is in the case of passive formed with *bli* plus a form formally like the perfect participle. In Norsource this construction is analyzed using the perfect participle rules (of which the one named in (6a) is an instance), while the account of the argument structure effects resides solely in a so-called *lexical rule*; in a tree, lexical rules sit in a different line than the inflection rule,

---

[21] 'nonfstr' stands for 'non-final stress', the vowel counting as the non-stressed final vowel here being the '-e' of the infinitival form of *forbause*.

and for the mapping purpose in question, they have to take precedence over the inflection rule entered. In the example (8), thus, for *(bli) straffet* ('be punished'), the relevant lexical rule is `pass-ncomps1-lrule`, which takes precedence over the inflection rule `ppart-nonfstr-et_infl_rule` in deciding the GLOSS tag to be assigned, namely PASS.PTCP:

```
(8)
ppart-nonfstr-et_infl_rule
  pass-ncomps1-lrule
    straffe_tv_vlxm
     straffet
```

GLOSS Specification for Non-inflected Words

TC GLOSS tags are defined also for features of non-inflected lexical items, there being 560 such entry-to-GLOSS tag mappings. Examples are given in (9), (9a, b, c) reflecting an essential aspect of the content of the words concerned, and (9d, e) reflecting recognized grammatical features; the format in (9f) is used whenever a set of items are characterized for a common semantics, the entry-id suffix `_dirtel-end-p`, for instance, is used for prepositions and adverbs expressing endpoint of directional movement:

```
(9)
a. fordi_comp = CAUS
b. idet_prep-time = TEMP
c. mer_cmpar-reg = CMPR
d. seg_refl = 3P.REFL.ACC
e. en_indef-art = SG.MASC.INDEF
f. dirtel-end-p = DIR.ENDPNT
```

Norwegian is among the languages where a possessive pronoun can on the one hand, by its reference, encode meaning features belonging in the space of grammatical features, and on the other, have *agreement* features relative to the head noun, and there are cases where a discrepancy can arise between these features. For instance, for the possessive pronoun *vårt* 'our(s)', the referent can be characterized as '1. person plural', whereas its agreement based features are *neuter singular* (in agreement with a neuter singular noun). A possible conflict of features could here be resolved in TC by defining SG.NEUT.DEF as GLOSS tags whereas 1.person plural could be used as a specification for the *Meaning* level, as reflected in the IGT snippet (10a). At the current state, though, the mapping for the word *vårt* is simply (10b), addressing only the agreement relation (and avoiding a morph split):

```
(10)
a.
vårt hus 'our house'
MORPH|vår|t|hus
MEANING 1P.PL|house
GLOSS|SG.NEUT.DEF
POS|PNposs|N
```

```
b.
vårt_postposs = SG.NEUT.DEF
```

POS Specification

For the tag level for POS, the mappings from Norsource partly reflect suffixes of entry identifiers, such as (11a-c) below for the major word classes, and (11d, e) for more restricted sets of words, and partly specific lexical identifiers in full, as in (11f), the latter type constituting most of the 472 mappings to POS tags.

```
(11)
a. nlxm = N
b. alxm = ADJ
c. vlxm = V
d. dirtel-end-p = PREPdir
e. reg-p-loc = PREPplc
f. mer_cmpar-mass = QUANT
```

Assessment of GLOSS and POS Mapping

It is the first time that a full grammar has been mapped for GLOSS and POS tag specifications to TC, and for TC this is an interesting situation of testing its annotation schemata. Essentially all of the GLOSS and POS tags required for representing features and word classes in Norwegian were available in TC, so that there are very few cases where a tag had to be created for this specific task.

It is in turn a benefit for Norsource that its features can be displayed in the fashion provided by TC, thus, as a morphological tagger, its morphological analyses otherwise being barely interpretable from the grammar-native feature structures or tree structures like (1) coming with a syntactic parse.

3   Search for Valence Information and GLOSS/POS Information

The search facilities in the corpus follow the general search procedures defined within TypeCraft. The most relevant interface is at 'SearchPhrase'[22], where for 'Language' one selects 'Norwegian Bokmål'. Search relative to a given lexeme is specified at 'Morpheme level/ Exact baseform', and search relative to a specific valence frame level is done at 'Phrase level/ Phrase description'. Combining, for instance, the verb *si* ('say') and the CL label 'tr-obDECL' yields a list of sentences as shown in Fig. 3, where each sentence is a hyperlink taking one to a display with the general format of Fig. 1 (except that there is no English translation).

All of the cases here shown are in passive form, and some with a word between *si* and the declarative clause. This search is thus equivalent to a search for *si* + the complementizer *at*, with one possible intervening word. Not tractable in terms of lexical items or morphemes, in contrast, is a search term like *scObNrgRes*, for 'resultative secondary object predicates', a construction type widely used in Norwegian and encoded in valence frames. Figure 4 shows the editor view of the sentence *Disse stoffene gir*

---

[22] https://typecraft.org/tc2wiki/Special:TypeCraft/SearchPhrase/.

Det sies forøvrig at Ghanas eldste koran ligger i denne moskéen.
Det sies gjerne at grunnboka har positiv og negativ troverdighet.
Det sies ofte at det gjeldende dyret reflekterer det tilhørende menneskets karakter.
Det sies at den som finner bøssen vil få stor fremgang i livet.
Det ble sagt at han omga sitt palass med syv konsentriske murer av ulike farger.
Det ble sagt at hver dag reiste Ra seg etter frokosten, og bordet sin hellige båt.

Fig. 3. Samples of return for search for *si* 'say' with the frame *tr-obDECL*; 64 examples.

veldig lett fra seg elektronene ('these materials very easily give from themselves the electrones'), found among the 646 results from a search:[23]

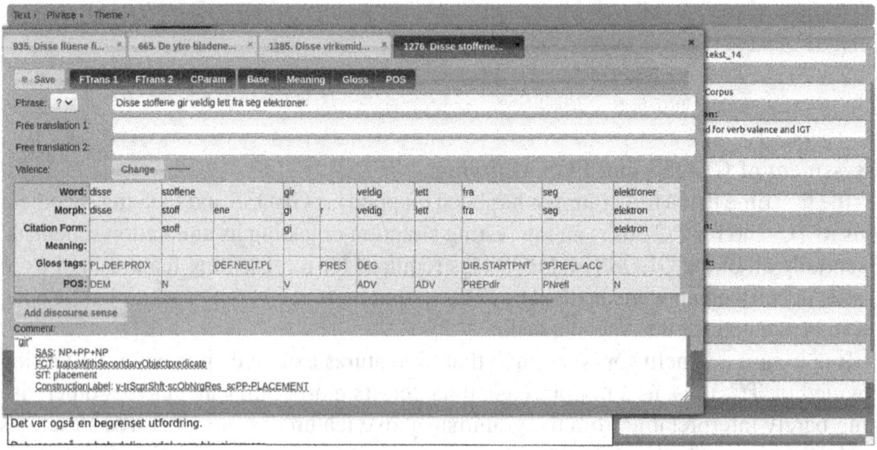

Fig. 4. TC editor view of the sentence *Disse stoffene gir veldig lett fra seg elektronene* ('these materials very easily give from themselves the electrones'), for the search string *scObNrgRes*.

None of the words or substrings among them could serve as a key to the valence frame belonging to this type for the sentence in question.[24]

Given the match between the valence labels in the valence repository and in the corpus, one can investigate the frequency of occurrence in the corpus of the various frame types. We illustrate with frames including *light reflexive pronouns*, instantiated pervasively throughout the Indo-European languages and well represented also in Norwegian. Among the 264 verb construction types represented in the valence repository, 36 have the simple reflexive as a distinctive element, and relative to the valence file

---

[23] with url: https://typecraft.org/tc2/ntceditor.html#3858,621088 (the screenshot also shows part of the TC metadata for the sentence). 'Predicative' here is a notion subsuming caused directional movement, reflected in the value 'Placement' under the label 'SIT', for 'situation type', which is used for some verbs in the grammar; for discussion of this parameter of specification, see Hellan (2019, 2020).

[24] One should add: with resources presently available. See Quasthoff et al. (2020) for an outline of a possible approach combining POS signatures and valence profiles for frame recognition in a text.

repository with its 12,884 entries, 1578 entries have one of these frames. The Table 1 below summarizes occurrence frequencies for some of the frame types:[25]

Table 1. Corpus frequencies for some frame types

Frame type, in CL terms[a]	Minimal example[b]	Entry instances	Instances in corpus
tr-obRefl	hun undrer seg	667	319
trObl-obRefl	hun befatter seg med dem	333	242
trObl-obRefl-oblEqObInf	hun beflitter seg på å lese	79	40
trObl-obRefl-oblLoc	hun oppholder seg her	15	47
trPath-obRefl-obDir	hun smyger seg hit	76	5
trPrtcl-obRefl	de dummer seg ut	139	14
ditr-iobRefl-obEqIobInf	hun foresetter seg å komme	8	4

[a]Meaning of labels: tr – transitive, trObl – transitive plus an oblique, trPath – transitive plus a path expression, trPrtcl – transitive plus a particle, ditr – ditransitive; obRefl – (indirect) object is a light reflexive, oblEqObInf – the oblique contains an infinitive controlled by the object, obEqIobInf – the object is an infinitive controlled by the indirect object, oblLoc – the oblique has a locative meaning, obDir – the object has a directional movement or orientation.
[b]Respective free translations (*seg* being the light reflexive): she wonders, she deals with them, she applies herself to reading, she stays here, she slithers hereto, they make fools of themselves, she decides to come.

The combination of valence and IGT enables one to also investigate correspondences between morphological properties not encoded in the valence labels and valence. For instance, to find instances where a given frame is realized in *passive* form, one can specify PASS in the field 'Morpheme level/Gloss tag' in the search interface. A search for a plain transitive frame (in CL code marked as '*tr-*') in the 'Comment' field, combined with the information 'Morpheme level/Gloss tag: PASS' in the subcorpus *Norsk valenskorpus tekst_15*, which has about 1000 sentences, gives 60 instances, as opposed to far more if the transitive frame is left unspecified for possible further dependents. If one restricts the morphological specification as PASS.PTCP, meaning periphrastic passives with *bli* or *være*, 51 of the 60 appear, while the other 9, carrying the gloss PASS.PRES (cf. (6) and (8) above), are *s*-passives. The cohabitation of valence representation and IGT thus is most useful.[26] [27]

---

[25] Both with values of December 2019. We make no attempt here to interpret the values.

[26] As Noreen (2014) reports it is in comparison more difficult to search for similar information in a corpus which lacks annotation for morphological features or even valence.

[27] The search interface described does not offer the possibility of negation. It may also be noted that the valence information under 'Comment' cannot be aligned with a particular column in the IGT, which leaves as a possibility that if a sentence has more than one verb, a given IGT specification like those described could in principle match any of the verbs.

4 Aspects of Evaluation and Further Perspectives

4.1 Comparative Remarks

We noted in the discussion of the valence annotation in Fig. 1 that the style used in the 'SAS' notation is one that may be called *alignable*, since it names the constituents in the order in which they occur in a sentence, while the other styles simply name the frame type as such. The alignable style is the prevalent in other valence resources, and then also *aligned*, as we illustrate in the following figure, with a schematic view from the immediate user interface of *Vallex*.[28] Here the valence code is a compact string of tags – "ACT(1) PAT(4)" -, whose elements appear also in the annotated sentence, distributed over the sentence argument constituents and aligned in the order in which they occur in the valence code:

> blokovat
> blokovat12x,24x ACT(1) PAT(4)
> blokují integrační procesy
> Example (among 200):
> pdt KamiónyACT blokovaly silniciPAT u hranice

Fig. 5. Illustration of verb frame encoding from the user interface of *Vallex*.

Norsource being a parser, one could of course supplement the labeling demonstrated with a parse tree with node labels reflecting at least the content represented in the SAS notation; however, a valence corpus is not a treebank, so such a move would be counter to the purpose of the present enterprise.

Notably, the tags used in Fig. 5 – ACT and PAT – are semantic role labels rather than syntactic labels; in this respect the *Vallex* tags are like the tags used in *VerbNet* and maybe most of the other resources mentioned in Sect. 1. Norsource being the general source of valence information in the present corpus, it should be noted that its parses are by no means void of semantic information: for instance, information from what may be called Semantic Argument Structure (shown in the representation format Minimal Recursion Semantics ('MRS'; cf. Copestake et al. (2005)), in the form of attributes like ARG1, ARG2 etc., could, although they are not semantic roles,[29] be adapted to uses of labels like ACT, PAT and similar. Most information at this level is however baked into the valence labels used,[30] and so adding such specification may be in principle redundant.

4.2 Assessments and Conclusion

From its launch in 2005, when it was the first online IGT linguistic editor connected to an online multilingual databasee, TC has had IGT as its main area. In addition, TC allows

---

[28] https://lindat.mff.cuni.cz/services/PDT-Vallex/PDT-Vallex.html; from December 2019.

[29] As opposed to uses made of such labels in, e.g., PropBank; cf. discussion in Hellan (2019, 2020).

[30] For instance, the label part *scObNrgRes* deployed in Fig. 4 would correspond to an MRS where the ARG1 of the predicate "fra seg" ('away from itself'), is embedded in a causal structure, but is mapped syntactically to the direct object.

for the export of examples to a Media wiki connected to the database, and thus provides ample possibilities of descriptions around the data. As part of the stored annotation of phrases TC stores the content of the Comment field which appears relative to each sentence.

Regarding the valence information in the corpus view, the quality of the valence information per se depends on the quality of the deep parser from which it is derived, that is, our valence corpus is only as good as the parser is in handling the relevant phenomena. Moreover the quality of the corpus depends also on the conversion mechanisms, so that mistakes could arise, and, per the automatic design, 'infect' a large number of sentences in the corpus.

Yet an obvious advantage of the method is that, once analyses are deemed plausible, one can in relatively little time obtain a comprehensive valence corpus.

Addition of content to the corpus can be done by augmenting Norsource to include the relevant information, for subsequent import into the corpus (as is being done for the next version). An alternative is manual augmentation of the corpus as produced at a given stage – this is possible in the TC interface. At present there is no versioning control procedure which would allow for such additions to be carried on (if compatible) into subsequent automatic imports from the grammar. If at some point where the grammar or the conversion procedures are not, or less regularly, maintained, this may be a natural 'step off' point from the automatic production of corpus as now described. This possibility at the same time secures the likelihood of further maintenence and enrichment of the corpus, and thus its role in the overall NorVal architecture.

References

Beermann, D., Mihaylov, P.: Collaborative databasing and resource sharing for Linguists. In: Languages Resources and Evaluation, vol. 48, pp. 1–23. Springer, Dordrecht (2014)

Beermann, D., Hellan, L., Mihaylov, P., Struck, A.: Developing a Twi (Asante) Dictionary from Akan Interlinear Glossed Text. In: Proceedings of LREC 2020 (2020)

Calzolari, N., et al. (eds.): Proceedings of the 9th International Conference on LanguageResources and Evaluation (LREC 2014). ELRA, Reykjavík, Iceland

Copestake, A.: Implementing Typed Feature Structure Grammars. CSLI Publications, Stanford (2002)

Copestake, A., Flickinger, D., Sag, I., Pollard, C.: Minimal recursion semantics: An introduction. J. Res. Lang. Comput. **3**, 281–332 (2005)

Dakubu, M.E., Kropp Hellan, L.: A labeling system for valency: linguistic coverage and applications. In: Hellan, L., Malchukov, A., Cennamo, M. (eds.) Contrastive Studies in Verb Valency. John Benjamins Publishing Co, Amsterdam & Philadelphia (2017)

Dyvik, H., et al.: NorGramBank: A 'Deep' treebank for Norwegian. In: Proceedings of the Tenth International Conference on Language Resources and Evaluation (LREC'16), pp. 3355–3366 (2016)

Goldhahn, D., Eckart, T., Quasthoff, U.: Building large monolingual dictionaries at the Leipzig corpora collection: From 100 to 200 languages. In: Proceedings of the Eighth International Conference on Language Resources and Evaluation LREC'12 (2012)

Hellan, L.: Construction-based compositional grammar. J. Logic Lang. Inform. **28**(2), 101–130 (2019). https://doi.org/10.1007/s10849-019-09284-5

Hellan, L.: Interoperable semantic annotation. In: Proceedings of LREC 2020 (2020)

Hellan, L., Kropp Dakubu, M.E.: Identifying verb constructions cross-linguistically. In: Studies in the Languages of the Volta Basin 6.3. Legon: Linguistics Department, University of Ghana (2010)

Hellan, L., Beermann, D., Bruland, T., Dakubu, M.E.K., Marimon, M.: MultiVal: Towards a multilingual valence lexicon. In: Calzolari et al. (eds) (2014)

Hellan, L., Bruland, T., Aamot, E., Sandøy, M.H.: A grammar sparrer for Norwegian. In: Proceedings of NoDaLiDa (2013)

Hellan, L., Bruland, T.: A cluster of applications around a deep grammar. In: Vetulani et al. (eds.) Proceedings from the Language & Technology Conference (LTC) 2015, Poznan (2015)

Norén, R.: Finite English passives and their Norwegian correspondences: A study based on the English-Norwegian parallel Corpus. Master thesis, University of Oslo (2014)

Przepiórkowski, A., Hajnicz, E., Patejuk, A., Woliński M., Skwarski, F., Swidziński, M.: Walenty: Towards a comprehensive valence dictionary of Polish. In: Calzolari et al. (eds) (2014)

Osenova, P.: Localizing a core HPSG-based grammar for Bulgarian. In: Hedeland, H., Schmidt, T., Worner, K., (eds.) Multilingual Resources and Multilingual Applications, Proceedings of GSCL 2011, ISSN 0176-599X, Hamburg, pp. 175–180 (2011)

Patejuk, A., Przepiórkowski, A.: Integrating a rich external valency dictionary with an implemented XLE/LFG grammar. In: Arnold, D., Butt, M., Crysmann, B., King, T.C., Müller, S.t. (eds.) Proceedings of the Joint 2016 Conference on Head-driven Phrase Structure Grammar and Lexical Functional Grammar, pp. 520–540. CSLI Publications, Stanford (2016)

Pollard, C., Sag, I.A.: Head-Driven Phrase Structure Grammar. Chicago University Press, Chicago (1994)

Quasthoff, U., Hellan, L., Körner, E., Eckart, T., Goldhahn, D., Beermann, D.: Typical sentences as a resource for valence. In: Proceedings of LREC 2020 (2020)

Sag, I.A., Wasow, T., Bender, E.: Syntactic Theory. CSLI Publications, Stanford (2003)

How to Improve Optical Character Recognition of Historical Finnish Newspapers Using Open Source Tesseract OCR Engine – Final Notes on Development and Evaluation

Mika Koistinen, Kimmo Kettunen[✉], and Jukka Kervinen

The National Library of Finland, DH Projects, Saimaankatu 6, 50100 Mikkeli, Finland
{mika.koistinen,kimmo.kettunen,jukka.kervinen}@helsinki.fi

Abstract. The current paper presents work that has been carried out in the National Library of Finland (NLF) to improve optical character recognition (OCR) quality of the historical Finnish newspaper collection 1771–1910. Evaluation results reported in the paper are based mainly on a 500 000 word sample of the Finnish language part of the whole collection. The sample has three different parallel parts: a manually corrected ground truth version, original OCR with ABBYY FineReader v. 7 or v. 8, and an ABBYY FineReader v. 11 re-OCRed version for comparison with Tesseract's OCR. Using this sample and its page image originals we have developed a re-OCRing procedure using the open source software package Tesseract v. 3.04.01. Our method achieved initially 27.48% improvement vs. ABBYY FineReader 7 or 8 and 9.16% improvement vs. ABBYY FineReader 11 on document level. On word level our method achieved 36.25% improvement vs. ABBYY FineReader 7 or 8 and 20.14% improvement vs. ABBYY FineReader 11. Our final precision and recall results on word level show clear improvement in the quality: recall is 76.0 and precision 92.0 in comparison to GT OCR. Other measures, such as recognizability of words with a morphological analyzer and character accuracy rate, show also steady improvement after re-OCRing.

Keywords: Optical character recognition · Historical newspaper collections · Evaluation · Finnish

1 Introduction

The National Library of Finland has digitized historical newspapers and journals published in Finland between 1771 and 1929 and provides them online [1, 2]. This collection contains approximately 7.51 million freely available pages primarily in Finnish and Swedish. The total amount of all digitized pages on the web is over 18.7 million, slightly over half (52%) of them being in restricted use due to copyright reasons. Besides newspapers and journals this material contains e.g. books, maps and different small prints, ephemera. The National Library's Digital Collections are offered via the *digi.kansalliskirjasto.fi* web service, also known as *Digi*. An open data package of all

© Springer Nature Switzerland AG 2020
Z. Vetulani et al. (Eds.): LTC 2017, LNAI 12598, pp. 17–30, 2020.
https://doi.org/10.1007/978-3-030-66527-2_2

the collection's newspapers from period 1771 to 1910 was released in early 2017 [3]. The open web newspaper and journal collection has been quite popular, and we estimate that is has at least 100 000 active users: historians, genealogists, students, researchers and teachers, among others, have been identified as users [4].

Digitization is a process with many phases. When originally non-digital materials, e.g. old newspapers and books, are digitized, the process involves first scanning of the documents which results in image files. Out of the image files one needs to sort out texts and possible non-textual data, such as photographs and other pictorial representations. Texts are recognized from the scanned pages with Optical Character Recognition (OCR) software. OCRing for modern prints and font types is considered to a large extent a resolved problem that yields high quality results, but results of historical document OCRing are still far from that [5].

Newspapers of the 19th and early 20th century were mostly printed in the Gothic (Fraktur, blackletter) typeface in Europe. Fraktur is used heavily in our data, although also Antiqua is common and both fonts can be used in the same publication in different parts. It is well known that the Fraktur typeface is especially difficult to recognize for OCR software [5–7]. Other aspects that affect quality of OCR recognition are the following [5, 6]:

- quality of the original source and microfilm
- scanning resolution and file format
- layout of the page
- OCR engine training
- unknown fonts
- etc.

Due to these difficulties scanned and OCRed document collections have a varying number of errors in their content. A quite typical example is *The 19th Century Newspaper Project* of the British Library [8]: based on a 1% double keyed sample of the whole collection Tanner et al. report that 78% of the words in the collection are correct. This quality is not good, but quite realistic for a digitized historical newspaper collection.

OCR errors in the digitized newspapers and journals may have several harmful effects for users of the data. One of the most important effects of poor OCR quality – besides worse readability and comprehensibility - is worse on-line searchability of the documents in the collections. Also all kind of post processing of the textual data is harmed by bad quality. Thus improvement of OCR quality of digitized historical collections is an important step in improving overall usability of the collections.

This paper reports results of re-OCR for a historical Finnish newspaper collection. The re-OCR process consists of a combination of different image preprocessing techniques, and a new Finnish Fraktur model for Tesseract OCR enhanced with morphological recognition and some simple rules to weight the candidate words of the OCR process.

The paper is organized as follows. In the second chapter we describe the basics of our new OCR process. The third chapter reports our first results, and the fourth shows final results after improvements made to the re-OCR process. Finally, Sect. 5 discusses the results and lists lessons learned during the development work.

2   How to Improve OCR Quality

Ways to improve quality of OCRed texts are few, if total rescanning is out of question, as it usually is due to labor costs. Improvement can be achieved with three principal methods: manual correction with different aids (e.g. editing software, [9]), re-OCRing [5] or algorithmic post-correction [10]. These methods can also be mixed. One popular method to realize manual correction has been crowdsourcing. Although this method can be useful, if there is enough population to carry it out [11], the method does not suit to large collections of languages that don't have enough people to carry out massive correction. Kettunen and Pääkkönen [2] have approximated earlier, that about 25–30% out of 2.4 billion Finnish words in the data of 1771–1910 are wrong. This means about 600–800 million word tokens and millions of word types. Effective manual correction of this amount of data is impossible. An earlier crowdsourcing effort resulted in correction of only about 65 000 words [12], which shows clearly the futility of this approach with a large heavily erroneous collection of a small language.[1]

Algorithmic post-correction can improve quality of texts, but its capabilities are still limited with low quality original data [10, 13]. Thus we chose re-OCRing with open source OCR engine Tesseract v. 3.04.01 as our primary method for improving quality of the texts. Post-correction can be tried later or it can be attached to the process as there are now available tools for doing post-correction of historical Finnish [13, 14].

2.1   Our re-OCR Process

OCRing of historical Finnish documents is difficult mainly because of the varying quality newspaper images and lack of model(s) for Finnish Fraktur. However, the character set of Finnish is very similar to other common Fraktur fonts: Finnish has *ä, ö* and *å* characters, but no *ü*, and *β* like German Fraktur. Thus existing fonts can be used in producing a new Fraktur font for Finnish.

Another problem is quality of page images of OCRed data. Scanned historical document images have many times different types of noise, such as scratches, tears, ink spreading, low contrast, low brightness, and skewing etc. [5]. Smitha et al. [15], among others, state that document image quality can be improved by binarization, noise removal, deskewing, and foreground detection. We use a set of different image preprocessing techniques in our process to improve the original page images. The image processing methods used in our process are explained in detail in Koistinen et al. [16]. It suffices to mention here, that use of different image processing methods and their combinations has been essential to achieve improvement in re-OCR of our data.

Our re-OCRing process consists of four parts: 1) image preprocessing, 2) Tesseract OCR, 3) choosing of the best word candidate from Tesseract's output and 4) transformation of Tesseract's output to ALTO format. The process is shown in Fig. 1.

The process uses five different image preprocessing techniques before transferring the page images to Tesseract for OCRing. Different combinations of image preprocessing are tried and best combinations are chosen based on the *hOCR* confidence values and

---

[1] A typical minor success story in crowdsourcing is described e.g. in Clematide et al. [9], where 180 000 characters on about 21 000 pages were corrected in about 7 months.

OCR improvement process

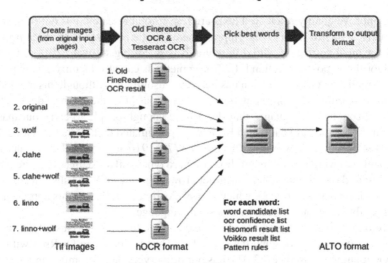

Fig. 1. The re-OCR process

results of morphological recognition of output words in phase three. After that the results are transformed to ALTO format.

After image preprocessing phase documents are OCRed using Tesseract OCR with font models *fin* and *fi_frak_mk41* that have been developed for the process. Our Finnish Fraktur model was developed using an existing German Fraktur model[2] as a starting point. The Fraktur model was improved iteratively. The characters that had most errors were improved in training data boxes (single letters and two letter combinations). Then Tesseract was run 1 to N times with the developed Finnish Fraktur model and already existing Finnish Antiqua model[3] in dual model mode, where the best alternative from Fraktur and Antiqua results was chosen.

The third phase of the process, **pick the best words,** selects the best word candidates. Tesseract uses hOCR as its output format[4]. hOCR is an open standard for presenting OCR results and it has a confidence value for each word produced by the used OCR tool [17]. Best words are selected by using hOCR word confidence values and morphological analysis software Omorfi[5] to check recognizability of the words. If candidate word is recognized by Omorfi, the hOCR confidence value of the word gets +10 points and if it is not recognized by Omorfi, it gets −2 points (on a scale of 0–100). If the word is a number, +10 extra points are not given, since there were multiple long number series errors among the first selected results if extra points were given.

---

[2] https://github.com/paalberti/tesseract-dan-fraktur.

[3] https://github.com/tesseract-ocr/langdata/tree/master/fin.

[4] https://kba.github.io/hocr-spec/1.2/.

[5] https://github.com/jiemakel/omorfi. We call this version HisOmorfi because it has some enhanced capability for recognition of 19th century Finnish words.

Frequency of characters in Finnish is also taken into consideration in the process. Rarely used characters like *c* and *f* are given −3 points for each occurrence in the word. Thus word candidate *kokonkfcsfa,* for example, would get −9 points, and *kokoukscssa* would get −3 points (*kokouksessa* would be right). This seems like a good rule for Finnish, but would not work for Swedish, the second major language of our collection, as Swedish texts contain lots of correct *f* and *c* characters. Similarly special characters ' *!; : _&"* are given minus points in the results. Also other special characters like *[] () /{}* *% # ? " &* etc. should be considered to be given minus points in the future.

The phase of combining the OCRed documents is run in steps. First documents 1 and 2 are combined, and then the combination of 1 and 2 is combined with document 3 and so on. The last phase, **Transform to output format,** transfers the documents into ALTO XML format[6]. ALTO is the format used by our production system docWorks[7], and the presentation system Digi.

3 Results – Part I

3.1 First Results

Koistinen et al. [16] reported **page level** evaluation results of the first developed re-OCR process with the 500 000 word GT sample comparing ABBYY FineReader v.7 and/or 8 (current OCR of the collection), ABBYY FineReader v.11 and Tesseract re-OCR with different image preprocessing methods and by using page level confidence as a measure.

The best Tesseract OCR result on page level was achieved by combining four image preprocessing methods: Linear Normalization + WolfJolion, Contrast Limited Adaptive Histogram Equalization + WolfJolion, original image and WolfJolion. Page level system improved the word level quality of OCR by 1.91% points (9.16%) against the best result of ABBYY FineReader 11 and by 7.21% points (27.48%) against ABBYY FineReader 7 and 8. Thus the method could correct at best about 84.6 million words in the 1.06 million Finnish newspaper page and 2.4 billion word collection (consisting of Finnish language) of the current OCR with ABBYY FineReader v. 7/8.

The developed method could still be improved. The method is 2.08% points from the optimal Oracle result, which is 16.94% word error rate. Oracle result is the result when the truly best document is always selected, instead of choosing the result based on the hOCR confidence value.

The character accuracy results for Fraktur model showed that characters *u, m* and *w* had less than 80% correctness even after re-OCRing. These letters are confused with partly overlapping letters such as *n* and *i*. It seems, however, that if accuracy for one of them is increased, accuracy of other characters will decrease. Also recognition of letter *ä* could possibly be improved, though it overlaps with letters *a* and *å*. From 20 most frequent errors in the character data only five characters had under 80% correctness rate at this phase of development.

---

[6] https://www.loc.gov/standards/alto/.
[7] https://content-conversion.com/#docworks-2.

3.2 Further Results

In the second **word level** evaluation document confidence was changed to select best single words from different images to make the method more accurate. In this method the original page image was transformed into five different page images using WolfJolion, Linear Normalization, Contrast Limited Adaptive Linear Normalization (CLAHE), Linear Normalization + WolfJolion, CLAHE + WolfJolion image preprocessing methods. Tesseract OCR was run on these six images and the best words were selected by the hOCR word accuracy value with Omorfi and rules c-f and special character detection to add/reduce points. The final result after the process is an ALTO format document for combined OCR results that contains the most accurate content and alternative blocks for less accurate content. On word level our method achieved a 9.43% unit improvement vs. ABBYY FineReader 7 or 8 and a 4.18% unit improvement vs. ABBYY FineReader 11.

For further analysis of the results we used a parallel version of the 500K collection with ground truth, old OCR and Tesseract OCR, and performed a detailed quality analysis for the results using different ways of evaluation [18]. Kettunen and Pääkkönen [2] have earlier estimated the quality of the whole historical newspaper collection with morphological analysis. We applied this method now with two morphological analyzers: original Omorfi v. 0.3[8] and HisOmorfi. Results of analyses are depicted in Table 1.

Table 1. Word recognition rates with two morphological analyzers

	Ground truth	Tesseract OCR	Current OCR
Omorfi 0.3	81.2%	76.1%	76.9%
HisOmorfi	94.0%	87.4%	80.7%

Figures show that the manually edited ground truth version is recognized clearly best, as it should be. Plain Omorfi recognizes words of the current OCR version slightly better than Tesseract words, the difference being 0.8% units. This is caused by the fact that HisOmorfi is used in the re-OCRing process and it favors w to v[9]. Plain Omorfi does not recognize most of the words that include w, but HisOmorfi is able to recognize them, which is shown in the high percentage of Tesseract's HisOmorfi result column.

As further evaluation measures we used standard measures of recall and precision and their combination, F-score [19]. These measures have been widely used in both post-correction and re-OCRing evaluations [13]. Other measures exist, too, but most of them, as for example correction rate used in Silfverberg et al. [13], are calculated only slightly differently from P/R figures.

---

[8] https://github.com/flammie/omorfi.

[9] Variation of w/v is one of the differences of 19th century Finnish in comparison to 20th century Finnish. *W* was many times used instead of *v* in 19th century, but now *w* is only used foreign names.

As the ground truth data is not wholly parallel with number of words varying from 459 942 to 500 604 in different versions of the data, we based our calculations on lines where there was character data in every column of the table consisting of GT, CurrOCR, and TesseractOCR words. Number of these lines was 459 930.

Table 2 shows basic P/R results and F-scores of the data and also correction rate. We show two results: one on the left column is achieved by comparing all the data without cleaning. The result on the right column shows the results with punctuation and all other non-alphabet and non-number characters removed from the lines. The removed character set is: ,;\':\'"\'\'_!@#%&*()+=<>[]{}?\V—~|'\'",,¡«©»®°¡. Variation of w/v is also neutralized.

Table 2. P/R results for Tesseract OCR vs. current OCR

Basic results	Results with cleaned data
Recall = 68.4	Recall = 71.0
Precision = 70.1	Precision = 71.0
F measure = 69.3	F measure = 71.0
Correction rate = 39.3	Correction rate = 43.0

The results achieved are clearly better than previous post-correction trial results in Kettunen [20], where F-scores of about 55–60 at best were reached with small test samples. As current results are also achieved with a more realistic sample of the data, they seem promising. It seems that our re-OCR has a satisfying recall of the errors, but it is not very precise. This is mainly due to new erroneous words introduced in the re-OCR.

We can additionally compare our re-OCRing results to some other correction results of data that originates from our newspaper material but where the data sample is only a part of our sample. Silfverberg et al. [13] have evaluated post-correction results of *hfst-ospell* software with the historical data using about 40 000 word pairs. They used *correction rate* as their measure, and their best result is 35.09 ± 2.08 (confidence value). Correction rate of our re-OCR process data in Table 2 is 39.3, which is slightly better than result of post-correction in Silfverberg et al. Besides, our result is achieved with a tenfold amount of word pairs.

Drobac et al. [14] have used neural network based software Ocropy to re-OCR a sample of historical Finnish newspaper materials. They have used two differently trained models, which they call DIGI and NATLIB. Besides these OCR models they use also post-correction with hfst-ospell. Drobac et al. use character accuracy (CAR) as their evaluation measure. Results reported in Drobac et al. and comparative results using CAR for our re-OCR data are shown in Table 3.

Figures reveal that plain Ocropy OCR is on the same level of performance as our re-OCR method. Post-correction brings some gain for the character accuracy with the NATLIB model, but not with the DIGI model. Version 11 of ABBYY FineReader performs slightly better than Ocropy, but is slightly beyond performance of NATLIB model and post-correction.

Table 3. Results of Drobac et al. compared with results of NLF's re-OCR results using character accuracy

	Drobac et al.	NLF re-OCR
Ocropy OCR	93.0	N/A
DIGI model+post corr.	93.3	N/A
NATLIB model+post corr.	95.2	N/A
NLF ReOCR	N/A	93.2
NLF FR11	N/A	94.5
NLF current OCR	N/A	90.9

4   Improvements to the re-OCR Process

The results we achieved with our initial re-OCR process were at least promising. They showed clear improvement of the quality in the GT collection, as was depicted in Tables 1 and 2. Slightly better OCR results were achieved by Drobac et al. [14] with Ocropy machine learning OCR system using character accuracy rate (CAR) as an evaluation measure. Post-correction results of Silfverberg et al. [13], however, were worse than our re-OCR results.

The main drawback of our first re-OCR system is that it is relatively slow. Image preprocessing and combining of images takes time, if it is performed to every page image as it was initially done. Execution time of the word level system was initially about 6 750 word tokens per hour when using a CPU with 8 cores in a standard Linux environment. With increase of cores to 28 the speed improved to 29 628 word tokens per hour. The speed of the process was still not very satisfying.

After modifications to the re-OCR process we have been able to improve the processing speed of re-OCR considerably. We have especially sped up the string replacements performed during the process, as they took almost as much time as the image processing. After recoding string replacements take only a fraction of the time they took earlier. However, image preprocessing cannot be sped up easily, but anyhow new processing takes about half of the time it used to take with the GT data. We are now able to process about 201 800 word tokens an hour in a 28 core system.

We improved also the process of the word candidate selection after re-OCR. The improved process uses now two morphological analyzers ((His)Omorfi and Voikko[10]), character trigrams and other character level data to be able to weight the word candidate suggestions given by the OCR process. We checked especially the trigram list we were using and removed the least frequent ones from it.

---

[10] https://voikko.puimula.org/.

4.1 Results – Part II

After improvements made to the re-OCR process we have been able to achieve also better results. The latest results are shown in Tables 4 and 5. Table 4 shows precision, recall and correction rate results and Table 5 shows results of character error rate, CER, word error rate, WER, and character accuracy rate CAR analyses using the ground truth data.

Table 4. Precision, recall, F-score and correction rate of the re-OCR after improvements: GT data

	Current result	Earlier result (cf. Table 2.)
Recall	76.0	68.4
Precision	92.0	70.1
F-score	83.0	69.3
Correction rate	69.0	39.3

Table 5. CER, WER and CAR of the re-OCR after improvements: GT data

	Re-OCR	Current OCR
CER	2.05	6.47
WER	6.56	25.30
WER (order independent)	5.51	23.41
CAR	97.64	92.62

Two other commonly used evaluation measures for OCR output are character error rate, CER, and word error rate, WER [21]. CER is defined as

$$CER = \frac{i + s + d}{n} \tag{1}$$

Calculation of CER employs the total number n of characters and the minimal number of character insertions i, substitutions s and deletions d required to transform the reference text into the OCR output.

Word error rate WER is defined as

$$WER = \frac{i_w + s_w + d_w}{n_w} \tag{2}$$

Here n_w is the total number of words in reference text, i_w is the minimal number of insertions, s_w is number of substitutions and d_w number of deletions on word level to obtain the reference text. Smaller WER and CER values mean better quality. Our CER and WER results for the OCR process are shown in Table 5. These results have been analyzed with the OCR evaluation tool described in Carrasco [21].

Results in Tables 4 and 5 show that the re-OCR process has improved clearly from the initial performance described in Sect. 3. Precision of the process has improved considerably, and although recall is still slightly low, F-score is now 83.0 (earlier 73.0). CER and WER have improved also distinctly. Our CAR is now also slightly better than Drobac's best value without post correction – ours is 97.6 and Drobac's 97.3 [14].

Evaluation of OCR results can be done experimentally either with or without ground truth. After finalizing development and evaluation of the re-OCR process with the GT data, we started testing of the re-OCR process with realistic newspaper data, i.e. without GT to avoid over fitting of the data by using only GT in evaluation. For analysis we use morphological recognition with HisOmorfi, a morphological analyzer that has been enhanced to process better historical Finnish. The results give an estimation of improvement in the word quality [2]. Recognition results of the latest re-OCR of Uusi Suometar are shown in Fig. 2. The data consists of years 1869–1918 of the newspaper with about 306 802 265 words and 86 000 pages.

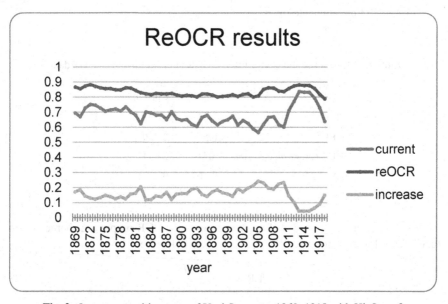

Fig. 2. Latest recognition rates of Uusi Suometar 1869–1918 with HisOmorfi

We can see that re-OCR is improving the quality of the newspaper text clearly and consistently for the whole period. The average improvement for 49 years is 15.3% units. The largest improvement is 20.5% units, and smallest 12% units.

5 Discussion and Conclusion

We have described in this paper results of a re-OCR process for a historical Finnish newspaper and journal collection. The developed re-OCR process consists of a combination of five different image preprocessing techniques, a new Finnish Fraktur model for Tesseract 3.04.01 OCR enhanced with morphological recognition and character level rules to weight the resulting candidate words. Out of the results we create new OCRed data in METS and ALTO XML format that can be used in our docWorks document presentation system.

We have shown that the re-OCRing process yields clearly better results than commercial OCR engine ABBYY FineReader v. 7/8, which is our current OCR engine. We have also shown that a 49 year time span of newspaper Uusi Suometar (86 000 pages and ca. 306 million words) gets significantly and consistently improved word recognition rates for Tesseract output in comparison to current OCR. Further we have shown that our results are either equal or slightly better than results of a machine learning OCR system Ocropy in Drobac et al. [14]. Our results outperform clearly post correction results of Silfverberg et al. [13].

Let us now turn to lessons learned during the creation of the re-OCR process so far. Our development cycle for a new re-OCR process has been relatively long and taken more time than we were able to estimate in advance. We started the process by first creating the GT collection for Finnish [18, 22]. The end result of the process was a ca. 525 000 word collection of different quality OCR data with ground truth. The size of the collection could be larger, but with regards to limited means it seems sufficient. In comparison to GT data used in OCR or post correction literature, it fares also well, being a mid-sized collection. The GT collection has been the cornerstone of our quality improvement process: effects of the changes in the re-OCR process have been evaluated with it[11].

The second time consuming part in the process was creation of a new Fraktur font model for Finnish. Even if the font was based on an existing German font model, it needed lots of manual effort in picking letter images from different newspapers and finding suitable Fraktur fonts for creating synthesized texts. This was, however, crucial for the process, and could not be bypassed.

A third lesson in our process was choice of the actual OCR engine. Most of the OCR engines that are used in research papers are different versions of latest machine learning algorithms. They may show nice results in the narrowly chosen evaluation data, but the software are usually not really production quality and could probably not be used in an industrial OCR process that processes 1–2 million page images in a year. Thus our slightly conservative choice of open source Tesseract that has been around for more than 20 years is justifiable.

Another, slightly unforeseen problem have been modifications needed to the existing ALTO XML output of the whole process. As ALTO XML is a standard approved by the ALTO board, changes to it are not made easily. An easy way to circumvent this is to use two different ALTOs in the database of docWorks: one conforming to the existing standard and another one that includes the necessary changes after re-OCR. We have chosen this route by including some of the word candidates of the re-OCR in the database as variants.

The last thing that needs emphasis is the main reason for doing re-OCR. OCR errors in the digitized newspapers and journals may have several harmful effects for users

---

[11] Both the GT data and the Fraktur font model for Tesseract are available on the open data pages of the NLF, https://digi.kansalliskirjasto.fi/opendata/submit?set_language=en

of the data. One of the most important effects of poor OCR quality – besides worse readability and comprehensibility – is worse on-line searchability of the documents in the collections [23, 24]. Although information retrieval is quite robust even with corrupted data, IR works best with longer documents and long queries, especially when the data is of bad quality. Empirical results of Järvelin et al. [25] with a Finnish historical newspaper search collection, for example, show that even impractically heavy usage of fuzzy matching in order to circumvent effects of OCR errors will help only to a limited degree in search of a low quality OCRed newspaper collection, when short queries and their query expansions are used.

Weaker searchability of the OCRed collections is one dimension of poor OCR quality. Other effects of poor OCR quality may show in the more detailed processing of the documents, such as sentence boundary detection, tokenization, named entity recognition, and part-of-speech-tagging, which are important in higher-level natural language processing tasks [26]. Part of the problems may be local, but part will cumulate in the whole pipeline causing errors. Thus quality of the OCRed texts is the cornerstone for any kind of further usage of the material and improvements in OCR quality are welcome. And last but not least, user dissatisfaction with the quality of the OCR, as testified e.g. in Jarlbrink and Snickars [27], is of great importance. Digitized historical newspaper and journal collections are meant for users, both researchers and lay person. If the users are not satisfied with the quality of the content, improvements need to be made.

Acknowledgements. This work was funded by the European Regional Development Fund and the program Leverage from the EU 2014–2020.

References

1. Kettunen, K., Honkela, T., Lindén, K., Kauppinen, P., Pääkkönen, T., Kervinen, J.: Analyzing and improving the quality of a historical news collection using language technology and statistical machine learning methods. In: IFLA World Library and Information Congress, Lyon (2014). http://www.ifla.org/files/assets/newspapers/Geneva_2014/s6-honkela-en.pdf
2. Kettunen, K., Pääkkönen, T.: Measuring lexical quality of a historical finnish newspaper collection – analysis of garbled OCR data with basic language technology tools and means. In: Calzolari, N., et al. (ed.) Proceedings of the Tenth International Conference on Language Resources and Evaluation (LREC 2016) (2016). http://www.lrec-conf.org/proceedings/lrec2016/pdf/17_Paper.pdf
3. Pääkkönen, T., Kervinen, J., Nivala, A., Kettunen, K., Mäkelä, E.: Exporting Finnish digitized historical newspaper contents for offline use. D-Lib Mag. **22**, July/August 2016 (2016)
4. Pääkkönen, T., Kettunen, K.: Kansalliskirjaston sanomalehtiaineistot: käyttäjät ja tutkijat kesällä 2018. Informaatiotutkimus **37**(3), 15–19 (2018). https://doi.org/10.23978/inf.76067
5. Piotrowski, M.: Natural language processing for historical texts. Synthesis Lectures on Human Language Technologies. Morgan & Claypool Publishers, San Rafael (2012)
6. Holley, R.: How good can it get? Analysing and improving OCR accuracy in large scale historic newspaper digitisation programs. D-Lib Mag. **15**(3/4) (2009)
7. Springmann, U., Lüdeling, A.: OCR of historical printings with an application to building diachronic corpora: a case study using the RIDGES herbal corpus. Digit. Humanit. Q. **11**(2) (2017)

8. Tanner, S., Muñoz, T., Ros, P.H.: Measuring mass text digitization quality and usefulness. Lessons learned from assessing the OCR accuracy of the british library's 19th century online newspaper archive. D-Lib Mag. **15**(8) (2009)
9. Clematide, S., Furrer, L., Volk, M.: Crowdsourcing an OCR gold standard for a german and french heritage corpus. In: Calzolari, N., et al. (eds.) Proceedings of the Tenth International Conference on Language Resources and Evaluation (LREC 2016) (2016). http://www.lrec-conf.org/proceedings/lrec2016/pdf/917_Paper.pdf
10. Reynaert, M.: Non-interactive OCR post-correction for giga-scale digitization projects. In: Proceedings of the 9th International Conference on Computational linguistics and Intelligent Text Processing, CICLing'08, pp. 617–630 (2008)
11. Holley, R.: Crowdsourcing: how and why should libraries do it? D-Lib Mag. **16**(3/4) (2010)
12. Chrons, O., Sundell, S.: Digitalkoot: making old archives accessible using crowdsourcing. In: Human Computation, Papers from the 2011 AAAI Workshop (2011). http://www.aaai.org/ocs/index.php/WS/AAAIW11/paper/view/3813/4246
13. Silfverberg, M., Kauppinen, P., Linden, K.: Data-driven spelling correction using weighted finite-state methods. In: Proceedings of the ACL Workshop on Statistical NLP and Weighted Automata, pp. 51–59 (2016). https://aclweb.org/anthology/W/W16/W16-2406.pdf
14. Drobac, S., Kauppinen, P., Lindén, K.: OCR and post-correction of historical Finnish texts. In: Tiedemann, J. (ed.) Proceedings of the 21st Nordic Conference on Computational Linguistics, NoDaLiDa, 22–24 May 2017, Gothenburg, Sweden, pp. 70–76 (2017)
15. Smitha, M.L., Antony, P.J., Sachin, D.J.: Document image analysis using imagemagick and tesseract-ocr. Int. Adv. Res. J. Sci. Eng. Technol. **3**(5), 108–112 (2016)
16. Koistinen, M., Kettunen, K., Pääkkönen, T.: Improving optical character recognition of finnish historical newspapers with a combination of fraktur & antiqua models and image preprocessing. In: Tiedemann, J. (ed.) Proceedings of the 21st Nordic Conference on Computational Linguistics, NoDaLiDa, 22–24 May 2017, Gothenburg, Sweden, pp. 277–283 (2017)
17. Breuel, T.: The hOCR microformat for OCR workflow and results. Document analysis and recognition, 2007. In: ICDAR 2007, Ninth International Conference on Document Analysis and Recognition (2007). http://ieeexplore.ieee.org/stamp/stamp.jsp?arnumber=4377078
18. Kettunen, K., Koistinen, M., Kervinen, J: Ground truth OCR sample data of finnish historical newspapers and journals in data improvement validation of a re-OCRing process. LIBER Q. **30**(1), 1–20 (2020). http://doi.org/10.18352/lq.10322
19. Manning, C.D., Schütze, H.: Foundations of Statistical Natural Language Processing. The MIT Press, Cambridge (1999)
20. Kettunen, K.: Keep, change or delete? Setting up a low resource OCR post-correction framework for a digitized old finnish newspaper collection. In: Calvanese, D., De Nart, D., Tasso, C. (eds.) IRCDL 2015. CCIS, vol. 612, pp. 95–103. Springer, Cham (2016). https://doi.org/10.1007/978-3-319-41938-1_11
21. Carrasco, R.C.: An open-source OCR evaluation tool. In: Proceeding DATeCH '14 Proceedings of the First International Conference on Digital Access to Textual Cultural Heritage, pp. 179–184 (2014)
22. Kettunen, K., Kervinen, J., Koistinen, M.: Creating and using ground truth OCR sample data for Finnish historical newspapers and journals. In: DHN2018, Proceedings of the Digital Humanities in the Nordic Countries 3rd Conference, pp. 162–169. http://ceur-ws.org/Vol-2084/)
23. Taghva, K., Borsack, J., Condit, A.: Evaluation of model-based retrieval effectiveness with OCR text. ACM Trans. Inf. Syst. **14**(1), 64–93 (1996)
24. Kantor, P.B., Voorhees, E.M.: The TREC-5 confusion track: comparing retrieval meth-ods for scanned texts. Inf. Retrieval **2**, 165–176 (2000)

25. Järvelin, A., Keskustalo, H., Sormunen, E., Saastamoinen, M., Kettunen, K.: Information retrieval from historical newspaper collections in highly inflectional languages: a query expansion approach. J. Assoc. Inf. Sci. Technol. **67**(12), 2928–2946 (2016)
26. Lopresti, D.: Optical character recognition errors and their effects on natural language processing. Int. J. Doc. Anal. Recogn. **12**, 141–151 (2009)
27. Jarlbrink, J., Snickars, P.: Cultural heritage as digital noise: nineteenth century newspapers in the digital archive. J. Doc. **73**(6), 1228–1243 (2017)

Fine-Tuning Tree-LSTM for Phrase-Level Sentiment Classification on a Polish Dependency Treebank

Tomasz Korbak[1,2]([✉])[ID] and Paulina Żak[3][ID]

[1] Institute of Philosophy and Sociology, Polish Academy of Sciences,
Nowy Świat 72, 00-330 Warsaw, Poland
tkorbak@ifispan.waw.pl
[2] Human Interactivity and Language Lab, Faculty of Psychology,
University of Warsaw, Krakowskie Przedmieście 26/28, 00-927 Warsaw, Poland
[3] Polish-Japanese Academy of Information Technology,
Koszykowa 86, 02-008 Warsaw, Poland
paulina.zak@pjwstk.edu.pl

Abstract. We describe a variant of Child-Sum Tree-LSTM deep neural network [16] fine-tuned for working with dependency trees and morphologically rich languages using the example of Polish. Fine-tuning included applying a custom regularization technique (zoneout, described by Krueger et al. [9], and further adapted for Tree-LSTMs) as well as using pre-trained word embeddings enhanced with sub-word information [2]. The system was implemented in PyTorch and evaluated on phrase-level sentiment labeling task as part of the PolEval competition.

Keywords: Tree-LSTM · Sentiment analysis · Dependency tree · Word embeddings

1 Introduction

In this article, we describe a variant of Tree-LSTM neural network [16] for phrase-level sentiment classification. The contribution of this paper is evaluating various strategies for fine-tuning this model for a morphologically rich language with relatively loose word order – Polish. We explored the effects of several variants of regularization technique known as zoneout [9] as well as using pre-trained word embeddings enhanced with sub-word information [2].

The system was evaluated in PolEval competition. PolEval is a SemEval-inspired evaluation campaign for natural language processing tools for Polish[1]. The task that we undertook was phrase-level sentiment classification, i.e. labeling the sentiment of each node in a given dependency tree. The dataset format was analogous to the seminal Stanford Sentiment Treebank[2] for English [14].

---

[1] http://poleval.pl.
[2] https://nlp.stanford.edu/sentiment/.

Tomasz Korbak was funded by the Ministry of Science and Higher Education (Poland) research Grant DI2015010945 as part of Diamentowy Grant program.

© Springer Nature Switzerland AG 2020
Z. Vetulani et al. (Eds.): LTC 2017, LNAI 12598, pp. 31–42, 2020.
https://doi.org/10.1007/978-3-030-66527-2_3

The source code of our system is publicly available at https://github.com/tomekkorbak/treehopper.

2   Phrase-Level Sentiment Analysis

Sentiment analysis is the task of identifying and extracting subjective information (attitude of the speaker or emotion she expresses) in text. In a typical formulation, it boils down to classifying the sentiment of a piece of text, where sentiment is understood as either binary (positive or negative) or multinomial label and where classification may take place on document level or sentence level. This approach, however, is of limited effectiveness in case of texts expressing multiple (possibly contradictory) opinions about multiple entities (or aspects thereof) [17]. What is needed is a more fine-grained way of assigning sentiment labels, for instance to phrases that build up a sentence.

Apart from aspect-specificity of sentiment labels, another important consideration is to account for the effect of syntactic and semantic composition on sentiment. Consider the role negation plays in the sentence "The movie was not terrible": it flips the sentiment label of the whole sentence around [14]. In general, computing the sentiment of a complex phrase requires knowing the sentiment of its subphrases and a procedure of composing them. Applying this approach to full sentences requires a tree representation of a sentence.

There are two broad families of formalism used to represent sentential syntactic structure: constituency grammars and dependency grammars. The choice of a formalism is highly dependent on peculiarities of language of interest. For instance, English can be nicely captured using a constituency grammar, therefore Stanford Sentiment Treebank represents sentences as binary constituency trees. Polish, on the other hand, has relatively loose word order and rich morphology, thus making dependency approaches more suitable.

PolEval dataset represents sentences as dependency trees. Dependency grammar models sentences in terms of tokens and (binary, directed) relations between them, with some additional constraint: there must be a single root node with no incoming edges and each non-root node must have a single incoming arc and a unique path to the root node. What this entails is that each phrase will have a single head that governs how its subphrases are to be composed [7].

PolEval dataset consisted of a 1200 sentence training set and 350 sentence evaluation test. Each token in a sentence is annotated with its head (the token it depends on), relation type (i.e. coordination, conjunction, etc.) and sentiment label (positive, neural, negative). For an example, consider Fig. 1.

3   LSTM and Tree-LSTM Neural Networks

3.1   Recurrent Neural Networks

Recurrent neural networks (RNNs) are a class of neural networks designed to handle sequential data. This includes EEG signals, protein sequences and natural

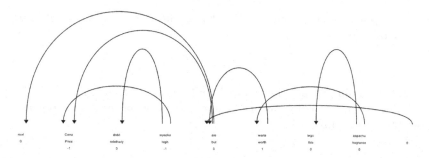

Fig. 1. An entry in Poleval dataset consists of (1) an ordered list of tokens, (2) dependency relations between them, (3) types of these relations (not used by our model, hence not shown) and (4) sentiment labels for each head (−1, 0, 1).

language sentences, modeled as linearly ordered sequences of words rather than tree structures. The power of recurrent neural networks lies in their ability to take advantage of previously seen samples in modeling subsequent ones.

Let us focus on the class of per-sample binary classification problems (also known as *sequence labeling* tasks). Let $\{x^{(t)}\}$ denote the sequence of vectors, where t ranges from 1 to τ, let $\{\hat{y}^{(t)}\}$ be the sequence of predicted labels and let $\{y^{(t)}\}$ be the sequence of ground truth labels. Then the cost function that undergoes maximization in the course of learning is

$$\prod_{i=1}^{\tau} P(\hat{y}^{(i)} = y^{(i)} \mid \theta; x^{(t)}, x^{(t-1)}, \dots, x^{(1)}) \tag{1}$$

where θ denotes all parameters (connection weights) of a model that are optimized.

The fundamental concept behind RNNs is sharing parameters across different parts of the model [5]. While a typical shallow neural network would have separate parameters for particular region of a fixed-size dataset, an RNN can generalize parameters across different time-steps of a sequence.

The idea of parameter sharing can be formalized in terms of unfolding computational graphs of RNNs. A computational graph of a neural networks is a directed graph $G = (N, E)$, where nodes $n_i \in N$ denote variables and edges $e_i \in E$ denote operations on variables (i.e. applying functions). Unfolded computational graphs for RNNs will contain cycles that can be interpreted as recurrent connections between variables. They enable the flow of information across time-steps during forward and backpropagation through time.

An RNN can be described as a dynamical system with transition function f:

$$h_t = f(h_t, x_t; \theta) \tag{2}$$

where h_t denotes hidden state at time-step t, x_t denotes t-th sample and θ denotes model parameters (weight matrices).

The output $\hat{y}_t$ is then a function of current hidden state h_t, current sample x_t and parameters θ:

$$\hat{y}^{(t)} = g(h^{(t)}, x^{(t)}; \theta) \tag{3}$$

In the most simple case (known as Vanilla RNN, or Elman network, cf. [4]), both f and g can be defined as an affine transformations of a concatenation of hidden states and inputs, $[h^{(t)}, x^{(t)}]$, that is:

$$f(h_t, x_t; \theta) = W_h[h_t, x_t] + b_h \tag{4}$$

$$g(h_t, x_t; \theta) = W_y[h_t, x_t] + b_y \tag{5}$$

for some $W_h, W_y, b_h, b_y \in \theta$. Importantly, none of these parameters depends on t; they are shared across time-steps.

3.2 LSTM Cells and Learning Long-Term Dependencies

Notice the recursion inherent in the definition of f and g. In principle, every RNN could be defined non-recursively by self-substitution of Eqs. 2 and 3 τ times, for instance

$$h^{(t)} = f(f(...f(h^{(1)}, x^{(1)}; \theta)..., x^{(t-1)}; \theta), x^{(t)}; \theta) \tag{6}$$

for some initial hidden state $h^{(1)}$. This re-formulation corresponds to unfolding a folded computation graph, yielding a directed acyclic graph.

Thanks to recurrent connections, RNNs are capable of maintaining a working memory (or short-term memory, as opposed to long-term memory captured in weights of forward connections) for storing information about earlier time-steps and use it for classifying subsequent ones. Linguistically, this corresponds to learning constraints that previous words in a sentence place on subsequent words or sentences. RNNs can theoretically handle arbitrarily complex and long-distance dependencies as they are proved to compute any function computable by a Turing machine [12]. One problem is that the distance between two time-steps has a huge effect on learnability of constraints they impose on each other. This particular problem with long-term dependencies is known as *vanishing gradient problem* [1].

Long short-term memory (LSTM) architecture [6] was designed address to the problem of vanishing gradient by enforcing constant error flow across time-steps. This is done by introducing a structure called *memory cell*; a memory cell has one self-recurrent connection with constant weight that carries short-term memory information through time-steps. Information stored in memory cell is thus relatively stable despite noise, yet it can be superimposed with each time-step. This is regulated by three gates mediating memory cell with inputs and hidden states: input gate, forget gate and output get.

For time-step t, let input gate i_t, forget gate f_t and output gate o_t be defined in terms of the following Eqs. (7–9):

$$i_t = \sigma(W^{(i)}x^{(t)} + U^{(i)}h_{t-1}) \tag{7}$$

$$f_t = \sigma(W^{(f)}x^{(t)} + U^{(f)}h_{t-1}) \tag{8}$$

$$o_t = \sigma(W^{(o)}x^{(t)} + U^{(o)}h_{t-1}) \tag{9}$$

where $W^{(i)}, W^{(f)}, W^{(o)}$ and $U^{(i)}, U^{(f)}, U^{(o)}$ denote weight matrices for input-to-cell (where input is x_t) and hidden-to-cell (where hidden layer is h_t) connections, respectively, for input gate, forget gate and output gate. σ denotes the sigmoid function.

Gates are then used for updating short-term memory. Let new memory cell candidate $\widetilde{c}_t$ at time-step t be defined as

$$\widetilde{c}_t = \tanh(W^{(c)}x_t + U^{(c)}h_{t-1}) \tag{10}$$

where $W^{(c)}, U^{(c)}$, analogously, are weight matrices for input-to-cell and hidden-to-cell connections and where tanh denotes hyperbolic tangent function.

Intuitively, $\widetilde{c}_t$ can be thought of as summarizing relevant information about word-token x_t. Then, $\widetilde{c}_t$ is used to update c_t, according to forget gate and input gate.

$$c_t = f_t \circ c_{t-1} + i_t \circ \widetilde{c}_t \tag{11}$$

where $A \circ B$ denotes the Hadamard product of two matrices, i.e. element-wise multiplication.

Forget gate, given current input and hidden state, decides which information from previous memory cell c_{t-1} can be dropped, while input gate, given current input and hidden state, decides which information from candidate memory cell should be incorporated into the memory cell c_t.

Finally, c_t is used to compute next hidden state h_t, again depending on output gate (defined in Eq. 9) that takes into account input and hidden states at current time-step.

$$h_t = o_t \circ \tanh(c_t) \tag{12}$$

In a sequence labeling task, h_t is then used to compute label $\hat{y}_t$ as defined by Eq. 5. The forward-propagation for a LSTM network is done by recursively applying Eqs. 7–12 while incrementing t.

3.3   Recursive Neural Networks and Tree Labeling

Recursive neural networks, or tree-structured neural networks, make a superset of recurrent neural networks, as their computational graphs generalize computational graphs of recurrent neural network from a chain to a tree. Whereas a recurrent neural networks hidden state h_t depends only on one previous hidden states, h_{t-1}, a hidden state of a recursive neural network depends on a set of descending hidden states $C(h_t)$, when $C(j)$ denotes a set of children of a node j.

Tree-structured neural networks have a clear linguistic advantage over chain-structured neural networks: trees make a very natural way of representing the syntax of natural languages, i.e. how more complex phrases are composed of simpler ones.[3] Specifically, in this paper we will be concerned with a tree labeling task, which is analogous generalization of sequence labeling to tree-structured inputs: each node of a tree is assigned with a label, possibly dependent on all of its children.

Finally, in this paper we will be concerned with a tree labeling task, which is analogous generalization of sequences labeling to tree-structured inputs: each node of a tree is assigned with a label, possibly dependent on all of its children.

3.4   Tree-LSTMs Neural Networks

A Tree-LSTM (as described by [16] is a natural combination of the approaches described in two previous subsections. Here we will focus on a particular variant of Tree-LSTM known as Child-Sum Tree-LSTM. This variant allows a node to have an unbounded number of children and assumes no order over those children. Thus, Child-Sum Tree-LSTM is particularly well-suited for constituency trees.[4]

Let $C(j)$ again denote the set of children of the node j. For a given node j, Child-Sum Tree-LSTM takes as inputs vector x_j and hidden states h_k for every $k \in C(j)$. The hidden state h_j and cell state c_j are computed using the following equations:

$$\widetilde{h}_j = \sum_{k \in C(j)} h_k \tag{13}$$

$$i_j = \sigma(W^{(i)}x_j + U^{(i)}\widetilde{h}_j + b_j) \tag{14}$$

$$f_{jk} = \sigma(W^{(f)}x_j + U^{(f)}\widetilde{h}_j + b_f) \tag{15}$$

$$o_j = \sigma(W^{(o)}x_j + U^{(o)}\widetilde{h}_j + b_o) \tag{16}$$

$$u_j = \tanh(W^{(u)}x_j + U^{(u)}\widetilde{h}_j + b_u) \tag{17}$$

$$c_j = i_j \circ u_j + \sum_{k \in C(j)} f_{jk} \circ c_k \tag{18}$$

---

[3] Although recursive neural networks are used primarily in natural language processing, they were also applied in other domains, for instance scene parsing [13].

[4] The other variant described by [16], N-ary Tree-LSTM assumes that each node has at most N children and that children are linearly ordered, making it natural for (binary) dependency trees. The choice between these two variant really boils down to the syntactic theory we assume for representing sentences. As PolEval dataset assumes dependency grammar, we decided to go along with Child-Sum Tree-LSTM.

$$h_j = o_j \circ \tanh(c_j) \tag{19}$$

Equations 14–19 are analogous to Eqs. 7–11; they correspond to applying input gate, forget gate, output gate, update gate and computing cell and hidden states.

In a tree labeling task, we will additionally have an output function

$$\hat{y}_j = W^{(y)} h_j + b_y \tag{20}$$

for computing a label of each node.

4   Experiments

We choose to implement our model in PyTorch[5] due to convenience of using a dynamic computation graphs framework.

We evaluated our model on tree labeling as described in Subsect. 3.3 using PolEval 2017 Task 2 dataset. (For an example entry, see Fig. 1).

4.1   Regularizing with Zoneout

Zoneout [9] regularization technique is a variant of dropout [15] designed specifically for regularizing recurrent connections of LSTMs or GRUs. Dropout is known to be successful in preventing feature co-adaptation (also known as overfitting) by randomly applying a zero mask to the outputs of a given layer. More formally,

$$h := d_t \circ h \tag{21}$$

where d_t is a random mask (a tensor with values sampled from Bernoulli distribution).

However, dropout usually could not be applied to recurrent hidden and cell states of LSTMs, since aggregating zero mask over a sufficient number of time-steps effectively zeros them out. (This is reminiscent of the vanishing gradient problem).

Zoneout addresses this problem by randomly swapping the current value of a hidden state with its value from a previous time-step rather than zeroing it out. Therefore, contrary to dropout, gradient information and state information are more readily propagated through time. Zoneout has yielded significant performance improvements on various NLP tasks when applied to cell and hidden states of LSTMs. This can be understood as substituting Eqs. 10 and 12 with the following ones:

$$c_t := d_t^c \circ c_t + (1 - d_t^c) \circ c_{t-1} \tag{22}$$

$$h_t := d_t^h \circ h_t + (1 - d_t^h) \circ h_{t-1} \tag{23}$$

---

[5] http://pytorch.org/.

where 1 denotes a unit tensor and d_t^c and d_t^h are random, Bernoulli-sampled masks for a given time-step.

Notably, zoneout was originally designed with sequential LSTMs in mind. We explored several ways of adapting it to tree-structured LSTMs. We will consider only hidden state updates, since cell states updates are isomorphic.

As Tree-LSTM's nodes are no longer linearly ordered, the notion of previous hidden states must be replaced with the notion of hidden states of children nodes. The most obvious approach, that we call "sum-child" will be randomly replacing the hidden states of node j with the sum of its children nodes' hidden states, i.e.

$$h_j := d_j^h \circ h_j + (1 - d_j^h) \circ \sum_{k \in C(j)} h_k \tag{24}$$

Another approach, called "choose-child" by us, is to randomly choose a single child to replace the node with.

$$h_j := d_j^h \circ h_j + (1 - d_j^h) \circ h_k \tag{25}$$

where k is a random number sampled from indices of the members of $C(j)$.

Apart from that, we explored different values for d^h and d^c as well as keeping a mask fixed across time-steps, i.e. d_t being constant for all t.

4.2   Using Pre-trained Word Embeddings

Standard deep learning approaches to distributional lexical semantics (e.g. word2-vec, [10]) were not designed with agglutinative languages, like Polish, in mind and cannot take advantage of compositional relation between words. Consider the example of "chodziłem" and "chodziłam" (Polish masculine and feminine past continuous forms of "walk", respectively). The model has no sense of morphological similarity between these words and has to infer it from distributional information itself. This poses a problem when the number of occurrences of a specific orthographic word form is small or zero and some Polish words can have up to 30 orthographic forms (thus, the effective number of occurrences is 30 times smaller than the number of occurrences when counting lemmas).

One approach we explore is to use word embeddings pre-trained on lemmatized data. The other, more promising approach, is take advantage of morphological information by enhancing word embeddings with subword information. We evaluate fastText word vectors as described by [2]. Their work extends the model of [10] with additional representation of morphological structure as a bag of character-level n-gram (for $3 \leq n \leq 6$). Each character n-gram has its own vectors representations and the resulting word embeddings is a sum of the word vector and its character vectors. Authors have reported significant improvements in language modeling tasks, especially for Slavic languages (8% for Czech and 13% for Russian; Polish was not evaluated) compared to pure word2vec baseline.

5   Results

We conducted a thorough grid search on a number of other hyperparameters (not reported here in detail due to spatial limitations). We found out that the best results were obtained with minibatch size of 25, Tree-LSTM hidden state and cell state size of 300, learning rate of 0.05, weight decay rate of 0.0001 and L2 regularization rate of 0.0001. No significant difference was found between Adam [8] and Adagrad [3] optimization algorithms. It takes between 10 and 20 epochs for the system to converge.

Here we focus on two fine-tunings we introduced: fastText word embeddings and zoneout regularization.

The following word embeddings model were used:

- word2vec [10], 300 dimensions, pre-trained on Polish Wikipedia and National Corpus of Polish [11] using lemmatized word forms. Lemmatization was done using Concraft morphosyntactic tagger [18].
- word2vec [10], same as above, but using orthographical word forms.
- fastText [2], 300 dimensions, pre-trained on Polish Wikipedia using orthographical word forms and sub-word information.

Table 1. Results of our faulty solution as evaluated by PolEval organizing committee. "Ensemble epochs" means the number of training epochs we averaged the weights over to obtain a snapshot-based ensemble model.

emb lr	Ensemble epochs	Accuracy
0.2	1	0.678
0.1	1	0.671
0.1	3	0.670

Table 2. A comparison of the effect of pre-trained word embedding on model's accuracy. "emb lr" means learning rate of the embedding layer, i.e. 0.0 means the layer was kept fixed and not optimized during training. "time" means wall-clock time of training on a CPU measured in minutes.

Word embeddings	emb lr	Accuracy	Time
word2vec, orthographic	0.0	0.7482	20:52
word2vec, orthographic	0.1	0.7562	20:26
word2vec, lemmatized	0.0	0.7536	20:01
word2vec, lemmatized	0.1	0.7737	20:09
fastText, orthographic	0.0	0.8011	20:04
fastText, orthographic	0.1	0.7993	20:17

Table 3. Results extracted from a grid search over zoneout hyperparameters. "Mask" denotes the moment mask vector is sampled from Bernoulli distribution: "common" means all node share the same mask, while "distinct" means mask is sampled per node. "Strategy" means zoneout strategy as described in Sect. 4.1. "d_j^c" and "d_j^h" mean zoneout rates for, respectively, hidden and cell states of a Tree-LSTM. No significant differences in training time were observed.

Mask	Strategy	d_j^c	d_j^h	Accuracy
Common	Choose-child	0.01	0.25	0.7931
Common	Sum-child	0.25	0.25	0.7954
Distinct	Sum-child	0.00	0.00	0.7957
Common	Choose-child	0.25	0.01	0.7958
Common	Sum-child	0.01	0.25	0.7961
Distinct	Sum-child	0.01	0.25	0.7963
Distinct	Choose-child	0.25	0.01	0.7970
Distinct	Choose-child	0.00	0.00	0.7970
Distinct	Sum-child	0.00	0.25	0.7972
Common	Choose-child	0.00	0.01	0.7979
Distinct	Sum-child	0.01	0.01	0.7984
Distinct	Choose-child	0.01	0.00	0.7984
Common	Choose-child	0.00	0.00	0.7986
Common	Sum-child	0.01	0.01	0.7990
Common	Choose-child	0.01	0.01	0.7990
Distinct	Choose-child	0.25	0.25	0.7990
Common	Sum-child	0.25	0.00	0.7995
Common	Sum-child	0.00	0.25	0.7995
Distinct	Choose-child	0.25	0.00	0.7996
Distinct	Sum-child	0.00	0.01	0.7997
n/a	n/a	0.00	0.00	0.8000
Distinct	Choose-child	0.00	0.25	0.8003
Common	Choose-child	0.00	0.25	0.8003
Common	Sum-child	0.25	0.01	0.8004
Common	Choose-child	0.25	0.25	0.8005
Distinct	Sum-child	0.25	0.25	0.8006
Distinct	Choose-child	0.00	0.01	0.8007
Common	Sum-child	0.01	0.00	0.8008
Common	Sum-child	0.00	0.01	0.8013
Distinct	Sum-child	0.01	0.00	0.8013
Common	Choose-child	0.01	0.00	0.8015
Distinct	Sum-child	0.25	0.00	0.8018
Distinct	Choose-child	0.01	0.01	0.8032
Distinct	Choose-child	0.01	0.25	0.8051
Common	Choose-child	0.25	0.00	0.8052
Distinct	Sum-child	0.25	0.01	0.8070

Our results for different parametrization of pre-trained word embeddings and zoneout are shown in Tables 2 and 3, respectively. The effects of word embeddings and zoneout were analyzed separately, i.e. results in Table 2 were obtained with no zoneout and results in Table 3 were obtained with best word embeddings, i.e. fastText.

Note that these results differ from what is reported in official PolEval benchmark. Our results as evaluated by organizing committee, reported in Table 1, left us behind the winner (0.795) by a huge margin. This was due to a bug in our implementation, which was hard to spot as it manifested only in inference mode. The bug broke mapping between word tokens and weights in our embedding matrix. All results reported in Tables 2 and 3 were obtained after fixing the bug (the model trained on training dataset and evaluated on evaluation dataset, after ground truth labels were disclosed). Note that these results beat the best reported solution by a small margin.

6   Conclusions

As far as word2vec embeddings are concerned, both training on lemmatized word forms and further optimizing embedding yielded small improvements; the two effects being cumulative. FastText vectors, however, beat all word2vec configurations by a significant margin. This result is interesting as fastText embeddings were originally trained on a smaller corpus (Wikipedia, as opposed to Wikipedia + NKJP in the case of word2vec).

When it comes to zoneout, it barely affected accuracy (improvement of about 0.6% point) and we did not found a hyperparameter configuration that stands out. More work is needed to determine whether zoneout could yield robust improvements for Tree-LSTM.

Unfortunately, our system did not manage to win the Task 2 competition, this being due to a simple bug. However, our results obtained after the evaluation indicate that it was very promising in terms of overall design and in fact, could beat other participants by a small margin (if implemented correctly). We intend to prepare and improve it for the next year's competition having learned some important lessons on fine-tuning and regularizing Tree-LSTMs for sentiment analysis.

References

1. Bengio, Y., Simard, P., Frasconi, P.: Learning long-term dependencies with gradient descent is difficult. IEEE Trans. Neural Netw. **5**(2), 157–166 (1994)
2. Bojanowski, P., Grave, E., Joulin, A., Mikolov, T.: Enriching word vectors with subword information. arXiv preprint arXiv:1607.04606 (2016)
3. Duchi, J., Hazan, E., Singer, Y.: Adaptive subgradient methods for online learning and stochastic optimization. J. Mach. Learn. Res. **12**, 2121–2159 (2011). http://dl.acm.org/citation.cfm?id=1953048.2021068
4. Elman, J.L.: Finding structure in time. Cogn. Sci. **14**(2), 179–211 (1990). https://doi.org/10.1207/s15516709cog1402_1

5. Goodfellow, I., Bengio, Y., Courville, A.: Deep Learning. MIT Press, Cambridge (2016)

6. Hochreiter, S., Schmidhuber, J.: Long short-term memory. Neural Comput. **9**(8), 1735–1780 (1997)

7. Jurafsky, D., Martin, J.H.: Speech and Language Processing: An Introduction to Natural Language Processing, Computational Linguistics, and Speech Recognition, 1st edn. Prentice Hall PTR, Upper Saddle River, NJ, USA (2000)

8. Kingma, D.P., Ba, J.: Adam: a method for stochastic optimization. CoRR Arxiv:1412.6980 (2014)

9. Krueger, D., et al.: Zoneout: Regularizing RNNS by randomly preserving hidden activations. CoRR ArXiv:1606.01305 (2016)

10. Mikolov, T., Chen, K., Corrado, G., Dean, J.: Efficient estimation of word representations in vector space. CoRR ArXiv:1301.3781 (2013)

11. Przepiórkowski, A., Górski, R.L., Lewandowska-Tomaszczyk, B., Łaziński, M.: Towards the national corpus of polish. In: Proceedings of the Sixth International Conference on Language Resources and Evaluation, LREC 2008. ELRA, Marrakech (2008)

12. Siegelman, H., Sontag, E.: Neural Nets are Universal Computing Devices. Sycon-91-08, Rutger University (1991)

13. Socher, R., Lin, C.C.Y., Ng, A.Y., Manning, C.D.: Parsing natural scenes and natural language with recursive neural networks. In: Proceedings of the 28th International Conference on International Conference on Machine Learning, ICML'11, Omnipress, USA, pp. 129–136 (2011). http://dl.acm.org/citation.cfm?id=3104482. 3104499

14. Socher, R., et al.: Recursive deep models for semantic compositionality over a sentiment treebank. In: EMNLP (2013)

15. Srivastava, N., Hinton, G., Krizhevsky, A., Sutskever, I., Salakhutdinov, R.: Dropout: a simple way to prevent neural networks from overfitting. J. Mach. Learn. Res. **15**(56), 1929–1958 (2014). http://jmlr.org/papers/v15/srivastava14a.html

16. Tai, K.S., Socher, R., Manning, C.D.: Improved semantic representations from tree-structured long short-term memory networks. CoRR ArXiv:1503.00075 (2015)

17. Thet, T.T., Na, J.C., Khoo, C.S.: Aspect-based sentiment analysis of movie reviews on discussion boards. J. Inf. Sci. **36**(6), 823–848 (2010). https://doi.org/10.1177/0165551510388123

18. Waszczuk, J.: Harnessing the CRF complexity with domain-specific constraints. the case of morphosyntactic tagging of a highly inflected language. In: Proceedings of COLING 2012. The COLING 2012 Organizing Committee, pp. 2789–2804 (2012). http://aclanthology.coli.uni-saarland.de/pdf/C/C12/C12-1170.pdf

Supervised Transfer Learning for Sequence Tagging of User-Generated-Content in Social Media

Sara Meftah[1], Nasredine Semmar[1(✉)], Othmane Zennaki[1], and Fatiha Sadat[2]

[1] Département Intelligence Ambiante et Systèmes Interactifs, Université Paris-Saclay, CEA, 91191 Gif-sur-Yvette, France
{sara.meftah,nasredine.semmar,othmane.zennaki}@cea.fr
[2] UQAM, Montreal, QC, Canada
fatiha.sadat@uqam.ca

Abstract. Neural-networks based approaches for Natural Language Processing (NLP) are effective when dealing with learning from large amounts of annotated data. However, these are only available for a limited number of languages and domains due to the cost of the manual annotation. Particularly, despite the valuable importance of Social Media's content for a variety of applications (*e.g.*, public security, health monitoring, or trends highlight), this important domain is still poor in terms of annotated data. In this work, we analyse the impact of supervised sequential transfer learning to overcome the sparse data problem in the *Tweets-domain* by leveraging the huge annotated data available for the *Newswire-domain*. We experiment our approach on three NLP tasks: part-of-speech tagging, chunking and named entity recognition.

Keywords: Transfer learning · Social media · Domain adaptation · Sequence tagging

1 Introduction

The past few years have witnessed the great success of the application of neural networks (NNs) models for NLP. These models extract a hierarchy of features, directly from data without any need for hand-crafted features. Indeed, several studies [14] have shown that NNs architectures are effective to extract morphological information (root, prefix, suffix, etc.) from words and encode it into neural representations, especially for morphological rich languages [4,16]. Nevertheless, these models are in most cases based on a supervised learning paradigm, i.e. trained from scratch on large amounts of labelled examples. Consequently, models with high performance often require huge volumes of manually annotated data to produce powerful results and prevent over-fitting.

Nowadays, sparse data problem is a serious challenge. Indeed, manual data annotation is labour intensive, and thus, most languages varieties are under-resourced in terms of annotated datasets [9]. According to Baumann et al. [2],

© Springer Nature Switzerland AG 2020
Z. Vetulani et al. (Eds.): LTC 2017, LNAI 12598, pp. 43–57, 2020.
https://doi.org/10.1007/978-3-030-66527-2_4

only 20 of 6,909 languages are high-resourced. As a consequence, research in NLP focuses only on a small number of languages, specifically standard forms of languages, which are well-resourced.

We focus in this work on the low-resourced domain, User Generated Content (UGC) in Tweets that we call *Tweets-domain*. In fact, the application of models purely trained on well-structured corpora from the *Newswire-domain* falls to work on noisy text from the *Tweets-domain*. For instance, as reported in [10], the accuracy of the Stanford part-of-speech tagger [33] trained on News falls from 97% on standard English to 85% on Twitter. The main reason for this drop in accuracy is that Tweets contain a lot of Out-Of-Vocabulary words compared to standard text. This is due to the conversational nature of the text, the lack of conventional orthography, the noise, linguistic errors and the idiosyncratic style. Also, Twitter poses an additional issue by imposing 280 characters limit for each Tweet.

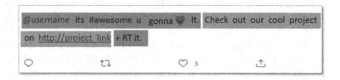

Fig. 1. An example of a Tweet [15]. Gray segments show expressions that are similar to formal texts and red ones show expressions that are specific to *Tweets-domain* (informal texts). (Color figure online)

Transfer Learning (TL) [24,26] is a promising method to workaround the problem of the lack of annotated data. Indeed, languages, tasks and domains may share common knowledge about language (e.g. linguistic representations, structural and semantic similarities, etc.). And thus, relevant knowledge previously learned in a source problem can be leveraged to help solving a new target problem.

The overall contribution of this paper is based on the intuition that *Tweets-domain* is an informal variety of the *Newswire-domain*. As illustrated in Fig. 1, in the same Tweet, we can find a part which is formal and the other which is informal. Therefore, the main objective of this work is to study the forcefulness of supervised sequential transfer learning to overcome the sparse data problem in the *Tweets-domain* by leveraging the huge available annotated data in the *Newswire-domain*. Our experiments show significant improvements on three NLP tasks: Part-of-Speech tagging, Chunking and Named Entity Recognition.

The remainder of the paper is organised as follows: Section 2 describes the base neural model architecture used for sequence tagging tasks. In Sect. 3, we introduce our approach for transferring knowledge from the *Newswire-domain* to the *Tweets-domain*. Section 4 describes the experimental setup. Section 5 reports our experimental results, followed by an in-depth analysis in Sect. 6.

2   Base Neural Model

As aforementioned, we perform experiments on three NLP sequence tagging (ST) tasks: Part-of-Speech tagging (POS), Chunking (CK) and Named Entity Recognition (NER). Given an input sentence $S = [w_1, \ldots, w_n]$ of n successive tokens, ST aims to predict the tag $c_i \in C$ of every w_i, with C being the tag-set.

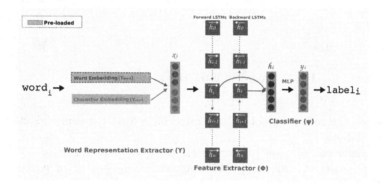

Fig. 2. Illustrative scheme of neural model architecture for sequence tagging tasks.

We use a common ST neural model, illustrated in Fig. 2. It includes three main components. First, the **Word Representation Extractor (WRE)** computes for each token w_i, two hybrid representations; a word-level embedding (denoted Υ^{word}) and character-level embedding based on a bidirectional-Long Short-Term Memory (biLSTMs) encoder (denoted Υ^{char}), and concatenates them to get a final representation x_i. Second, this representation is fed into a **Features Extractor (FE)** (denoted Φ) based on a single-layer BiLSTMs network, to produce a hidden representation h_i which is fed into a **Classifier (Cl)**, a fully-connected (FC) layer (denoted Ψ), for classification. Formally, given w_i, the logits are obtained using the following equation: w_i: $\hat{y}_i = (\Psi \circ \Phi \circ \Upsilon)(w_i)$. With Υ being the concatenation of Υ^{char} and Υ^{word}.

3   Transfer Learning Approach

We use a common sequential TL method to transfer knowledge from the *Newswire-domain* to the *Tweets-domain* [19]. It consists on learning a source model on the source task with enough data from the *Newswire-domain*, then transferring a part of its learned weights to initialise the target model, which is further fine-tuned on the target task with few training examples from the *Tweets-domain*.

As illustrated in Fig. 3, given a source neural network N_s with a set of parameters θ_s split into two sets: $\theta_s = (\theta_s^1, \theta_s^2)$, and a target network N_t with a set of parameters θ_t split into two sets: $\theta_t = (\theta_t^1, \theta_t^2)$. Our method includes three simple yet effective steps:

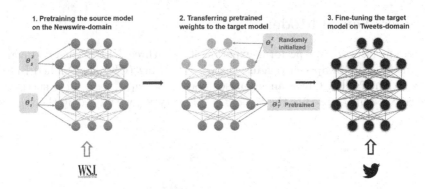

Fig. 3. Overview of transfer learning process.

1. We learn the source model on annotated data from the source domain on a source dataset D_s.
2. We transfer the first set of parameters from the source network N_s to the target network N_t: $\theta_t^1 = \theta_s^1$. Whereas the second set θ_t^2 of parameters is randomly initialised.
3. Then, the target model is further fine-tuned on the small target data-set D_t.

4 Experimental Setup

4.1 Data-Sets

We conducted experiments on 5 English Tweets datasets: TPoS, ArK and Twee-Bank datasets for POS task, TChunk for CK task and WNUT for NER task. We summarise Tweets datasets statistics in Table 1[1].

Table 1. Statistics of the used Tweets datasets.

Datasets		English				
		POS			CK	NER
		TPoS	Ark	TweeBank	TChunk	WNUT
#Classes		40	25	17	18	6
#Tokens	Train	10,5k	26,5k	24,7k	10,6k	62,7k
	Validation	2,3k	n/a	11,7k	2,2k	15,7k
	Test	2,9k	7,7k	19k	2,2k	23,3k
Metrics		Accuracy (Acc.)				F1

In the following, we describe briefly each task and the used datasets for both source (*Newswire*) and target (*Tweets*) domains:

---

[1] SeqEval package were used to calculate F1 metric.

Part-of-Speech Tagging assigns an adequate and unique grammatical category, POS tag, to each word in a text, e.g. noun, verb, etc. Note that tag-sets may differ from one data-set to another. The most common tag-sets are: 1) Treebank's tag-sets, where POS tags vary between languages due to cross-lingual differences. For example, the English Penn TreeBank (PTB) [17] comprises 36 tags[2]. 2) Universal Dependencies 2.0 [23] POS tag-set[3], contains a set of 17 POS tags common between all languages. For the *source data-set*, we used the PTB portion of **WSJ**, a large English dataset (formal texts) from the *Newswire-domain*, annotated with the PTB tag-set. Regarding the *target datasets*, we used three English Tweets data-sets:

- **TPoS** [29] uses the PTB tag-set plus four Twitter special tags: *URL* for web addresses, *HT* for hashtags, *USR* for username mentions and *RT* for *Retweet* signifier. For our experiments on TPoS, we used the same data splits used by [6]; 70:15:15 into training, validation and test sets.
- **ARK** [25]: published in two parts, Oct27 (training-set) and Daily547 (test-set), using a novel and coarse grained tag-set comprising 25 tags[4]. We split the Oct27 dataset into training-set and validation-set (90:10)[5] and used Daily547 as a test set.
- **TweeBank** [15]: a recent dataset annotated with the UD 2.0 POS tag-set.

Chunking (or shallow parsing) is an "intermediate step from POS towards dependency parsing"[6], extracting high-order syntactic spans (chunks) from texts, e.g. noun phrases, verbal phrases or adverbial phrases, etc. CK datasets use generally the BIO annotation scheme that categorises tokens as either being outside the syntactic span (O), the beginning of the syntactic span (B_X) or inside the syntactic span (I_X), here X refers to one of the 11 syntactic chunk types like NP, VP, PP, ADJP, ADVP, etc. For the *source dataset*, we used the **CONLL2000** shared task's English data-set [32] that uses Sect. 15–18 from the WSJ corpus for training and Sect. 20 for testing. Regarding *target dataset*, we used **TChunk** Tweets data-set [29] (the same corpus as TPoS). Both source and target datasets use the BIO format annotation scheme.

Named Entity Recognition consists on classifying named entities that are present in a text into pre-defined categories like names of persons, organisations or locations. Regarding the *source domain*, we make use of the English *Newswire* dataset **CONLL-03** from the CONLL 2003 shared task [31], tagged with four different entity types (Persons, Locations, Organisations and MISC). For the *target domain*, we conducted our experiments on the Emerging Entity Detection shared task of **WNUT** [5], that includes six entity types, three of which

---

[2] https://www.ling.upenn.edu/courses/Fall_2003/ling001/penn_treebank_pos.html.
[3] https://universaldependencies.org/u/pos/.
[4] http://www.cs.cmu.edu/~ark/TweetNLP/annot_guidelines.pdf.
[5] data splits portions are not mentioned in original papers.
[6] https://www.clips.uantwerpen.be/conll2000/chunking/.

are common with CONLL-03's types: *Persons*, *Organisations* and *Locations*. While three are emergent: *Products*, *Groups* and *Creative Works*. Similarly to CK datasets, both NER datasets use the BIO format.

4.2 Comparison Methods

We compare our results to the following best State-Of-The-Art (SOTA) works: **POS:** We compare our results to 2 recent works. 1) **TPANN** [12] is a model that uses adversarial pre-training to leverage, in addition to supervised pretraining (on WSJ) from the *Newswire-domain*, huge amounts of unlabelled Tweets. 2) **PretRand** [21] improves standard fine-tuning by reducing the negative bias occurring in the pretrained weights.
CK: We compare our results to **GATE** [29], a CRF-based model.
NER: We compare our results to **Flairs** [1], a biLSTM-CRF model fed with the Pooled Contextual Embeddings.

4.3 Implementation Details

The hyper-parameters we used are as follows. For **WRE:** The dimensions of character embedding $= 50$, hidden states of the character-level biLSTM $= 100$ and word-level embeddings $= 300$ (these latter are pre-loaded from Glove pre-trained vectors [27] and fine-tuned during training). For **FE:** We use a single-layer biLSTM (token-level feature extractor), with dimension $= 200$. In all experiments, SGD was used for pre-training and fine-tuning with early stopping and mini-batches were set to 16 sentences.

5 Experimental Results

In the following, we report our experimental results on supervised transfer learning from the *Newswire-domain* to the *Tweets-domain*. First, in Sect. 5.1, we report the main results when transfer is performed between the same NLP tasks. Second, in Sect. 5.2, we take further experiments to analyse layer-per-layer transferability. Third, in Sect. 5.3, we investigate the transferability between different NLP tasks.

5.1 Overall Performance

Here, we discuss the results of the main experiment of our approach described in Sect. 3, where the pretraining task is the same as the fine-tuning one. Precisely, for TPoS, ArK and TweeBank datasets, the pretrained weights are learned on WSJ dataset; for TChunk, the pretrained weights are learned on CONLL2000 dataset; and for WNUT dataset, the pretrained weights are learned on CONLL-03 dataset.

As shown in Sect. 4.1, source and target datasets may have different tag-sets, even within the same NLP task. Therefore, in this experiment, WRE's layers

(Υ) and FE's layers (Φ) are pre-trained on the *source-dataset* and the classifier (Ψ) is randomly initialised. Then, the three modules are further jointly trained on the *target-dataset* by minimising a Softmax Cross-Entropy (SCE) loss using the SGD algorithm.

Table 2. Main results of supervised transfer learning on Tweets datasets (Acc (%) for POS and CK and F1 (%) for NER), compared to best SOTA works (Gray lines) and the reference training from scratch. Best score for each dataset is highlighted in bold.

Task	POS (Acc.)					CK (Acc.)		NER (F1)
Method	Dataset							
	TPoS		ARK	Tweebank		TChunk		WNUT
	Dev	Test	Test	Dev	Test	Dev	Test	Test
GATE	n/a	88.69	n/a	n/a	n/a	n/a	87.5	n/a
TPANN	n/a	<u>90.92</u>	92.8	n/a	n/a	n/a	n/a	n/a
PretRand	**91.56**	**91.46**	**93.77**	**94.51**	**94.95**	n/a	n/a	n/a
FLAIRS	n/a	n/a	n/a	n/a	n/a	n/a	n/a	**49.59**
From scratch	88.52	86.82	90.89	91.61	91.66	87.76	85.83	36.75
Transfer Learning	90.95	89.79	<u>92.09</u>	<u>93.04</u>	<u>93.29</u>	**90.71**	**89.21**	<u>41.25</u>

Results are reported in Table 2. In gray lines, we report best SOTA results for each dataset (Sect. 4.2). Then, we report the results of our reference *training from scratch*, followed by the results of supervised transfer learning approach, which provides much better results compared to the reference. Specifically, transfer learning exhibits an improvement of $\sim$+3% for TPoS, $\sim$+1.2% for ArK, $\sim$+1.6% for TweeBnak, $\sim$+3.4% for TChunk and $\sim$+4.5% for WNUT. Compared to SOTA, our results are competitive despite the simplicity of the method compared to recent methods.

5.2 Layer-per-Layer Transferability

In this experiment, we investigate the transferability of each layer of our model. We start by transferring from the bottom-most layers (Υ) up to the top-most layers (Φ). In addition, we experiment two settings: 1) 🔒: pretrained layers are frozen; and 2) 🔓: pretrained layers are fine-tuned. As illustrated in Fig. 4, we define 4 transfer schemes[7]:

- **Scheme A**: Only WRE (Υ) layers; i.e. word-level embedding and BiL-STMs character embedding; are initialised with pretrained weights from the *Newswire-domain* whereas the FE (Φ) and the classifier (Ψ) are randomly initialised. The pretrained layers are frozen (🔒) during training on *Tweets-domain* dataset.

---

[7] Note that, Sect. 5.1 experiments correspond to scheme D.

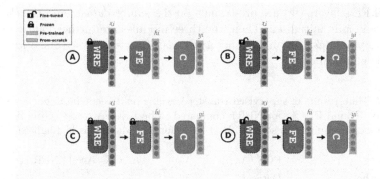

Fig. 4. Overview of the experimental schemes of transferring pretrained parameters. **Scheme A & B**: Only WRE's layers are initialised with pretrained weights from the *Newswire-domain* whereas FE's layers and the classifier are randomly initialised. **Scheme C & D**: In addition to WRE's layers, FE's layers are initialised with pretrained weights from the *Newswire-domain* whereas the classifier is randomly initialised. The pretrained layers are frozen in schemes A and C and tuned in schemes B and D during fine-tuning on *Tweets-domain*.

- **Scheme B**: The same as Scheme A, except that the pretrained layers are tuned (🔓) during fine-tuning on *Tweets-domain* dataset.
- **Scheme C**: In addition to WRE layers, FE layers are initialised with pretrained weights from the *Newswire-domain*, whereas the classifier is randomly initialised. The pretrained layers are frozen (🔒) during fine-tuning on *Tweets-domain*.
- **Scheme D**: The same as Scheme C, except that the pretrained layers are tuned (🔓) during fine-tuning on *Tweets-domain*.

Results are shown in Table 3. First, as expected, the best performance across all tasks and datasets is yielded by the transfer scheme D. Indeed, since transfer is performed between the same NLP tasks, transferring both low-most and top-most layers is beneficial. Second, we can observe that the transferability of each layer depends on whether the pretrained parameters are frozen or tuned:

- *When pretrained layers are frozen* (🔒) (schemes A and C) only the low-most layers are transferable, with a little improvement (+0.64% in average) compared to the reference *training from scratch* (scheme O). Whereas, when transferring top-most layers as well (scheme C), the performance degrades dramatically (−4.26% in average) compared to the reference. Which is explained by the fact that top-most layers are grossly domain specific, and thus need to be updated during fine-tuning to learn new patterns that are specific to the *Tweets-domain*.
- *When pretrained layers are tuned* (🔓) (schemes B and D): both pretrained low-most and top-most layers are beneficial across all tasks and datasets. Specifically, transferring embeddings layers that are updated during fine-tuning (Scheme B) yields a slight gain (+1.62% in average) compared to the

Table 3. Layer-per-layer transferability analysis results on Tweets datasets *TPoS, ARK, TweeBank, TChunk* and *WNUT*. Scheme O consists on training target models from scratch (random initialisation) on small Tweets training-sets. Transfer schemes A,B, C and D are illustrated in Fig. 4. Scores marked with ◇ are higher than the reference training from scratch (scheme O). First best scores by dataset are highlighted in bold, second best scores are underlined. The last column (avg.) gives the average score of each scheme across all datasets.

Scheme	POS (Acc.)					CK(Acc.)		NER (F1)	Avg.
	TPoS		ArK	TweeBank		TChunk		WNUT	
	Dev	Test	Test	Dev	Test	Dev	Test	Test	
O	88.52	86.82	90.89	91.61	91.66	87.76	85.83	36.75	82.47
A	88.87◇	87.48◇	90.85	92.05◇	92.48◇	87.85◇	86.10◇	39.27◇	83.11◇
B	90.17◇	88.66◇	91.55◇	92.31◇	92.65◇	89.22◇	87.19◇	40.97◇	84.09 ◇
C	86.10	86.91◇	85.39	87.48	87.92	82.35	81.75	27.83	78.21
D	**90.95◇**	**89.79◇**	**92.09◇**	**93.04◇**	**93.29◇**	**90.71◇**	**89.21◇**	**41.25◇**	**85.04◇**

reference. Moreover, transferring the feature extractor layers as well enhances the performance further (+2.57% in average).

5.3 Inter Tasks Transferability

Through the precedent experiments, we analysed transfer learning from the *News-domain* to the *Tweets-domain* in a scenario where pretraining (source) task is the same as fine-tuning (target) one. Here, we carry further experiments to analyse the transferability between different NLP tasks. Results are shown in Table 4. The results of transferring pretrained models from the same NLP task are illustrated in red cells, while the results of transferring from different NLP task are illustrated in Gray cells.

In the first line of Table 4, we report the results of the reference training from scratch. The second set of lines reports the results when the weights are pretrained on the POS dataset from the *Newswire-domain* (WSJ). The third set of lines reports the results when the weights are pretrained on the CK dataset from the *Newswire-domain* (CONLL2000). Finally, the fourth set of lines reports the results when the weights are pretrained on the NER dataset from the *Newswire-domain* (CONLL-03).

From Table 4, we draw the following observations:

1. **Fine-Tuning *vs* Freezing Pretrained Weights**: Across all transfer schemes, fine-tuning pretrained parameters yields better performance compared to their freezing. An expected observation since pretrained parameters need to be updated to better match *Tweets-domain* specificities. Specifically, when transferring only the low-most layers, the damage brought by freezing (scheme A) is slight compared to fine-tuning (scheme B); ∼-1% in average. In the other hand, when transferring top-most layers as well, we observe

Table 4. Main results (%) of neural transfer learning from *Newswire*-domain to *Tweets*-domain on Tweets datasets *TPoS*, *ARK*, *TweeBank*, *TChunk* and *WNUT*. The first column (pretraining) shows the pretraining task on the *Newswire*-domain. Scheme O represents training target models from scratch (random initialisation) on small Tweets train-sets. Transfer schemes A, B, C and D are illustrated in Fig. 4. Red cells show results on transfer from the same NLP task and gray ones represent transfer from different NLP task. Scores marked with ◇ are higher than the baseline training from scratch scheme. First best score by dataset are highlighted in bold, second best scores are underlined.

Pretraining	Scheme	POS					CK		NER
		TPoS		ArK	TweeBank		TChunk		WNUT
		Dev	Test	Test	Dev	Test	Dev	Test	Test
n/a	O	88.52	86.82	90.89	91.61	91.66	87.76	85.83	36.75
POS	A	88.87°	87.48°	90.85	92.05°	92.48°	88.29°	87.41°	35.31
	B	90.17°	88.66°	91.55°	92.31°	92.65°	88.82°	87.72°	38.51°
	C	86.10	86.91°	85.39	87.48	87.92	79.67	79.39	20.17
	D	**90.95°**	**89.79°**	**92.09°**	**93.04°**	**93.29°**	89.79°	88.99°	34.98
CK	A	87.66	86.65	90.27	91.39	92.03°	87.85°	86.10°	37.12°
	B	89.82°	87.70°	91.44°	92.07°	92.54°	89.22°	87.19°	37.35°
	C	82.94	79.14	81.04	83.31	83.84	82.35	81.75	19.02
	D	90.08°	87.70°	91.33°	92.57°	92.66°	**90.71°**	**89.21°**	34.76
NER	A	87.27	86.26	89.98	91.33	91.87°	86.93	84.61	39.27°
	B	89.43°	87.57°	90.75	91.73°	92.21°	88.56°	87.06°	40.97°
	C	69.77	66.27	67.03	70.00	70.13	60.92	57.89	27.83
	D	89.35°	88.31°	90.90°	91.62°	92.05°	88.03°	86.54°	**41.25°**

that freezing pretrained parameters (scheme C) dramatically hurts the performances compared to fine-tuning (scheme D); ∼-13.35% in average. A plausible explanation is that, generally, the lowest layers of NNs tend to represent domain-independent features and thus encode information that could be useful for all tasks and domains, whereas top-most ones are more domain specific, and thus, should be updated. These observations are not surprising and confirm many works in the literature [22,34].

2. **Pretraining Task**: We can observe that the best average score is obtained when using parameters pretrained on POS tagging task. Specifically, the first best score is obtained using the scheme D and the second by scheme B. Which confirms the fact that the information encoded by POS task is universal and important for higher-level NLP tasks and applications [3].

    For POS→POS and POS→TChunk, we can observe that both low-most and top-most layers play an important role. Specifically, transferring low-most layers (scheme B) yields an average improvement of ∼+1.13% for POS→POS and ∼+1.27% for POS→TChunk. In addition, transferring top-most layers as well (scheme D) yields and improvement of $\sim$+1.8% for POS$\rightarrow$POS and $\sim$+2.6% for POS$\rightarrow$TChunk. However, only low-most layers are transferable from POS to NER. As illustrated in the results, for POS$\rightarrow$WNUT transfer scenario, scheme B yields an improvement of $\sim$+1.75%, while scheme D degrades the performance by $\sim$-1.8% compared to training from scratch. A plausible explanation is that the transferability decreases as the dissimilarity between source and target tasks increases, and since NER task is less similar to POS, only low-level features learned in embedding layers are beneficial for NER.

6 Analysis

In this section, we analyse the impact of transfer learning from the *Newswire-domain* to the *Tweets-domain*, when transfer is performed between the same NLP tasks (Sect. 5.1 experiments). First, we analyse the impact of our approach on extremely low-resources scenarios (Sect. 6.1). Then, we investigate the impact of the pretraining state on the transfer performance (Sect. 6.2).

6.1 Target Data-Sets Size Impact

In this experiment we analyse the impact of supervised transfer learning in extremely low-resources scenarios. For this, we evaluate, in Fig. 5, the gain in performance brought by transfer compared to the baseline *training from scratch*, according to different target training-sets' sizes. From the results, we can observe that transfer learning has desirably a bigger gain with small target-task datasets, which clearly means that, unsurprisingly, the less target training-data we have, the more interesting transfer learning will be.

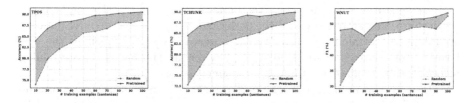

Fig. 5. Performance (on validation-sets of TPoS, TChunk and WNUT) according to different Tweets training-set sizes. Transparent orange highlights the gain brought by transfer learning approach compared to the reference training from scratch.

Specifically, we can observe that for TPoS, TChunk and WNUT datasets, the gain is, respectively, about 10%, 11% and 15% when only 10 annotated sentences are available. However, when 100 annotated sentences are available, the gains decrease to 3%, 2.5% and 1%, respectively.

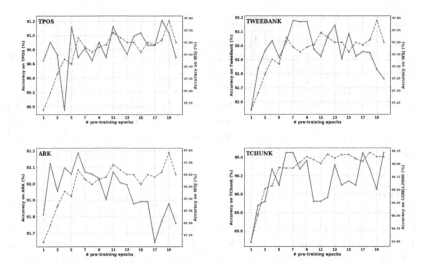

Fig. 6. Accuracy curves (Gray) on source-datasets from the Newswire domain according to different pretraining epochs. And accuracy curves (green) on target-datasets from Tweets-domain using the pretrained parameters at each pretraining epoch. Note that different scales of the y-axis are used for source and target curves. (Color figure online)

6.2   Pretraining State Impact

So far in our experiments we used the pretrained parameters from the best model trained on the source dataset. In simple words, we picked the model at the epoch with the highest performance on the source validation-set. In this analysis, we study when pretrained parameters are ready to be transferred. Specifically, we pick the pretrained weights at different pretraining epochs; that we call the pretraining states. Then, we assess the performance when transferring each.

In Fig. 6, we plot in Gray lines the performance' curves on the source datasets according to pretraining epochs. We can observe that, unsurprisingly, the performance on source datasets increase rapidly on the first epochs of pretraining before reaching a plateau with a slight augmentation.

Then, we plot in green lines the performance on target datasets when using pretrained weights from different pretraining epochs. We can observe that, globally, the best performance on target-datasets is yielded when using pretrained weights from early pretraining epochs. Specifically, for POS tagging, the best performance on *TweeBank* and *ArK* target-datasets is obtained when using the weights from the 7th and 6th pretraining epochs, respectively. Whereas the best performance on the *WSJ* source-dataset is not obtained until the 19th epoch. Interestingly, the performance on both datasets, *TweeBank* and *ArK* degrades gradually at the last pretraining epochs. However, for *TPoS* target-dataset, we can observe that the performance follows the performance on *WSJ* source-dataset. This phenomena could be explained by the fact that, *TPoS* shares the same

PTB tag-set as *WSJ*, whereas *TweeBank* and *ArK* use different tag-sets. And thus, at the last states of pretraining, the pretrained parameters become well-tuned to the source dataset and specific to the source tag-set.

7 Related Work

TL has become omnipresent in NLP works in the last years. However, due to limited space, we focus only on TL for social media texts. The majority of work studied the impact of pretrained language models (unsupervised pretraining). For instance, Dirkson and Verberne [8] proposed to fine-tune the pretrained language model ULMFiT [13] on Twitter health data for adverse drug response extraction. Tian et al. [7] used pretrained BERT model for early detection of rumours in Tweets. And in [30], the authors used pretrained ELMo representations [28] for hate speech detection in Social Media. As far as we are aware, there has been little work studying supervised TL from the Newswire domain to Social Media domain. For instance, supervised pretraining [18,20] and adversarial pretraining [12] were applied to transfer from the POS tagging of News to POS tagging of Tweets. And in [11], a dynamic conversion neural networks based method was proposed to generate different parameters to separately model similarities and dissimilarities between News and Tweets for POS tagging task.

8 Conclusion

In this paper, we analysed the impact of supervised transfer learning from the high-resourced *Newswire-domain* to the low-resourced *Tweets-domain*. Our experiments on three NLP tasks confirm the efficacy of our approach, especially in extremely low-resources scenarios. For future work, first, we plan to study the flexibility of our approach for transfer between less-related languages and domains, such as Arabic dialects. Second, we aim to study the impact of incorporating pretrained contextualised representations, like ELMo, BERT and Flairs, to our method.

References

1. Akbik, A., Bergmann, T., Vollgraf, R.: Pooled contextualized embeddings for named entity recognition. In: Proceedings of the 2019 Conference of the North American Chapter of the Association for Computational Linguistics: Human Language Technologies. Long and Short Papers, vol. 1, pp. 724–728 (2019)
2. Baumann, P., Pierrehumbert, J.B.: Using resource-rich languages to improve morphological analysis of under-resourced languages. In: LREC, pp. 3355–3359 (2014)
3. Changpinyo, S., Hu, H., Sha, F.: Multi-task learning for sequence tagging: an empirical study. In: Proceedings of the 27th International Conference on Computational Linguistics, pp. 2965–2977 (2018)
4. Chiu, J.P., Nichols, E.: Named entity recognition with bidirectional LSTM-CNNs. arXiv preprint arXiv:1511.08308 (2015)

5. Derczynski, L., Nichols, E., van Erp, M., Limsopatham, N.: Results of the wnut2017 shared task on novel and emerging entity recognition. In: Proceedings of the 3rd Workshop on Noisy User-generated Text, pp. 140–147 (2017)
6. Derczynski, L., Ritter, A., Clark, S., Bontcheva, K.: Twitter part-of-speech tagging for all: overcoming sparse and noisy data. Proceedings of the International Conference Recent Advances in Natural Language Processing RANLP **2013**, 198–206 (2013)
7. Devlin, J., Chang, M.W., Lee, K., Toutanova, K.: Bert: Pre-training of deep bidirectional transformers for language understanding. In: Proceedings of the 2019 Conference of the North American Chapter of the Association for Computational Linguistics: Human Language Technologies,Long and Short Papers, vol. 1, pp. 4171–4186 (2019)
8. Dirkson, A., Verberne, S.: Transfer learning for health-related twitter data. In: Proceedings of the Fourth Social Media Mining for Health Applications (# SMM4H) Workshop & Shared Task, pp. 89–92 (2019)
9. Duong, L.: Natural language processing for resource-poor languages. Ph.D. thesis, University of Melbourne (2017)
10. Gimpel, K., et al.: Part-of-speech tagging for twitter: annotation, features, and experiments. In: Proceedings of the 49th Annual Meeting of the Association for Computational Linguistics: Human Language Technologies: short papers-Volume 2. pp. 42–47. Association for Computational Linguistics (2011)
11. Gui, T., Zhang, Q., Gong, J., Peng, M., Liang, D., Ding, K., Huang, X.J.: Transferring from formal newswire domain with hypernet for twitter pos tagging. In: Proceedings of the 2018 Conference on Empirical Methods in Natural Language Processing, pp. 2540–2549 (2018)
12. Gui, T., Zhang, Q., Huang, H., Peng, M., Huang, X.: Part-of-speech tagging for twitter with adversarial neural networks. In: Proceedings of the 2017 Conference on Empirical Methods in Natural Language Processing, pp. 2411–2420 (2017)
13. Howard, J., Ruder, S.: Universal language model fine-tuning for text classification. In: Proceedings of the 56th Annual Meeting of the Association for Computational Linguistics, Long Papers, vol. 1, pp. 328–339 (2018)
14. Jozefowicz, R., Vinyals, O., Schuster, M., Shazeer, N., Wu, Y.: Exploring the limits of language modeling. arXiv preprint arXiv:1602.02410 (2016)
15. Liu, Y., Zhu, Y., Che, W., Qin, B., Schneider, N., Smith, N.A.: Parsing tweets into universal dependencies. In: Proceedings of the 2018 Conference of the North American Chapter of the Association for Computational Linguistics: Human Language Technologies, Long Papers, vol. 1, pp. 965–975 (2018)
16. Ma, X., Hovy, E.: End-to-end sequence labeling via bi-directional LSTM-CNNs-CRF. In: Proceedings of the 54th Annual Meeting of the Association for Computational Linguistics, Long Papers, vol. 1, pp. 1064–1074 (2016)
17. Marcus, M., Santorini, B., Marcinkiewicz, M.A.: Building a large annotated corpus of English: the penn treebank. Comput. Linguist. **19**(2), 313–330 (1993)
18. März, L., Trautmann, D., Roth, B.: Domain adaptation for part-of-speech tagging of noisy user-generated text. In: Proceedings of the 2019 Conference of the North American Chapter of the Association for Computational Linguistics: Human Language Technologies, Long and Short Papers, vol. 1, pp. 3415–3420 (2019)
19. Meftah, S., Semmar, N.: A neural network model for part-of-speech tagging of social media texts. In: Proceedings of the Eleventh International Conference on Language Resources and Evaluation (LREC 2018) (2018)

20. Meftah, S., Semmar, N., Sadat, F., Raaijmakers, S.: Using neural transfer learning for morpho-syntactic tagging of South-Slavic Languages tweets. In: Proceedings of the Fifth Workshop on NLP for Similar Languages, Varieties and Dialects (VarDial 2018), pp. 235–243 (2018)

21. Meftah, S., Tamaazousti, Y., Semmar, N., Essafi, H., Sadat, F.: Joint learning of pre-trained and random units for domain adaptation in part-of-speech tagging. In: Proceedings of the 2019 Conference of the North American Chapter of the Association for Computational Linguistics: Human Language Technologies, Long and Short Papers, vol. 1, pp. 4107–4112 (2019)

22. Mou, L., et al.: How transferable are neural networks in NLP applications? In: Proceedings of the 2016 Conference on Empirical Methods in Natural Language Processing, pp. 479–489 (2016)

23. Nivre, J., et al.: Universal dependencies v1: a multilingual treebank collection. In: Proceedings of the Tenth International Conference on Language Resources and Evaluation (LREC 2016), pp. 1659–1666 (2016)

24. Oquab, M., Bottou, L., Laptev, I., Sivic, J.: Learning and transferring mid-level image representations using convolutional neural networks. In: Proceedings of the IEEE Conference on Computer Vision and Pattern Recognition, pp. 1717–1724 (2014)

25. Owoputi, O., O'Connor, B., Dyer, C., Gimpel, K., Schneider, N., Smith, N.A.: Improved part-of-speech tagging for online conversational text with word clusters. In: Proceedings of the 2013 Conference of the North American Chapter of the Association for Computational Linguistics: Human Language Technologies, pp. 380–390 (2013)

26. Pan, S.J., Yang, Q., et al.: A survey on transfer learning. IEEE Trans. Knowl. Data Eng. **22**(10), 1345–1359 (2010)

27. Pennington, J., Socher, R., Manning, C.: Glove: global vectors for word representation. In: Proceedings of the 2014 Conference on Empirical Methods in Natural Language Processing (EMNLP), pp. 1532–1543 (2014)

28. Peters, M.E., et al.: Deep contextualized word representations. In: Proceedings of NAACL-HLT, pp. 2227–2237 (2018)

29. Ritter, A., Clark, S., Etzioni, O., et al.: Named entity recognition in tweets: an experimental study. In: Proceedings of the Conference on Empirical Methods in Natural Language Processing, pp. 1524–1534. Association for Computational Linguistics (2011)

30. Rizoiu, M.A., Wang, T., Ferraro, G., Suominen, H.: Transfer learning for hate speech detection in social media. arXiv preprint arXiv:1906.03829 (2019)

31. Sang, E.F., De Meulder, F.: Introduction to the CoNLL-2003 shared task: language-independent named entity recognition. arXiv:cs/0306050 (2003)

32. Sang, T.K., Erik, F., Buchholz, S.: Introduction to the CoNLL-2000 shared task: chunking. In: Proceedings of CoNLL-2000, Lisbon, Portugal, pp. 127–132 (2000)

33. Toutanova, K., Klein, D., Manning, C.D., Singer, Y.: Feature-rich part-of-speech tagging with a cyclic dependency network. In: Proceedings of the 2003 Conference of the North American Chapter of the Association for Computational Linguistics on Human Language Technology, vol. 1, pp. 173–180. Association for Computational Linguistics (2003)

34. Yosinski, J., Clune, J., Bengio, Y., Lipson, H.: How transferable are features in deep neural networks? In: Advances in Neural Information Processing Systems, pp. 3320–3328 (2014)

Investigating the Lack of Consensus Among Sentiment Analysis Tools

Marco A. Palomino[1]([⊠]) [iD], Aditya Padmanabhan Varma[2],
Gowriprasad Kuruba Bedala[3], and Aidan Connelly[1]

[1] University of Plymouth, Plymouth, UK
{marco.palomino,aidan.connelly}@plymouth.ac.uk
[2] Vellore Institute of Technology, Chennai, India
adityapadmanabhan.2016@vitstudent.ac.in
[3] Saveetha Institute of Technical and Medical Sciences, Chennai, India
bedalagowriprasad16@saveetha.com

Abstract. Sentiment analysis, the classification of human emotion expressed in text, has the potential to enhance our ability to analyse the ever growing amount of information published each day on social media. Thus, we compare here seven of the most well-regarded sentiment analysis tools, and conclude that none of them is sufficiently reliable to be used on its own. Combining them and relying on their results only when various tools reach an agreement seems to be a better option. The pros and cons of such an approach are discussed in this paper, while providing recommendations related to the usability of the tools in question. Our work is of particular relevance to small and medium-sized enterprises (SMEs), which constitute a large and integral part of the economy. SMEs seem to be ideal candidates to turn data derived from sentiment analysis into business opportunities.

Keywords: Sentiment analysis · Twitter · Social media · scikit-learn · Sentiment140 · SentiStrength · uClassify · VADER · TextBlob · SMEs

1 Introduction

According to the *European Union*, and other international organisations, such as the World Bank and the United Nations, *small and medium-sized enterprises* (SMEs) are businesses whose personnel falls below 250 employees, and whose annual turnover does not exceed EUR 50 million [13]. In the UK, 5.8 million small businesses were in operation at the start of 2019 [47]. Indeed, small businesses accounted for 99.3% of all private sector businesses—these were 5.82 million businesses with 0 to 49 employees—and SMEs accounted for three fifths of the employment and around half of the turnover in the UK private sector at the start of 2019 [47]. Considering that the total employment in SMEs across the UK is currently 16.6 million, which equates to 60% of all private sector employment, supporting the needs of SMEs has become a critical issue.

© Springer Nature Switzerland AG 2020
Z. Vetulani et al. (Eds.): LTC 2017, LNAI 12598, pp. 58–72, 2020.
https://doi.org/10.1007/978-3-030-66527-2_5

The software designed to support the operation of SMEs is meant to help them to run operations, cut costs and replace paper processes [27]. We are particularly interested in the software choices available for a specific application that has been gaining interest and popularity: *sentiment analysis*, the process of computationally categorising opinions [14].

Sentiment analysis—sometimes known as *opinion mining*—aims to systematically identify, extract, quantify, and study opinions about specific topics, and attitudes towards particular entities [4,14]. Sentiment analysis has a great potential as a technology to enhance the capabilities of customer relationship management and recommendation systems—for example, showing which features customers are particularly happy about, or excluding from recommendations items that have received negative feedback. Sentiment analysis can also be exploited for troll-filtering and spam detection [5]. Intelligence applications able to monitor surges in hostile communications are examples of non-commercial systems employing sentiment analysis [18,29].

The basic tasks of sentiment analysis are *emotion recognition* [41] and *polarity detection* [21]. While the first task focuses on identifying a variety of emotional states, such as "anger", "sadness" and "happiness", the second one is either a binary classification task—whose outputs are 'positive' versus 'negative', 'thumbs up' versus 'thumbs down', or 'like' versus 'dislike'—or a ternary classification task—whose outputs are 'positive', 'neutral' or 'negative'. Several sentiment analysis tools have been developed lately—both Feldman [14] and Ribeiro *et al.* [39] claim that 7,000 articles on sentiment analysis had been written up by 2016, while dozens of start-ups are developing sentiment analysis solutions.

Despite the interest in the subject, it is still unclear which sentiment analysis tool is more adaptable to different domains, or cheaper and easier to manage. Therefore, the goal of this paper is to help SMEs to evaluate off-the-shelf tools for the purpose of sentiment analysis, and ascertain which tool is better for specific needs that businesses may encounter. Little is known about the relative performance of the various tools available [39]; thus, comparative studies such as this one are needed.

Our initial evaluation suggested that sentiment analysis can be severely biased, depending on which tool is used [8]. Consequently, we launched a larger investigation in 2020, where we have added new sentiment analysis tools to the analysis and used a much larger corpus as a testbed for our experimentation. We can now confirm that considering the consensus among a selection of tools is a better alternative than choosing one and using it in isolation.

The remainder of this paper is organised as follows: Sect. 2 introduces the corpora for our experiments—we have gathered two different datasets to compare sentiment analysis tools: one in 2017 and one in 2020. Section 3 describes the tools we have compared: *Sentiment140* [16], *SentiStrength* [9], *scikit-learn* [35], *TextBlob* [23], *Treebank* [46], *uClassify* [53] and *VADER* [37]. Section 4 presents the results yielded by the tools we compared and discusses our analysis. Finally, Sect. 5 offers our conclusions.

2 Experimental Corpus

While large companies can afford time and resources to look into the best sentiment analysis tools for their purposes—for example, IBM acquired *AlchemyAPI*
in 2015, before replacing it with the *Watson Natural Language Understanding
Service* [11]—most SMEs would find it unreasonable to invest significantly on
such an activity. Hence, we decided to launch an investigation of sentiment analysis tools in 2017, as a means to inform SMEs about the features, strengths and
drawbacks of popular off-the-shelf sentiment analysis tools.

As a testbed for our experiments, we have chosen *Twitter* [52], the microblog-
ging service that enables people to publish short messages—namely, *tweets*—
expressing interests and attitudes they are willing to share [3]. Twitter users
employ *hashtags*—words or phrases preceded by a hash sign '#'—to categorise
tweets topically, so that others can follow conversations on a particular topic. A
more detailed description of Twitter and its jargon can be found in [26].

Twitter is a valuable source of opinions and sentiments [31]—for example,
manufacturing companies are always interested in how positive or negative the
opinions tweeted about their products are. Companies across the world have
embraced Twitter as a powerful way to connect with their customers and grow
their businesses [7]. Twitter is now indispensable in marketing, sales and customer service. Thus, we have used Twitter for the evaluation of sentiment analysis tools since our first study in 2017.

In 2017, we worked with a corpus consisting of 40,912 tweets collected at
the beginning of the year, when people tend to make New Year resolutions.
Such resolutions are commonly associated with weight loss and dietary regimes.
Hence, this gave us an opportunity to monitor tweets related to nutritional,
detox and dietary products. We began the retrieval of the 2017 corpus on 26th
January 2017, and we ended it 20 d later—14th February 2017. To guarantee
that we gathered a good sample of tweets, a professional in the field provided
a list of hashtags and phrases relevant to the subject, which are displayed in
Table 1. Such hashtags and phrases captured conversations related to health and
disease connected with nutritional and dietary products. Table 1 also displays
the number of tweets we collected for each hashtag and phrase. While some of
the hashtags seem unintelligible to a layman, they are all sensible in the context
of dietary products. For example, *irritable bowel syndrome*—referred to by the
hashtag #IBS; see row 3 in Table 1—is a condition of the digestive system that is
frequently mentioned in dietary conversations. In fact, #IBS was the third most
popular hashtag in our 2017 corpus.

As explained in [8], the study we carried out in 2017 showed significant differences in the number of tweets classified as positive, negative or neutral, depending on the tool chosen for the classification. Figure 1 displays the polarity of the
tweets according to the different tools involved in the study. Such contrasting
results led us to undertake further investigation. We are currently studying a
greater number of sentiment analysis tools, and we are employing a much larger
corpus.

Table 1. Number of tweets per hashtag and phrase in the 2017 corpus.

Hashtag or Phrase	Number of tweets
#healthy #food	11,267
#cleaneating	7,853
#IBS	3,974
#foodallergy	3,817
#gluten	3,652
#superfoods	3,556
#lowfodmap	867
#fodmap	829
#natural #diet	546
detox diet nutrition	320
#detoxdiet	224
#diet #research	58
#lowgi	56
#nutraceutical	29
#medicalfood	19
#cleansing #diet	12
#diet #scam	7
food is your medicine	0

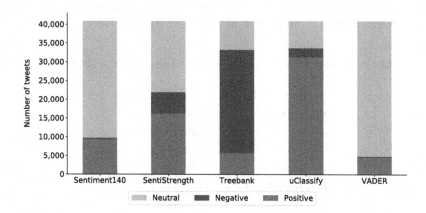

Fig. 1. Polarity per tool.

Our new 2020 corpus consists of 1,525,050 tweets gathered during 13 continuous hours, starting on Sunday 2nd February 2020 at 20:47:00 (GMT)—hereafter, all times are GMT times. The first tweet was captured at 20:47:03, and the last one at 09:25:08 on Monday 3rd March 2020. Our corpus consists of publicly available tweets referring to the *Super Bowl*, the annual championship game of the *National Football League* (NFL), which was played on 2nd February 2020.

We chose the Super Bowl as the subject of our corpus, because it is not only a sporting event, but also a marketing event. Many large firms air their commercial campaigns during the televised broadcast of the Super Bowl at great expense [24]. The Super Bowl is also an entertainment event, as internationally known artists perform during the Halftime Show. There were 32.2 million interactions across Facebook, Instagram and Twitter during the Halftime Show in 2019 [1].

We retrieved all the tweets in Plymouth (UK), where the game started at 23:30. We began the retrieval two hours before the game, because this allowed us to capture the start of the televised broadcast, when a spike of Twitter activity became evident. We collected the corpus using *Tweepy* [40], an open-source, Python library for retrieving tweets in real time. Tweepy makes it easier to use the *Twitter Streaming API* by handling authentication and connection [28,50].

Rather than retrieving tweets comprising a certain collection of hashtags, we looked for tweets containing specific keywords and phrases directly associated with the Super Bowl. Such keywords and phrases are displayed in Table 2, along with the number of tweets we collected for each of them. Note that the Twitter Streaming API is case insensitive, which guarantees the retrieval of any tweets containing the keywords in Table 2, regardless of case—for instance, the use of the keyword `superbowl` guarantees the retrieval of any tweets containing the terms `superbowl`, `SUPERBOWL`, `Superbowl`, `SuperBowl`, and any other possible case variation. The hashtag `#SuperBowl`, and all its case variations, are also retrieved by including the keyword `superbowl` in our study. Similarly, the phrase `American Football` guarantees the retrieval of any tweets including the terms `American` and `Football`, regardless of order and ignoring case.

Table 2. Number of tweets per keyword and phrase in the 2020 corpus.

Keywords and phrases	Number of tweets
`superbowl`	856,240
`nfl`	280,766
`football`	198,899
`touchdown`	37,421
`American Football`	4,478
`americanfootball`	257

The figures reported on Table 2 do not sum to give the total number of tweets available in the 2020 corpus. This is because there are many tweets which include two or more of the keywords listed in Table 2. Also, the text of some of the tweets in the 2020 corpus may not include explicit occurrences of the keywords and phrases listed in Table 2; yet, the Streaming API would provide us with such tweets if the keywords appear as part of URLs or metadata, such as user names, associated with those tweets [50]. A total of 252,678 tweets in the 2020 corpus fall into this case.

Figure 2 on page 6 shows the number of tweets captured per hour during the collection of our corpus—on average, we captured 117,311 tweets per hour. Each tweet was retrieved as a `status object`—in the context of Twitter development tools, tweets are also known as *status updates* [51]. The Streaming API provided the tweets and their corresponding metadata in Java Script Object Notation (JSON) format, and we produced a Python parser to extract the text of the tweets and other relevant information, such as the time when the tweets were published and the identifiers of the users who published those tweets. To store and manage the tweets that we collected, we uploaded them into a MySQL database, which we are using to analyse the corpus and generate statistics.

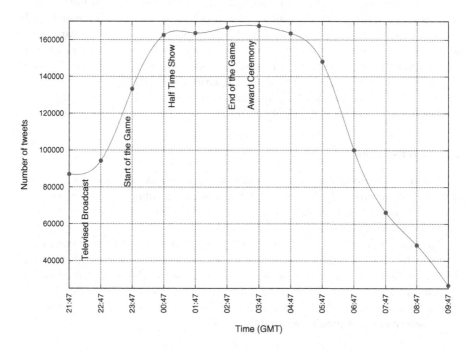

Fig. 2. Number of tweets per hour.

The Super Bowl Halftime Show began at around 01:00, which is a time when Fig. 2 reaches a peak in the number of tweets collected. Plenty of people engage in conversation on Twitter during the Halftime Show. Indeed, the pinnacle of Twitter activity in the 2019 Super Bowl was logged at 01:23, when 171,000 interactions were recorded in a single minute in reaction to the Halftime Show [1]. Between 02:47 and 03:47, this is during the last quarter of the game, we gathered the largest number of tweets in the 2020 corpus: 167,492. As indicated in Fig. 2, the volume of tweets started to decrease on 3rd March 2020 at 04:30, approximately—this is after the Award Ceremony had concluded.

3   Sentiment Analysis Tools

We can divide the main techniques used in sentiment analysis into *machine learning*, *lexicon-based* and *hybrid* techniques [21,32]. Whereas machine learning uses linguistic features, the lexicon-based techniques rely on a sentiment lexicon—a collection of known and pre-compiled sentiment terms. The lexicon-based techniques are separated into *dictionary-based* and *corpus-based* approaches, which use statistical or semantic methods to determine the sentiment expressed. Hybrid techniques combine both machine learning and lexicon-based approaches with sentiment lexicons playing a critical role [25].

Traditionally, product reviews have constituted the source of data for sentiment analysis. Product reviews are important to businesses, because they can make decisions based on the analysis of the opinions about their products. However, research looking into the sentiment analysis of tweets has been widely published recently [54]: Reis *et al.* used SentiStrength to measure the negative-ness or positive-ness of news headlines [12]; O'Connor *et al.* suggested that tweets with sentiment can potentially serve as votes and substitute traditional polling [30]; and Tamersoy *et al.* explored the utilisation of VADER's lexicon to study patterns of smoking and drinking abstinence in social media [17]. We will briefly outline below the main features of the tools chosen for our evaluation.

3.1   scikit-Learn

scikit-learn is a freely-available machine learning library for the Python programming language [34]. While scikit-learn does not offer specific support on sentiment analysis, it provides all that is needed to build a classifier capable of determining the polarity of tweets.

The main reason why we chose scikit-learn over other existing alternatives is that it focuses on making machine learning available to non-specialists. Good documentation and ease of use make scikit-learn approachable and powerful. It is ideal for SMEs, which require affordable software, but it is also amply used by multinationals, such as JPMorgan, which considers scikit-learn part of its toolkit for classification and predictive analytics [43].

3.2   Sentiment140

Sentiment140 [15], formerly known as *Twitter Sentiment*, started as a student project at Stanford University, where research in sentiment analysis used to focus on large pieces of text, as opposed to tweets, which are meant to be more casual and limited to 140 characters[1]. A key contribution made by Sentiment140 at the time of its creation was the use of machine learning classifiers, rather than the then traditional lexicon-based approach.

---

[1] The maximum length of a tweet used to be 140 characters. Although Twitter doubled its character length in 2017, only 1% of tweets reach the new 280-character limit, and only 12% of tweets are longer than 140 characters [20].

Given the wide range of topics discussed on Twitter, it would be too difficult to manually annotate sufficient data to train a sentiment classifier for all sorts of tweets; thus, the developers of Sentiment140 applied a technique called *distant supervision* [15], where the training data consists of tweets with emoticons. This approach was introduced by Read [38], and utilises the emoticons as "noisy" labels—for instance, :) in a tweet indicates that the tweet refers to a positive sentiment and :(indicates that the tweet expresses a negative sentiment.

Since it is relatively easy to extract several tweets containing emoticons, distant supervision is potentially a major improvement over the cost and resources that may otherwise be involved in hand-labelling training data.

3.3   SentiStrength

SentiStrength was specifically implemented to determine the strength of sentiment in informal English text, using methods to exploit the de-facto grammars and spelling styles of the informal communication that regularly takes place in social media, blogs and discussion forums [48]. Applied to *MySpace* comments, SentiStrength was able to predict positive emotion with 60.6% accuracy and negative emotion with 72.8% accuracy, both based upon numerical strength scales. SentiStrength's prediction of positive emotion has been found to be better than general machine learning approaches [49].

To assess the results of the different tools included in this paper on the same basis, we used SentiStrength as a trinary sentiment classification tool, which means that we employed it to identify the polarity of tweets as positive, negative or neutral, though SentiStrength can also work as a binary classification tool—positive or negative.

3.4   TextBlob

TextBlob is a Python library for processing text. It offers an API to perform a number of *natural language processing* (NLP) tasks, such as noun phrase extraction, language translation and spelling correction [22]. While the most commonly known Python library for NLP is the *Natural Language Toolkit* (NLTK) [2], we favoured the selection of TextBlob in our study because it is simpler and more user-friendly than the NLTK.

With respect to sentiment analysis, TextBlob provides two options for polarity detection: `PatternAnalyzer`, which is based on the data mining *Pattern* library developed by the *Centre for Computational Linguistics and Psycholinguistics* (CLiPS) [10], and a `NaiveBayesAnalyzer` classifier, which is an NLTK classifier trained on movie reviews [36].

The default option for sentiment analysis in TextBlob is `PatternAnalyzer`, and that is precisely the option we favoured, because we are not working with movie reviews, which is the specialty of the `NaiveBayesAnalyzer` classifier. We may consider the use of the `NaiveBayesAnalyzer` classifier in the future, provided we can train it suitably for the domain of our corpus.

3.5   Treebank

Most lexicon-based sentiment analysis tools work by looking at words in isolation: giving positive points for positive words, negative points for negative words,
and then summing up those points. Hence, the order of the words that compose a
sentence is ignored in such tools. In contrast, Treebank, the deep learning tool for
sentiment analysis developed at Stanford University, builds up a representation
based on the structure of the sentences [45].

Roughly speaking, Stanford University's deep learning model computes sentiment based on how words contribute to the meaning of longer phrases. The
underlying technology is based on a new type of recursive neural network that
is built on top of grammatical structures.

3.6   uClassify

uClassify was launched as a Web service in 2008, by a group of machine learning enthusiasts based in Stockholm [53]. Developers can utilise such a service to
create text classifiers for various tasks, such as sentiment analysis and language
detection. The uClassify sentiment classifier is trained on a corpus of 2.8 million entries comprising tweets, Amazon product evaluations and movie reviews.
Hence, it can cope with both short and long texts–including tweets, Facebook
statuses, blog posts and product reviews.

The uClassify API can serve a maximum of 500 requests for free on a daily
basis [53]. Therefore, we would have needed several days to test uClassify with
the 2020 corpus. However, the providers of this API service kindly permitted us
to undertake the whole testing at once, by granting us an academic license for a
limited period of time [19].

3.7   VADER

VADER, *Valence Aware Dictionary and sEntiment Reasoner*, is a rule-based tool
that is specifically adapted to identify sentiments expressed in social media [17].
Using a combination of qualitative and quantitative methods, the developers of
VADER built a gold-standard list of lexical features, along with their associated
sentiment intensity measures. Such features are combined with consideration
for five general rules, comprising grammatical and syntactical conventions for
expressing and emphasising sentiment intensity.

The simplicity of VADER carries several advantages. First, it is both fast
and computationally economical. Second, the lexicon and rules used by VADER
are available to everyone [17]—they are not hidden within a black-box.

By exposing both the lexicon and rule-based model, VADER makes the inner
workings of its sentiment analysis engine accessible—and thus, interpretable—to
a broader audience beyond the scientific community.

4 Results

Figures 3, 4, 5, 6, 7, 8, 9, 10 to 10 display how the polarity of the tweets in the 2020 corpus evolved per hour, according to the different tools described in Sect. 3.

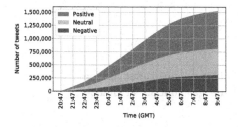

Fig. 3. scikit-learn

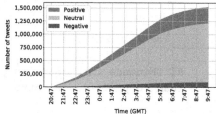

Fig. 4. Sentiment140

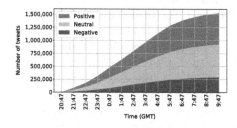

Fig. 5. SentiStrength

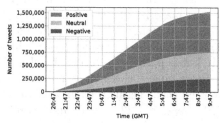

Fig. 6. TextBlob

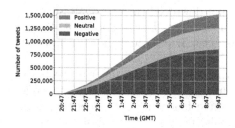

Fig. 7. Treebank

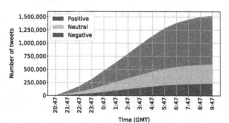

Fig. 8. uClassify

We start the analysis with Sentiment140, because it categorised as neutral more tweets than any other tool in our previous (2017) and current (2020) studies. The developers of Sentiment140 use the following litmus test to determine the polarity of a tweet: "if a tweet could ever appear as a front-page newspaper headline, or as a sentence in Wikipedia, then it is neutral" [15]. For example, the following tweet is considered neutral, because it could have been a newspaper headline, though it projects an overall negative feeling about General Motors:

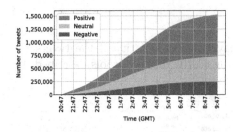
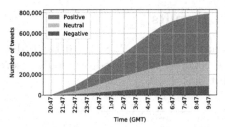

Fig. 9. VADER   **Fig. 10.** Agreement

```
RT @Finance Info Bankruptcy filing could put GM on
road to profits (AP) http://cli.gs/9ua6Sb #Finance
```

The original training and test set used to develop Sentiment140 did not comprise neutral tweets—only positive and negative [15]. While not having a neutral class may have contributed to the success of the tool in previous research, it is clearly a limitation. We think this is the reason why Sentiment140 concludes the vast majority of our tweets are neutral. Indeed, Sentiment140 concludes 73% of the 2020 corpus—that is, 1,120,405 tweets—is neutral. This is so different to the results yielded by other tools that we recommend to employ Sentiment140 only if it is possible to retrain its classifier.

Retraining a classifier is precisely what we did in the case of scikit-learn. We used scikit-learn to train a linear classifier with *stochastic gradient descent* (SGD) learning [42]. To obtain suitable training and test sets, we employed a corpus of tweets gathered by Sinha *et al.* [44], which includes 290,879 tweets published by the general public at the end of NFL regular season games played in 2012. We refer to this corpus as the *2012 postgame corpus*.

Although the tweet identifiers for the 2012 postgame corpus are available at www.ark.cs.cmu.edu/football, we could not download the text of all them, as some of those tweets have already been removed from Twitter, or are no longer public. Thus, the total number of tweets we actually used to train our sentiment classifier was 100,996.

To train our classifier, we determined the polarity of the 100,996 tweets we were able to download, and then check which of them have exactly the same polarity according to two or more of the three following tools: SentiStrength, TextBlob and VADER. We chose these tools, because our analysis identified them as the ones with the greatest amount of consensus among them. SentiStrength, TextBlob and VADER agree on the classification of 52% of the 2020 corpus—that is, 790,529 tweets. Figure 10 shows how the agreement on tweet polarity changed per hour among SentiStrength, TextBlob and VADER during the retrieval of the 2020 corpus.

From the 100,996 tweets we were able to download from the 2012 postgame corpus, two or more of the chosen tools—SentiStrength, TextBlob and VADER—agree on the classification of 86,278 tweets. A total of 75% of these 86,278 tweets became our training set, whereas the remaining 25% became the test set. Our classifier achieved 90.48% accuracy, and when we used it to determine the polarity of the entire 2020 corpus, it reached a 50.23% consensus with SentiStrength, TextBlob and VADER.

Treebank returned the largest number of negative tweets in both our previous (2017) and current (2020) study. Treebank computes the sentiment based on how words compose the meaning of longer phrases [45]. For instance, Treebank recognises words such as "funny" and "witty" as positive; yet, a sentence can still be negative, regardless of the presence of positive words. Hence, the following sentence is classified as negative overall, despite the occurrences of the words "funny" and "witty" in the text:

`This movie was actually neither that funny, nor super witty.`

Treebank was trained using 11,855 sentences extracted from movie reviews [45]. While movie reviews are widely used in sentiment analysis [33], they form part of a completely different domain, which is beyond the scope of our corpus. Therefore, we plan to conduct further investigation on Treebank, but we will do it after retraining it with a corpus which is closer to the domain of our study: Twitter and social media in general.

Table 3 displays the consensus between any pair of tools. For example, the cell corresponding to Sentiment140 and SentiStrength indicates the percentage of tweets classified with the same polarity by both Sentiment140 and SentiStrength.

Table 3. Consensus between any pair of tools.

	scikit-learn	Sentiment140	SentiStrength	TextBlob	Treebank	uClassify	VADER
scikit-learn	--	49.01%	68.31%	81.95%	40.02%	52.25%	72.31%
Sentiment140	49.01%	--	53%	46.71%	33.42%	39.90%	48.08%
SentiStrength	68.31%	53%	--	59.70%	38.58%	49.40%	70.71%
TextBlob	81.95%	46.71%	59.70%	--	37.37%	51.96%	68.55%
Treebank	40.02%	33.42%	38.58%	37.37%	--	33.28%	40.42%
uClassify	52.25%	39.90%	49.40%	51.96%	33.28%	--	54.50%
VADER	72.31%	48.08%	70.71%	68.55%	40.42%	54.50%	--

5 Conclusions

In a landscape where little is known about the relative performance of the various sentiment analysis tools available [39], we have presented a study that aims at comparing and contrasting a selection of well-known tools. Our work, based on two different studies carried out using different corpora, reveals that the choice of sentiment analysis tool has a considerable impact on the evaluation of a corpus. Consensus among certain tools is so small that the analysis of the sentiment expressed in Twitter can be severely biased, depending on which tool is used.

We suggest considering the consensus among a number of tools as a better alternative than choosing one tool and using it in isolation.

Although we started our research largely interested in supporting the needs of SMEs, our work is also of relevance to the scientific community and anyone involved in building applications using the tools discussed here.

While research on sentiment analysis continues to make progress, it remains evident that further investigation is still necessary, especially given the number of NLP problems that need to be solved first to achieve human-like performance in sentiment analysis [6]—namely, word-sense disambiguation, anaphora resolution, sarcasm detection and metaphor understanding, among others.

Acknowledgements. The authors gratefully acknowledge the free academic license provided by *uClassify* to support their experiments. We are thankful to Martin Lavelle for reading our manuscript and providing insightful comments.

References

1. Adgate, B.: What you should know about super bowl LIV advertising and broadcast. Forbes (2020)
2. Bird, S., Klein, E., Loper, E.: Natural Language Processing with Python: Analyzing Text with the Natural Language Toolkit. O'Reilly Media Inc., Sebastopol (2009)
3. Boyd, D., Golder, S., Lotan, G.: Tweet, Tweet, Retweet: Conversational Aspects of Retweeting on Twitter. In: Proceedings of the 43rd Hawaii International Conference on System Sciences (HICSS), 2010, pp. 1–10. IEEE, Honolulu, HI (2010)
4. Cambria, E.: Affective computing and sentiment analysis. IEEE Intell. Syst. **31**(2), 102–107 (2016)
5. Cambria, E., Das, D., Bandyopadhyay, S., Feraco, A.: Affective computing and sentiment analysis. In: Ahmad, K. (ed.) A Practical Guide to Sentiment Analysis, vol. 45, pp. 1–10. Springer, Dordrecht (2017). https://doi.org/10.1007/978-94-007-1757-2
6. Cambria, E., Poria, S., Gelbukh, A., Thelwall, M.: Sentiment analysis is a big suitcase. IEEE Intell. Syst. **32**(6), 74–80 (2017)
7. Collins, B.: More than 80% of SMEs Recommend Twitter for Business. Twitter Blog (2014)
8. Connelly, A., Kuri, V., Palomino, M.: Lack of consensus among sentiment analysis tools: a suitability study for SME firms. In: Proceedings of the 8th Language and Technology Conference, pp. 54–8. Poznań, Poland (2017)
9. CyberEmotions: SentiStrength (2020). http://sentistrength.wlv.ac.uk/
10. De Smedt, T., Daelemans, W.: Pattern for python. J. Mach. Learn. Res. **13**(66), 2063–2067 (2012)
11. Devarajan, D.: Retirement of AlchemyAPI service (2017). https://www.ibm.com/cloud/blog/announcements/bye-bye-alchemyapi
12. Dos Rieis, J.C.S., de Souza, F.B., de Melo, P.O.S.V., Prates, R.O., Kwak, H., An, J.: Breaking the news: first impressions matter on online news. In: 9th International AAAI Conference on Web and Social Media, pp. 357–366. The AAAI Press, Oxford, UK (2015)

13. European Commission: Commission Recommendation of 6 May 2003 concerning the Definition of Micro, Small and Medium-Sized Enterprises. Official Journal of the European Union, pp. 36–41 (2003). https://eur-lex.europa.eu/legal-content/EN/ALL/?uri=CELEX:32003H0361
14. Feldman, R.: Techniques and applications for sentiment analysis. Commun. ACM **56**(4), 82–89 (2013)
15. Go, A., Bhayani, R., Huang, L.: Twitter Sentiment Classification Using Distant Supervision. CS224N Project report, Stanford 1(12) (2009)
16. Go, A., Bhayani, R., Huang, L.: Sentiment140 (2020). http://www.sentiment140.com/
17. Hutto, C.J., Gilbert, E.: VADER: a parsimonious rule-based model for sentiment analysis of social media text. In: 8th International AAAI Conference on Weblogs and Social Media, pp. 216–225. The AAAI Press, Ann Arbor, Michigan (2014)
18. Keyvanpour, M.R., Javideh, M., Ebrahimi, M.R.: Detecting and investigating crime by means of data mining: a general crime matching framework. Proc. Comput. Sci. **3**, 872–880 (2011)
19. Kågström, J.: RE: Academic Licence (Personal Communication) – E-Mail (2020)
20. Lee, K.: The Proven Ideal Length of Every Tweet, Facebook Post, and Headline Online, pp. 1–14 . Fast Company (2014)
21. Liu, B.: Sentiment Analysis and Opinion Mining. Synthesis Lectures on Human Language Technologies, vol. 5, pp. 1–167. Morgan & Claypool, San Rafael (2012)
22. Loria, S.: TextBlob Documentation. Release 0.15 2 (2018)
23. Loria, S.: TextBlob: Simplified Text Processing (2020). https://textblob.readthedocs.io/en/dev/index.html
24. Matheson, V.A., Baade, R.A.: Padding required: assessing the economic impact of the super bowl. Eur. Sport Manage. Q. **6**(4), 353–374 (2006)
25. Medhat, W., Hassan, A., Korashy, H.: Sentiment analysis algorithms and applications: a survey. Ain Shams Eng. J. **5**(4), 1093–1113 (2014)
26. Milstein, S., O'Reilly, T.: The Twitter Book. O'Reilly Media, Sebastopol (2009)
27. Mohamed, A.: The Best Software for Small Businesses (SMEs) - Essential Guide. Computer Weekly (2009). https://www.computerweekly.com/feature/The-best-software-for-small-businesses-SMEs-Essential-Guide. TechTarget
28. Morstatter, F., Pfeffer, J., Liu, H., Carley, K.M.: Is the sample good enough? comparing data from twitter's streaming API with twitter's Firehose. In: 7th International AAAI Conference on Weblogs and Social Media, pp. 400–408. The AAAI Press, Cambridge, MA (2013)
29. Nath, S.V.: Crime data mining. In: Elleithy, K. (ed.) Advances and Innovations in Systems, Computing Sciences and Software Engineering, vol. xxx, pp. 405–409. Springer, Dordrecht (2007). https://doi.org/10.1007/978-1-4020-6264-3_70
30. O'Connor, B., Balasubramanyan, R., Routledge, B.R., Smith, N.A.: From tweets to polls: linking text sentiment to public opinion time series. In: 4th International AAAI Conference on Weblogs and Social Media, pp. 122–129. Washington, DC (2010)
31. Pak, A., Paroubek, P.: Twitter as a corpus for sentiment analysis and opinion mining. In: International Conference on Language Resources and Evaluation (LREC), pp. 1320–1326 (2010)
32. Pang, B., Lee, L., et al.: Opinion mining and sentiment analysis. Found. Trends ® Inf. Retrieval **2**(1–2), 1–135 (2008)
33. Parkhe, V., Biswas, B.: Sentiment analysis of movie reviews: finding the most important movie aspects using driving factors. Soft. Comput. **20**(9), 3373–3379 (2016)

34. Pedregosa, F., et al.: Scikit-learn: machine learning in python. J. Mach. Learn. Res. **12**, 2825–2830 (2011)
35. Pedregosa, et al.: scikit-learn: machine learning in python (2020). https://scikit-learn.org/stable/index.html
36. Perkins, J.: Python 3 Text Processing with NLTK 3 Cookbook. Packt Publishing Ltd. (2014)
37. Połtyn, M.: VADER Sentiment Analysis (2020). https://pypi.org/project/vader-sentiment/
38. Read, J.: Using emoticons to reduce dependency in machine learning techniques for sentiment classification. In: Proceedings of the ACL Student Research Workshop, pp. 43–48. Association for Computational Linguistics, Ann Arbor, MI (2005)
39. Ribeiro, F.N., Araújo, M., Gonçalves, P., Gonçalves, M.A., Benevenuto, F.: Sentibench—a benchmark comparison of state-of-the-practice sentiment analysis methods. EPJ Data Sci. **5**(1), 1–29 (2016)
40. Roesslein, J.: Tweepy Documentation (2020). http://docs.tweepy.org/en/v3.5.0/
41. Schuller, B., Batliner, A., Steidl, S., Seppi, D.: Recognising realistic emotions and affect in speech: state of the art and lessons learnt from the first challenge. Speech Commun. **53**(9–10), 1062–1087 (2011)
42. scikit-learn Developers: sklearn.linear_model.SGDClassifier (2019). `https://scikit-learn.org/0.15/modules/generated/sklearn.linear_model.SGDClassifier.html`
43. scikit-learn developers: Who is using scikit-learn? (2019). https://scikit-learn.org/stable/testimonials/testimonials.html
44. Sinha, S., Dyer, C., Gimpel, K., Smith, N.A.: Predicting the NFL using twitter. In: ECML/PKDD Workshop on Machine Learning and Data Mining for Sports Analytics. Prague, Czech Republic (2013)
45. Socher, R., Perelygin, A., Wu, J., Chuang, J., Manning, C.D., Ng, A.Y., Potts, C.: Recursive deep models for semantic compositionality over a sentiment treebank. In: Conference on Empirical Methods in Natural Language Processing, pp. 1631–1642. Association for Computational Linguistics, Seattle, WA (2013)
46. Stanford NLP Group: Sentiment Treebank (2020). https://nlp.stanford.edu/sentiment/treebank.html
47. The Federation of Small Businesses (FSB): UK Small Business Statistics (2020). https://www.fsb.org.uk/uk-small-business-statistics.html
48. Thelwall, M., Buckley, K., Paltoglou, G.: Sentiment strength detection for the social web. J. Am. Soc. Inf. Sci. Technol. **63**(1), 163–173 (2012)
49. Thelwall, M., Buckley, K., Paltoglou, G., Cai, D., Kappas, A.: Sentiment strength detection in short informal text. J. Am. Soc. Inf. Sci. Technol. **61**(12), 2544–2558 (2010)
50. Twitter Inc.: Filter Realtime Tweets (2020). https://developer.twitter.com/en/docs/tweets/filter-realtime/overview
51. Twitter Inc.: Tweet Object (2020). https://developer.twitter.com/en/docs/tweets/data-dictionary/overview/tweet-object
52. Twitter Inc.: Twitter. It's what's happening (2020). https://twitter.com
53. uClassify: uClassify – Free Text Classification (2020). https://www.uclassify.com/
54. Zhou, X., Tao, X., Yong, J., Yang, Z.: Sentiment analysis on tweets for social events. In: 17th International Conference on Computer Supported Cooperative Work in Design (CSCWD), pp. 557–562. IEEE, Whistler, Canada (2013)

Automated Normalization and Analysis
of Historical Texts

Paweł Skórzewski$^{(\boxtimes)}$ ⃝, Krzysztof Jassem⃝, and Filip Graliński⃝

Adam Mickiewicz University, Poznań, Poland
{pawel.skorzewski,jassem,filipg}@amu.edu.pl

Abstract. The paper presents an original method for processing historical texts. A historical text is converted into its modernized equivalent by a tool called diachronic normalizer, embedded into a linguistic toolkit. The solution has a few merits. Firstly, the toolkit architecture allows for imposing the morphological constraints on diachronization rules. Secondly, the diachronic normalizer may be launched in the pipeline together with other NLP tools, such as parsers or translators. Lastly, the toolkit makes it possible to efficiently apply, in the diachronic normalization, a long list of diachronic pairs, found out with the aid of word distribution vectors in historical corpora.

Keywords: Natural language processing · Diachronic normalization · NLP toolkits

1 Introduction

Historical texts are becoming available in the digitized form, which opens up opportunities to analyze them by means of natural language processing tools, like POS-taggers [22] or named entity recognizers [17]. Some aspects of creating tools dedicated for processing historical texts are discussed in [20]. However, it is not always necessary to create dedicated tools for this purpose – historical texts may also be successfully treated with NLP tools designed for contemporary languages: a tool may be re-trained on historical text, or a historical document may be pre-processed by transforming words to their contemporary forms. In our approach we use the last method, which we call diachronic normalization.

As stated in [26], text normalization remains one of few text processing tasks that still achieves better results with rule-based approach than with machine learning. This is due to the lack of training resources and the sparsity of tokens that undergo changes. The normalization rules may be hand-crafted or acquired from text corpora semi-automatically. In our experiment we combine both methods.

A standard procedure for automatized text analysis applies modular approach, where the disambiguated output of one tool (e.g. POS-tagger) becomes the input of another (e.g. syntax analyzer). Diachronic normalization does not fully submit to such treatment, as in some cases disambiguation decisions need to

Z. Vetulani et al. (Eds.): LTC 2017, LNAI 12598, pp. 73–86, 2020.
https://doi.org/10.1007/978-3-030-66527-2_6

be postponed to later stages. An example for the Polish language is the obsolete joint spelling of the particle *nie* ('not') with verbs, which should be normalized to contemporary disjoint spelling. The decision whether to separate the prefix *nie* depends on the morphological characteristics of the remaining word (is it a verb?), which may be verified only **after** the separation. The process is illustrated in Table 1, where the treatment of the word *niemódz* (anachronistic spelling of 'not to be able to') is contrasted to that of the word *niedźwiedź* ('a bear').

Table 1. Delayed decision in diachronic normalization

Input	Spelling change	Separation	Confirmation	Accept change?
niedźwiedź	*niedźwiedź*	↗ *nie dźwiedź* ↘ *niedźwiedź*	is *dźwiedź* a verb?	No. Reject: *nie dźwiedź* Accept: **niedźwiedź**
niemódz	*niemóc*	↗ *nie móc* ↘ *niemóc*	is *móc* a verb?	Yes. Accept: **nie móc** Reject: *niemóc*

The paper proposes a solution to the task – the diachronic normalization is inserted to a pipeline of NLP tools. Such an approach makes it possible to overcome difficulties exemplified above.

The paper is structured as follows: In Sect. 2, we give the historical insight into diachronic normalization and sketch the methods applied in our solution. In Sect. 3 we sketch the semi-automated method that searches for diachronic variants in historical corpora. Section 4 describes the insertion of the diachronic normalizer into an NLP toolkit. In Sect. 5 we evaluate the efficiency of our solution. In Sect. 6, we give a recipe on how to apply the presented method to customized text normalization (not necessarily diachronic normalization).

2 Diachronic Normalization of Historical Texts

The first attempts at rule-based diachronic normalization used for historical text in English were described by [2,22]. Similar studies were conducted for German [1]. There, context rules operated at the level of letters instead of words. The normalization rules may be derived from corpora, as [3,10] showed for German. Diachronic normalization may be also performed using a noisy channel model, as described by [18] on the example of Old Hungarian texts. The research on diachronic normalization was also conducted for Swedish [19], Slovene [24], Spanish [21] and Basque [5].

Normalization of Polish historic texts was first tackled by Waszczuk in 2012[1] for the sake of the SYNAT project[2] (a reference to the SYNAT project may be found in [16], where the authors report on the paper-into-electronic conversion

---

[1] https://hackage.haskell.org/package/hist-pl-transliter.
[2] http://synat.nlp.ipipan.waw.pl.

of an Old Polish dictionary). The solution, called *hist-pl-transliter*, is designed for the creation of transliteration rules, which allow for the usage of character classes within context constraints. The rule set is not large and consists of rules for transliteration of diacritic letters, and a few context-dependent rules.

The most recent effort in the area has been carried out within the KORBA project[3] [4]. The task consists in the transliteration of Polish texts originating from the 17th and 18th centuries. The solution uses the Ameba Supertool[4], which applies a large set of context-dependent rules defined by means of regex-based formalism.

Our approach differs from the two previous efforts for Polish in the computational aspect – it applies finite-state transducers, as well in the linguistic background – the set of rules for diachronic normalization is based on the historical description of Polish orthographic changes gathered in [13]. We have designed a machine-readable, Thrax-compliant [23,27] rule formalism, into which we rewrote the human-readable rules described by Malinowski [11]. Details on the formalism may be found in [11]. An example rule is given here:

```
Infinitive_z = ("e" | "ó") ("dz" : "c");

Rule1 = CDRewrite[Infinitive_z,
                  NonEmptyString,
                  End,
                  Any*];
```

The first line of the above code is the declaration of a transducer `Infinitive_z` that replaces *dz* with *c* if it follows *e* or *ó*. The second line is the declaration of transducer `Rule1` that performs a context-dependent rewrite (`CDRewrite`) using the transducer `Infinitive_z` on any location in the text where the left-handed constraint (non-empty string) and the right-handed constraint (end of the word) are satisfied. The rule has been created to change the infinitive forms of verbs such as *biedz* or *módz* into their contemporary forms: *biec* ('to run'), *móc* ('to be able to'). We named our solution `iayko`, for an archaic word *iayko*, whose contemporary form is *jajko* ('an egg').

3   Searching for Diachronic Spelling Variants

3.1   Mining for Diachronic Pairs

The idea (described at length in [8]) consists in applying the Word2vec model [14] on the Odkrywka[5] corpus. For a given lemma w we assume that its potential diachronic spelling variant v:

---

[3] http://clip.ipipan.waw.pl/KORBA.

[4] https://bitbucket.org/jsbien/pol.

[5] Odkrywka [9], contains 40 billion tokens and consists of Polish publications (mostly newspapers and books) originating mainly from the years 1810–2013.

- is similar to w according to the Word2vec model,
- is close to w orthographically,
- does not belong to a contemporary vocabulary,
- and is unlikely to result from erroneous OCR processing.

The fully-automatic approach yielded over 42 000 diachronic pairs (differing by exactly one edit) resulting from close to 1800 edit types. After the deletion of less frequent edit types and the human inspection of the remaining variants 5729 diachronic lemma pairs were selected as the result of the experiment. The automated generation of all inflected forms from the lemmas resulted in 104 064 pairs of inflected forms.

The procedure for searching diachronic pairs that differ by more than one edit required the development of a special iteration mechanism. This resulted in another 461 lemma pairs (7127 inflected forms).

3.2   Mining Sub-word Diachronic Variants

We used a similar idea in searching for sub-word diachronic variants, i.e. character n-grams which are distinctive enough to be safely substituted with their modern counterparts within any word. For instance, *hypno* could be substituted by *hipno* as no modern Polish word has the former prefix/infix.

This idea resulted in 654 sub-word spelling variants. The variants have been manually inspected in order to improve the hand-crafted normalization rules. Two types of adjustments were made:

- new rules were defined,
- more strict constraints were put on existing rules.

Here is a new rule, defined from a sub-word spelling variant:

```
Replace_gs_with_s = (a | e | u) ("rgsk" : "rsk");
```

```
Rule2 = (AnyString
         Replace_gs_with_s
         AnyString);
```

Rule2 deletes the consonant g inside some adjectives referring to cities, e.g. *hamburgski* → *hamburski*, *petersburgski* → *petersburski*, *norymbergski* → *norymberski*.

Rule3 and Rule4 emerged as a result of precising the contexts for the replacement $y \rightarrow j$. Before the adjustment the rule had indicated the replacement between any vowel and any consonant (e.g. *bayka* → *bajka*). After the adjustment the replacement was both expanded and limited to specific contexts:

```
LeftContext_y_j = ( "ac" | "as" | "az" | "dza" |
                    "ec" | "es" | "ez" |
                    "ic" | "is" | "iz" |
                    ...
                  );
```

```
Replace_y_with_j = "y" : "j";

Rule3 = CDRewrite[Replace_y_with_i,
                  LeftContext_y_j,
                  Vowel,
                  Any*];
```

Rule3 (only a part of the rule is listed here) expanded the replacement to some contexts between a consonant and a vowel.

```
Rule4 = AnyString (
    "a" Replace_y_with_j ("d"|"k"|"s") |
    "c" Replace_y_with_j ("al"|"an"|"en"|"ol"|"on"|"ow") |
    "e" Replace_y_with_j ("d"|"k"|"s") |
    ...
    ) AnyString;
```

Rule4 (only a part of the rule is listed here) limited the replacement between a vowel and a consonant to specific contexts – confirmed by sub-word spelling variants.

3.3 PSI-Toolkit

A distinct feature of the PSI-Toolkit is that all its processors operate on a common lattice-based data structure, called PSI-lattice. As described in [6] PSI-lattice is defined as a graph where vertices represent the intra-character points in input string and edges represent input characters and subsequent annotations. This feature has proved crucial for our approach to diachronic normalization.

The current list of PSI-Toolkit processors is available on the framework's documentation page[6]. The list consists of data readers, text segmenters, lemmatizers, POS-taggers, bilingual lexicons, syntactical parsers, machine translators and data writers in various formats (both textual and graphical).

The PSI-lattice allows for easy extension of the toolkit. For example, some external processors, such as Link Grammar parser [25] or Morfologik lemmatizer[7] [15, 28] have been incorporated. This feature facilitated the embedding of our new tool, diachronic normalizer, into the toolkit.

PSI-Toolkit allows users to provide custom linguistic resources. This will enable to apply our solution for normalization tasks, other than diachronic normalization.

A PSI-Toolkit command may be formed as a pipeline of tools. A pipeline of processors may be shortened by means of aliases – one-word alternative names for processor sequences.

PSI-Toolkit commands may be called from the command line in Linux-type operating systems. In this scenario, PSI-tools may be pipe-lined with Linux filtering programs.

---

[6] http://psi-toolkit.wmi.amu.edu.pl/help/documentation.html.
[7] http://morfologik.blogspot.com.

4   Embedding the Diachronic Normalizer into PSI-Toolkit

It is our intention to allow users create customized normalizers, either based on our set of rules, or totally independent, possibly intended for other normalization tasks.

4.1   Docker Technology

In order to improve the system portability we applied a container technology, called Docker[8]. A container is a virtualization engine that allows for running processes in an isolated environment. In contrast to virtual machines, containers do not contain a full operating system, but only libraries and settings required for the process to run. As containers can share operating system components (such as the kernel), they are more lightweight than virtual machines. The Docker's infrastructure makes PSI-Toolkit portable and flexible.

4.2   Running PSI-Toolkit with Docker

To install Docker under Ubuntu Linux distribution, you should follow the instructions provided on the project's website[9].

After Docker's installation, the dockerized PSI-Toolkit can be used as if it were installed and configured on user's machine. Docker will download the relevant container if not found in the local Docker cache and run its contents (the first attempt may take some time because Docker needs to download the container image):

```
echo "text to process" | docker run -i skorzewski/psi-toolkit
    psi_toolkit_pipeline
```

Here is an example of a command that uses our diachronic normalization tool, iayko.

```
echo 'iayko czy nie yaiko' | docker run -i
    skorzewski/psi-toolkit iayko
```

As a result, the normalized (contemporary) text will appear:

```
jajko czy nie jajko
```

Docker can also be installed on Microsoft Windows 10 Professional or Enterprise 64-bit. The installer and the installation instructions can be found at the project's website[10].

4.3   Examples of Usage

Here are some examples of a basic usage of iayko:

A historic text is normalized to its modern version, using the default set of finite-state rules:

---

[8] https://www.docker.com.
[9] https://docs.docker.com/engine/installation/linux/docker-ce/ubuntu.
[10] https://store.docker.com/editions/community/docker-ce-desktop-windows.

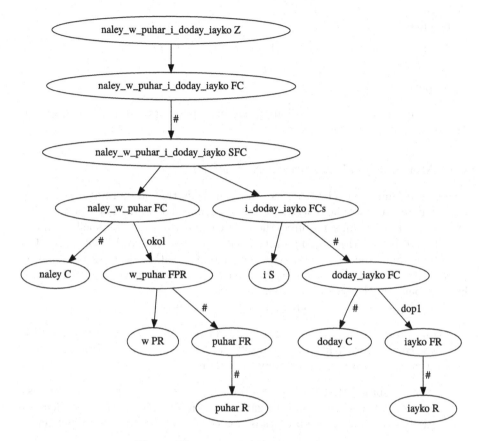

Fig. 1. The parse tree for a historic text

- pipeline:

  ```
  iayko --lang pl
  ```

- input: *naley w puhar i doday iayko*
- output: *nalej w puchar i dodaj jajko*

Normalized text may be used as an input for further processing, e.g. machine translation:

- pipeline:

  ```
  iayko --lang pl ! parse ! bilexicon --lang pl
    --trg-lang en ! transferer --lang pl --trg-lang en
  ```

- input: *naley w puhar i doday iayko*
- output: *pour in the cup and add the egg*

The normalization is used to obtain the parse tree for a historic text:

- pipeline:

```
iayko --lang pl ! parse ! draw-parse-tree
```

- input: *naley w puhar i doday iayko*
- output: see Fig. 1

A comprehensive set of various usage examples may also be found at the processor's documentation site[11].

4.4 Morphological Diachronic Normalizer

One of the limitations of `iayko` as a normalizer is that its normalization rules are indifferent to morphological features of normalized words. For example, the archaic rules of spelling required the particle *nie* ('not') to be spelled jointly with the verb, and the contemporary spelling of *nie* with verb is separate. On the other hand, the contemporary spelling of *nie* with adjectives or nouns is joint. So the normalizer should insert a space between the particle and the verb, but not between the particle and an adjective.

To address this issue, we created two new processors:

- a conditional diachronic normalizer, called `niema` (for an archaic word *niema*, whose contemporary form is *nie ma* 'there is no'),
- a selector annotator, called simply `selector`.

The lattice-based structure of PSI-Toolkit allows us to perform the process of morphological diachronic normalization by simply adding relevant edges to the initial lattice. The whole process is carried out by the following pipeline:

```
iayko ! niema ! lemmatizer ! selector
```

There is an alias `diachronizer` provided for this pipeline to facilitate its use.

Let's walk through all four steps of the pipeline on the example of input text *niemódz*:

1. `iayko` applies unconditional normalization rules to the input text. The rule $dz \rightarrow c$ is applied and the edge *niemóc* (transient form) is appended to the lattice.
2. `niema` applies conditional normalization rules on the output of `iayko`. The set of rules includes (among others) the separation of the particle *nie* before the verb. The edge sequence *nie móc* is appended to the lattice with the attached information about the constraint that the second part should be a verb.
3. A morphological analyzer (lemmatizer) adds morphological information for token edges in the lattice. In particular, the word *móc* is annotated as a verb.
4. `selector` finds that the constraint of *móc* being a verb is satisfied, and this edge is marked as selected. If the constraint weren't satisfied, a fallback edge (*niemóc*, in this case) would be marked as selected instead.

---

[11] http://psi-toolkit.wmi.amu.edu.pl/help/processor.psis?name=iayko.

Like in the case of **iayko**, the rules for the conditional diachronic normalizer **niema** apply the Thrax formalism. However, they are supplemented by an additional line, starting with %, which specifies the morphological constraint. Here are a few examples:

```
    separate_nie = ("nie" ("" : " ") NonEmptyString);
%*+verb
    Rule5 = CDRewrite[separate_nie,
                      Beginning,
                      End,
                      Any*];
```

Rule5 sends to **selector** the request to accept the change only if the outcome consists of two tokens: the particle *nie* (*** denotes that any morphological form is accepted) and a verb.

```
    InfinitiveEnding = ("y" | "eś") ("dź" : "ć") |
                       ("e" | "y") ("śc" : "źć") |
                       ("ą" | "e" | "ó") ("dz" : "c");
%inf
    Rule6 = CDRewrite[InfinitiveEnding,
                      "",
                      End,
                      Any*];
```

Above, the change is accepted only if it occurs at the end of the **infinitive** form of the verb. It should not be applied e.g. to the verb *kładź*, which is the imperative form of the verb *kłaść* ('to place').

```
    Ending_zki = ("zk" : "sk");
%adj
    Rule7 = CDRewrite[Ending_zki,
                      ("ą" | "i" | "ę" | "e"),
                      ("a" | "ą" | "i" | "ą" | "e"),
                      Any*]];
```

Rule7 is applied to adjectives derived from nouns, whose endings turned unvoiced in the first half of twentieth century. The morphological constraint prevents e.g. the unwanted change in the word *walizka* ('suitcase').

```
    add_y = "" : "y";
%verb&aspect=imperf&number=pl&person=pri
    Rule8 = CDRewrite[add_y,
                      AnyPrefix,
                      End,
                      Any*];
```

Rule8 sends to **selector** the request to accept the change only if the outcome consists of an adequate verb form (imperfective aspect, plural number, first person). This works for verbs such as *zrobim* → *zrobimy* ('we will do').

4.5   Dealing with Diachronic Homographs

A diachronic homograph is a word, whose obsolete spelling is identical to the spelling of an existing word of a different meaning. An example might be the obsolete word *niema* (currently written as *nie ma*), which coincides with the existing adjective *niema* ('dumb'). In order to prevent the diachronizer from unwanted change in the spelling of contemporary words we verify the existence of the word in the current language before launching the iayko processor. The pipeline for the diachronizer, improved this way, is following:

```
lemmatizer ! antiselector ! iayko ! niema ! lemmatizer !
   selector
```

Note that a lemmatizer is used here twice: in the first run, it serves as a lexicon, in the second run – as a POS-tagger.

Additionally, we introduced here a new processor called antiselector. It can be seen as a processor that is complementary to the selector. Like selector, the antiselector processor is used after a lemmatizer and runs over the edges of the lattice, but then it selects and marks all the edges that has *not* been tagged with the specified tags.

The sequence of steps in this pipeline is following:

1. A morphological analyzer (lemmatizer) is used for checking which input words are contemporary, and therefore should not undergo normalization.
2. The antiselector processor selects all tokens that had not been lemmatized (implying they are absent from the contemporary lexicon) and marks them as "needing to be normalized".
3. The iayko processor applies unconditional normalization rules to the tokens marked in the previous step.
4. The niema processor applies conditional normalization rules on the output of iayko, marking select edges with information about the conditions they should meet.
5. A lemmatizer is run again to add morphological information to the tokens.
6. The selector processor marks the edges meeting the conditions supplied by niema as *selected*, and these edges are used to form the final output.

The alias better-diachronizer (or diachronizer-2018) serves as a short-cut for this pipeline of processors.

4.6   Dealing with Diachronic Pairs

The procedure simply substitutes words in diachronic pairs before any further processing. Using pure Thrax mechanism for the task would require building huge transducers, as the list is over 100 000 items long. A simple normalization processor – simplenorm – executes the task. Then, antiselector can be used to filter out these already normalized tokens, thus preventing them from being normalized again by iayko.

Here is the final pipeline:

```
simplenorm ! lemmatizer ! antiselector ! iayko ! niema !
    lemmatizer ! selector
```

5 Evaluation

The normalization pipeline described here was test on the DiaNorm test set, which is a collection of 100 texts (each one comprising 500 words) that were normalized manually [12]. The test is realistic as it contains a number OCR mistakes (but OCR post-processing is out of scope from the point of view of spelling normalization). We used CharMatch, introduced in [12], as the evaluation metric. CharMatch compares the output against both the input of the normalizer and the expected output. The similarity to the expected output is rewarded, but, at the same time, unnecessary changes (when compared to the input) are penalized. The evaluation results, given in Table 2, show a clear improvement over the previous version of PSI-Toolkit diachronizer.

Table 2. Evaluation results of the previous and current version of the PSI-Toolkit diachronizer. Reference codes to repositories stored at Gonito.net [7] are given in curly brackets. Such a repository may be also accessed by going to http://gonito.net/q and entering the code there.

Diachronizer version	CharMatch dev	CharMatch test	Reference
2020	0.2045	0.2934	{e8b699}
2018	0.1375	0.1951	{361f2a}

6 Creating a Customized Normalizer

The capabilities of PSI-Toolkit's normalizers iayko and niema can be extended by writing customized normalization rules. The normalization rules can be created using Thrax – a language for defining finite state grammars[12].

In fact, iayko and niema can be used not only for diachronic normalization, but with the right set of Thrax rules, they can act as any desired normalizers.

Suppose you have PSI-Toolkit installed as a Docker container, as described in Sect. 4.2 You can write a finite-state grammar using Thrax, and save it to a file, let's say *my_ grammar.grm*. This file can be stored in any location, but you should remember to provide the whole file path when you use it:

```
docker run -i skorzewski/psi-toolkit iayko
    --grm path/to/my_grammar.grm
```

Grammars for iayko and niema can be also written in Markdown format, which allows for easier commenting, as well as for adding niema constraints:

---

[12] http://www.openfst.org/twiki/bin/view/GRM/Thrax.

```
docker run -i skorzewski/psi-toolkit niema
  --md path/to/my_grammar.md
```

The processing can be limited to a selected transducer:

```
docker run -i skorzewski/psi-toolkit iayko
  --grm path/to/my_grammar.grm --fst MyTransducer
```

The normalization will run much slower compared to using the embedded grammars, because the system needs to compile the text grammar into the FAR archive. If you intend to use your grammar more than once, you can ask the system to save the compilation result to a FAR file for future use:

```
docker run -i skorzewski/psi-toolkit iayko
  --grm path/to/my_grammar.grm --fst MyTransducer
  --save-far path/to/compiled.far
```

Then, you can use the compiled grammar to normalize texts as fast as with the embedded normalizer:

```
docker run -i skorzewski/psi-toolkit iayko
  --far path/to/compiled.far --fst MyTransducer
```

7   Conclusions

The paper reports on an attempt to process historical texts with contemporary NLP tools. The approach consists in applying diachronic normalization executed by means of the Thrax platform. Normalization rules are obtained both manually and semi-automatically. Hand-crafted rules are based on existing linguistic research as well as the manual examination of historical text corpora. The semi-automated approach applies the Word2vec word representations in searching for diachronic pairs. The normalization process is executed inside a pipeline of text annotators, included in PSI-Toolkit. The solution enables to reject unwanted changes, thus increasing accuracy on the one hand, and to process the historical input with NLP tools designed for contemporary texts on the other. The solution may be easily applied for other normalization tasks.

References

1. Archer, D., Ernst-Gerlach, A., Kempken, S., Pilz, T., Rayson, P.: The identification of spelling variants in English and German historical texts: manual or automatic? In: Digital Humanities 2006, CATI, Université Paris-Sorbonne, Paris, France, pp. 3–5 (2006)
2. Baron, A., Rayson, P., Archer, D.: Automatic standardization of spelling for historical text mining. In: Digital Humanities 2009 (June 2009)
3. Bollmann, M., Petran, F., Dipper, S.: Rule-based normalization of historical texts. In: Proceedings of the International Workshop on Language Technologies for Digital Humanities and Cultural Heritage, Hissar, Bulgaria, pp. 34–42 (2011)

4. Bronikowska, R., Modrzejewski, E.: The enrichment of the lexical information and the corpus resources by using the results of the morphological analysis of historical texts (2017). https://ijp.pan.pl/images/konferencje/elex-budapeszt-2017.pdf
5. Etxeberria, I., Alegria, I., Uria, L., Hulden, M.: Evaluating the noisy channel model for the normalization of historical texts: Basque, Spanish and Slovene. In: Proceedings of the 10th International Conference on Language Resources and Evaluation, LREC 2016 (2016)
6. Graliński, F., Jassem, K., Junczys-Dowmunt, M.: PSI-toolkit: a natural language processing pipeline. In: Przepiórkowski, A., Piasecki, M., Jassem, K., Fuglewicz, P. (eds.) Computational Linguistics, Studies in Computational Intelligence. Studies in Computational Intelligence, vol. 458, pp. 27–39. Springer, Heidelberg (2013). https://doi.org/10.1007/978-3-642-34399-5_2
7. Graliński, F., Jaworski, R., Borchmann, Ł., Wierzchoń, P.: Gonito.net - open platform for research competition, cooperation and reproducibility. In: Branco, A., Calzolari, N., Choukri, K. (eds.) Proceedings of the 4REAL Workshop: Workshop on Research Results Reproducibility and Resources Citation in Science and Technology of Language, pp. 13–20 (2016)
8. Graliński, F., Jassem, K.: Mining historical texts for diachronic spelling variants. Poznan Stud. Contemp. Lingustics (2019). https://ai.wmi.amu.edu.pl/wp-content/uploads/2020/02/gralinski2019mining-2.pdf. Accepted 6 Mar 2019
9. Graliński, F., Wierzchoń, P.: Odkrywka, czyli leksykografia diachroniczna live. In: Bańko, M., Karaś, H. (eds.) Między teorią a praktyką. Metody współczesnej leksykografii, vol. 1, pp. 59–69. Wydawnictwa Uniwersytetu Warszawskiego, Warszawa (2018)
10. Hauser, A.W., Schulz, K.U.: Unsupervised learning of edit distance weights for retrieving historical spelling variations. In: Proceedings of the 1st Workshop on Finite-State Techniques and Approximate Search, Borovets, Bulgaria, pp. 1–6 (2007)
11. Jassem, K., Graliński, F., Obrębski, T., Wierzchoń, P.: Automatic diachronic normalization of Polish texts (2017, to appear)
12. Jassem, K., Graliński, F., Obrębski, T.: Pros and cons of normalizing text with Thrax. In: Proceedings of the 8th Language and Technology Conference, Poznań, pp. 230–235 (2017)
13. Malinowski, M.: Ortografia polska od II połowy XVIII wieku do współczesności. Kodyfikacja, reformy, recepcja. Ph.D. thesis, Uniwersytet Śląski w Katowicach, Katowice (2011)
14. Mikolov, T., Sutskever, I., Chen, K., Corrado, G.S., Dean, J.: Distributed representations of words and phrases and their compositionality. In: Burges, C.J.C., Bottou, L., Ghahramani, Z., Weinberger, K.Q. (eds.) 27th Annual Conference on Neural Information Processing Systems 2013. Advances in Neural Information Processing Systems, Lake Tahoe, Nevada, United States, 5–8 December 2013, vol. 26, pp. 3111–3119 (2013)
15. Miłkowski, M.: Developing an open-source, rule-based proofreading tool. Softw. Pract. Exp **40**(7), 543–566 (2010)
16. Mykowiecka, A., Rychlik, P., Waszczuk, J.: Building an electronic dictionary of Old Polish on the base of the paper resource. In: Osenov, P., Piperidis, S., Slavcheva, M., Vertan, C. (eds.) Proceedings of the Workshop on Adaptation of Language Resources and Tools for Processing Cultural Heritage at LREC 2012, pp. 16–21. European Language Resources Association (ELRA) (2012)

17. Nissim, M., Matheson, C., Reid, J.: Recognising geographical entities in Scottish historical documents. In: Proceedings of the Workshop on Geographic Information Retrieval at SIGIR ACM 2004, Sheffield, UK (2004)
18. Oravecz, C., Sass, B., Simon, E.: Semi-automatic normalization of Old Hungarian codices. In: Proceedings of the ECAI 2010 Workshop on Language Technology for Cultural Heritage, Social Sciences, and Humanities, LaTeCH 2010, Lisbon, Portugal, pp. 55–59 (2010)
19. Pettersson, E., Megyesi, B., Nivre, J.: Rule-based normalisation of historical text - a diachronic study. In: Empirical Methods in Natural Language Processing: Proceedings of the 11th Conference on Natural Language Processing, KONVENS 2012, Vienna, Austria, pp. 333–341. Österreichische Gesellschaft für Artificial Intelligence (ÖGAI) (2012)
20. Piotrowski, M.: Natural Language Processing for Historical Texts. Morgan & Claypool, San Rafael (2012). https://doi.org/10.2200/S00436ED1V01Y201207HLT017
21. Porta, J., Sancho, J.L., Gómez, J.: Edit transducers for spelling variation in Old Spanish. In: Proceedings of the Workshop on Computational Historical Linguistics at NODALIDA 2013; NEALT Proceedings, Oslo, Norway, pp. 70–79. No. 87 in 18, Linköping University Electronic Press; Linköpings Universitet (2013)
22. Rayson, P., Archer, D., Baron, A., Smith, N.: Tagging historical corpora - the problem of spelling variation. In: Dagstuhl Seminar Proceedings. Schloss Dagstuhl-Leibniz-Zentrum fr Informatik (2007)
23. Roark, B., Sproat, R., Allauzen, C., Riley, M., Sorensen, J., Tai, T.: The OpenGrm open-source finite-state grammar software libraries. In: Proceedings of the ACL 2012 System Demonstrations, Jeju Island, Korea, pp. 61–66. Association for Computational Linguistics (July 2012). http://www.aclweb.org/anthology/P12-3011
24. Scherrer, Y., Erjavec, T.: Modernizing historical Slovene words with character-based SMT. In: 4th Biennial Workshop on Balto-Slavic Natural Language Processing, BSNLP 2013 (2013)
25. Sleator, D., Temperley, D.: Parsing English with a link Grammar. In: 3rd International Workshop on Parsing Technologies (1993)
26. Sproat, R., Jaitly, N.: RNN approaches to text normalization: A challenge (2016)
27. Tai, T., Skut, W., Sproat, R.: Thrax: an open source grammar compiler built on OpenFst. In: IEEE Automatic Speech Recognition and Understanding Workshop, ASRU 2011, Waikoloa Resort, Hawaii, vol. 12 (2011)
28. Woliński, M., Miłkowski, M., Ogrodniczuk, M., Przepiórkowski, A., Szałkiewicz, Ł.: PoliMorf: a (not so) new open morphological dictionary for Polish. In: Proceedings of the 8th International Conference on Language Resources and Evaluation, LREC 2012, Istanbul, Turkey, pp. 860–864. European Language Resources Association (ELRA) (2012)

PADI-web: An Event-Based Surveillance System for Detecting, Classifying and Processing Online News

Sarah Valentin[1,2,3]($\boxtimes$)(iD), Elena Arsevska[2,3](iD), Alize Mercier[2,3](iD),
Sylvain Falala[2], Julien Rabatel[3], Renaud Lancelot[2,3](iD),
and Mathieu Roche[1,3](iD)

[1] UMR TETIS, Univ Montpellier, AgroParisTech, CIRAD, CNRS, INRAE,
Montpellier, 34398 Montpellier, France
sarah.valentin@cirad.fr
[2] UMR ASTRE, Univ Montpellier, CIRAD, INRAE, 34398 Montpellier, France
[3] CIRAD, Montpellier, France

Abstract. The Platform for Automated Extraction of Animal Disease Information from the Web (PADI-web) is a multilingual text mining tool for automatic detection, classification, and extraction of disease outbreak information from online news articles. PADI-web currently monitors the Web for nine animal infectious diseases and eight syndromes in five animal hosts. The classification module is based on a supervised machine learning approach to filter the relevant news with an overall accuracy of 0.94. The classification of relevant news between 5 topic categories (confirmed, suspected or unknown outbreak, preparedness and impact) obtained an overall accuracy of 0.75. In the first six months of its implementation (January–June 2016), PADI-web detected 73% of the outbreaks of African swine fever; 20% of foot-and-mouth disease; 13% of bluetongue, and 62% of highly pathogenic avian influenza. The information extraction module of PADI-web obtained F-scores of 0.80 for locations, 0.85 for dates, 0.95 for diseases, 0.95 for hosts, and 0.85 for case numbers.
PADI-web allows complementary disease surveillance in the domain of animal health.

Keywords: Epidemic intelligence · Animal health · Web monitoring · Text mining · Classification · Information extraction

1 Introduction

Until the early 2000s, the surveillance of diseases has been essentially dependent on the principles of traditional, indicator-based surveillance (IBS). The IBS mainly involves reporting of known diseases based on sets of rules for verification and confirmation of cases, from clinicians to laboratories and health officials [14].

However, the rapid growth of the Internet and the connectivity of the users to the World Wide Web addressed the need of a supplementary disease surveillance,

© Springer Nature Switzerland AG 2020
Z. Vetulani et al. (Eds.): LTC 2017, LNAI 12598, pp. 87–101, 2020.
https://doi.org/10.1007/978-3-030-66527-2_7

i.e. event-based surveillance (EBS). Compared to the IBS, the EBS has been proven to be more flexible to detect both known and new diseases through the use of multiple different sources, languages and geographic coverage [8,10,18].

ProMED-mail, one of the first EBS systems, is based on the sharing of sanitary information between users and manual Web searches implemented by human analysts [11]. In contrast, the HealthMap [6] and MedISys [16] systems automatically detect disease-related contents on the Web, extract, and visualise events on interactive maps to monitor trends. The current EBS systems mainly focus on diseases of public health interest [7]. As a consequence, these systems have a limited value for animal health authorities due to the inconsistent coverage of animal health topics.

This paper describes the contributions towards the development and implementation of an EBS system dedicated to the detection and monitoring of new and emerging animal infectious diseases occurring worldwide. We present the Platform for Automated Extraction of Animal Disease Information from the Web (PADI-web), a text mining platform for automatic detection, translation, classification and extraction of disease (outbreak) information from news articles published on the Web (further on referred to as "news").

PADI-web was primarily developed for the French Epidemic Intelligence System (FEIS, or *Veille sanitaire internationale* in French), which is part of the French animal health epidemiological surveillance Platform (ESA Platform[1]). It is now publicly available[2] through an online platform, in both English and French. Currently, PADI-web scans the Web for nine animal infectious diseases, i.e. African swine fever, classical swine fever, avian influenza, foot-and-mouth disease, bluetongue, Schmallenberg virus infection, West Nile, lumpy skin disease and Rift Valley fever. In order to detect the emergence of potentially new or unknown infectious diseases, PADI-web monitors the Web for eight syndromes, i.e. general, respiratory, digestive, locomotion/neurologic, skin/mucous, haemorrhagic, reproductive and postnatal/congenital in five hosts, i.e. avian, bovine, ovine, caprine and porcine animals.

2   PADI-web Approaches

Figure 1 shows the main steps implemented in PADI-web, i.e. data collection, data processing, data classification, and extraction of epidemiological information (features of an outbreak).

2.1   Data Collection

PADI-web collects news articles in near-real time through Really Simple Syndication (RSS) feeds. Once a day, 7 days a week, RSS feeds from Google News are processed. Similarly to other EBS systems, we have chosen Google News for

---

[1] https://www.plateforme-esa.fr/.

[2] https://padi-web.cirad.fr/en/.

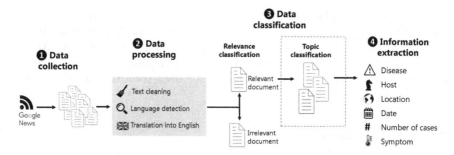

Fig. 1. Pipeline implemented in PADI-web. PADI-web scans the Web through customised RSS feeds (step 1). Once collected and stored in a database, the news contents are cleaned and translated (step 2). Then, the news are classified as relevant or irrelevant (step 3). The topic classification, presented in this paper, is not yet integrated into the PADI-web pipeline. Using a combined method for information extraction, epidemiological indicators are extracted from the news content (step 4).

to its international coverage and flexible RSS feeds. In order to detect relevant disease-related news articles, the RSS feeds use specific terminology based on: i) disease names - to detect news which describe outbreaks of known diseases, and ii) associations of terms on hosts and clinical signs - to detect news related to the occurrences of unknown diseases and syndromes in animals.

The terminology is automatically extracted using text mining techniques from a corpus of relevant news articles. We apply additional text mining techniques to automatically obtain associations between terms describing the clinical signs and hosts. Finally, using a Delphi approach and through a consensus, a group of animal health experts validates the extracted terms and associations and, if necessary, complements the list with additional terms. Detailed description of our methods for extraction of disease-related terminology is described elsewhere [2].

All the RSS feeds have an English version. In addition, several RSS feeds are adapted into additional languages, including Chinese, Turkish, French or Arabic. The RSS feed keywords (disease, hosts and symptoms) are translated using two vocabularies, i.e. UMLS [4] and Agrovoc[3], a controlled vocabulary developed by the Food and Agriculture Organization (FAO). The choice of languages is based on epidemiological expertise, in order to target specific high-risk areas. For instance, we are currently monitoring foot-and-mouth disease through RSS feeds in English, French and Arabic language, as it is endemic in Northern Africa.

2.2 Data Processing

Duplicated news (i.e. news for which the url already exists in the database) are filtered out. PADI-web retrieves the news content from its webpage. The textual content (title and text) is cleaned to remove irrelevant elements (e.g.

---

[3] http://aims.fao.org/vest-registry/vocabularies/agrovoc.

pictures, ads, hyperlinks) and the language is detected, using respectively the *BeautifulSoup* [17] and *langdetect* python libraries. All non-English news articles are translated using the Translator API of the Microsoft Azure system[4]. We use English as a bridge-language because the models for classification (Sect. 2.3) and information extraction (Sect. 2.4) modules have been trained with labeled data in this language.

2.3   Data Classification

First, news are classified as "relevant" or "irrelevant". Then, the relevant news are classified according to their topic.

Relevance Classification. We define a news article as relevant if it explicitly refers to a recent or current infectious animal health event. This definition includes several topics to capture all the available online information about an on-going event. It excludes topics such as research or general information about a disease.

In its first version, PADI-web categorized the collected news articles by using a list of 32 outbreak-related keywords, i.e. "positive keywords". More precisely, news articles were classified as relevant if they contained in the title or the body one of the text positive keywords related to an outbreak event (e.g. outbreak, cases, spread). This approach is called *keyword-based classification*.

To improve the accuracy of the classification, we integrated a classifier based on a supervised machine learning approach [22]. To create a learning dataset, a corpus of 800 annotated news labelled by an epidemiology expert (400 relevant news articles and 400 irrelevant news articles). To obtain a feature representation of the corpus, each document from the corpus was converted into a bag-of-words representation using the Term Frequency - Inverse Term Frequency (TF-IDF) as term weighing method [19]. The meaningless terms (stop-words) and the punctuation are removed, all the remaining terms are lowercased. Using the *scikit-learn* python library [15], a selection of model families is trained on the corpus (random forests, linear support vector classifier, neural networks, etc.). The model obtaining the highest mean accuracy score along the 5-fold cross-validation scheme is subsequently used to classify each newly retrieved article.

The PADI-web interface allows users to manually label the relevance of the retrieved news articles. Each manually labelled article is added to the initial training corpus. Thus, each training step is enriched with the user contribution.

Topic Classification. To go beyond the binary relevance classification, we defined more fine-grained categories for the relevant news. These categories were created in collaboration with the FEIS team, and aim at improving the news classification regarding two points. First, a part of the relevant news does not

---

[4] https://azure.microsoft.com/en-gb/services/cognitive-services/translator-text-api/.

directly refer to a disease outbreak, but rather describes a disease-free country in alert or the economic impacts on an affected area several days after an outbreak. Therefore, automatically extracting epidemiological information from their content can generate a number of false positive alerts, which has been identified as a significant limitation of PADI-web performances [3]. Secondly, the news declaring or suspecting an outbreak have a higher priority level than those describing outbreak consequences, for instance. In the context of daily monitoring of a continuous stream of news, it is therefore crucial to correctly identify the topic and prioritize the retrieved news. We present the topic categories as follows, in decreasing priority order:

– *Confirmed outbreak:* the news declares or provides updates about a current or a recent confirmed outbreak[5];
– *Suspected outbreak:* the news refers to current or a recent cases not yet diagnosed, associated with an explicit suspicion of an infectious disease;
– *Unknown outbreak:* the news refers to current or a recent cases not yet diagnosed, not associated with any suspicion;
– *Preparedness:* the news refers to the alert status of a country at risk of being affected by a disease spreading in a neighbouring area;
– *Impact:* the news refers to the economic, political or social consequences of an outbreak in an affected country or area.

2.4   Information Extraction

Once a new article is categorized as relevant, PADI-web uses a combined method for information extraction (IE). The combined IE method uses dictionaries and machine learning algorithms (Fig. 2). It allows the identification of key pieces of epidemiological information in the free text (epidemiological events), i.e. location and date of an outbreak, affected hosts, their numbers and encountered clinical signs.

Firstly, our method uses the previously defined dictionaries (Sect. 2.1) to identify relevant candidates for extraction of disease names, hosts, and clinical signs in a given free text. External resources, such as GeoNames [1] and Heidel-Time [20] allow the detection of the location and date of a given outbreak. The number of infected cases is recognized using regular expressions.

Secondly, as some of the candidates might be incorrect (e.g. not every date mentioned in the news is the date of an outbreak), each candidate is tested against a set of rules that distinguish correct from incorrect candidates. Such rules, which are at the core of the IE, are automatically extracted as association rules [21] from a corpus of 352 news articles where correct/incorrect candidates were manually annotated by two domain experts in epidemiology and health informatics (EA and JR). Finally, these rules are used as features feeding a Support Vector Machine to predict the relevance of a given candidate [13].

---

[5] This definition is only based in the news semantic, and do not take into account the official confirmation by a formal source.

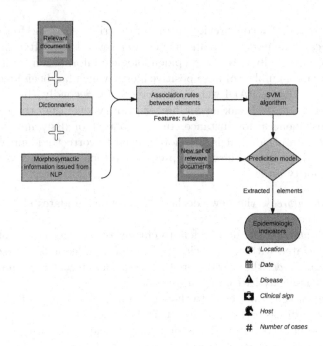

Fig. 2. Event extraction method implemented in PADI-web, based on dictionaries and SVM classification.

Finally, once each candidate has been processed, the interface of PADI-web permits users to visualise all extracted elements from a given news article and the location of a given event are associated with a link to Google Maps (Fig. 3). The interface of PADI-web offers users additional features such as trend analysis, i.e. monthly number of relevant news articles for a given topic. Users can also filter outbreak events by disease, hosts, clinical signs, date interval, and source of information. These events can then be downloaded in a structured format.

3   Experiments

We evaluated PADI-web in its integrity and for each step of its pipeline. The results of the first two steps of the method are detailed elsewhere [2]. In this work, we present the results from the evaluation of the classification step, the information extraction step and the overall performance of PADI-web.

3.1   Performance of the Classification

Relevance Classification. To evaluate the improvement of integrating a supervised classifier, we compared the performances of the *keyword-based classification* to the performances of three classifiers from different model families, i.e. Random Forest [5], Linear Support Vector Machine (Linear SVM) [9] and Multilayer Perceptron [12] on the learning dataset described in Sect. 2.3.

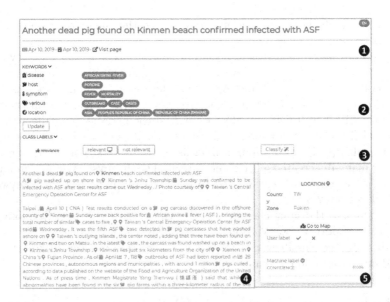

Fig. 3. Print screen of a user interface in PADI-web. The example shows a news article classified as relevant and related to African swine fever (1, 3). The automatically annotated candidates are coloured in green when they are estimated to be correct by the algorithm (4) and summarized in a keywords panel (2). The locations candidates are associated with complementary information: country, administrative zone, and a link to Google Maps (5) (Color figure online).

Topic Classification. To evaluate the performances of supervised classifiers for the topic classification, an epidemiologist first annotated the initial set of 400 relevant news from the learning dataset (Sect. 2.3). This first annotation phase led to a very imbalanced dataset, the class *confirmed outbreak* being over-represented among the other classes. Thus, excluding the *confirmed outbreak* class, we further increased the dataset by annotating additional relevant news extracted from PADI-web database until reaching balanced classes. The final dataset contained 631 news, distributed as follows: *confirmed outbreak*: 308 news, *suspected outbreak*: 86 news, *unknown outbreak*: 77 news, *preparedness*: 80 news, and *impact*: 80 news.

We evaluated two textual representations (Fig. 4):

- R_1: Bag-of-words representation
- R_2: R_1 enriched with a terms-count matrix

R_1 corresponds to the bag-of-words matrix obtained after the pre-processing steps described in Sect. 2.3. We further enriched this matricial representation with 5 features, i.e. the counts of diseases, hosts, symptoms, outbreak-related terms, and mystery-related terms in the news. More precisely, in each news, we counted the total number of terms belonging to each category listed here-above (using 5 lists of terms), thus obtaining a term-count matrix. This matrix was concatenated with the initial bag-of-words matrix, thus increasing the representation space by 5 columns.

As lists of terms, for diseases, hosts and symptoms, we used the previously defined dictionaries (Sect. 2.1). We further enriched the disease and host lists with synonyms from UMLS and Agrovoc. The disease list and host list contain 1,967 terms and 265 terms respectively. For symptoms, we enriched the vocabulary with the list of disorders from Agrovoc and we manually retrieved the clinical signs from the technical disease cards of the World Organisation for Animal Health (OIE)[6], obtaining a final list of 798 terms. For mystery-related terms, we manually created a list of terms related to the mysterious and unknown aspect of an event. Outbreak-related terms consists in the list of 32 keywords described in Sect. 2.3.

We compared the same classifiers as described previously (i.e. Random Forest, Linear Support Vector Machine and Multilayer Perceptron).

Both relevance classification and topic classification were evaluated through a 5-fold cross-validation scheme.

3.2   Performance of the Information Extraction

The IE step was evaluated on a set of 352 manually labelled news articles. These articles were acquired from Google News and covered content related to

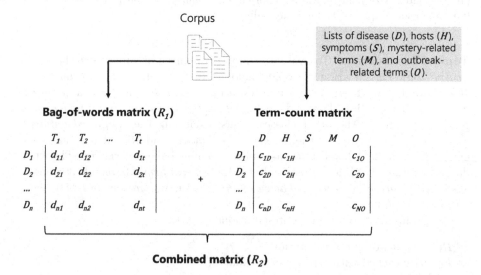

Fig. 4. Two types of textual representations used for the topic classification, where d_{ij} is the weight of the term j in the document i (bag-of-words matrix), and c_{ix} in the count of terms X in the document i (term-count matrix). R_2 is the contatenation of both representations.

---

[6] https://www.oie.int/en/animal-health-in-the-world/technical-disease-cards/.

a reporting of at least one disease outbreak in animals from 2014 to 2015. For each relevant document, the information about the candidates was automatically found, i.e. using the dictionaries (Sect. 2.1) for each type of information (disease, host, clinical sign, etc.). Next, the information about each candidate was annotated by two independent annotators (EA and JR) as: i) correct, when the corresponding candidate correctly described the desired piece of information, and ii) incorrect, when the candidate had no link to the corresponding event. We validated the machine learning approach, i.e. Support Vector Machine (SVM), by using a ten-fold cross-validation technique.

Both Data classification and Information extraction were evaluated using the following metrics:

- **Accuracy** measures the overall correctness of classification, i.e. the fraction of documents (resp. candidates) correctly classified from the total number of documents (resp. candidates).
- **Precision** indicates the correctness of classification, i.e. the fraction of documents (resp. candidates) correctly classified in a class i from the total number of documents (resp. candidates) classified in class i.
- **Recall** indicates the completeness of classification, i.e. the fraction of documents (resp. candidates) correctly classified in a class i from the total number of documents (resp. candidates) from class i.
- **F-score** is the harmonic mean of precision and recall.

3.3   Overall Performance of PADI-web

The general performance of PADI-web was evaluated during a six-month period, from 1st January to 28th June 2016. We evaluated the performance in terms of precision, sensitivity and timeliness of PADI-web to alert signals of emergence of African swine fever (ASF), foot-and-mouth disease (FMD), bluetongue (BTV), and highly pathogenic avian influenza (HPAI) at international level.

We considered as a gold standard all official reports for these diseases, freely available from the Empres-i database of the Food and Agriculture Organization (FAO). This database consists of verified information about new epidemiological events (immediate notifications) and ongoing outbreaks (follow-up reports) from a list of obligatory notifiable animal infectious diseases.

Precision is the capacity of PADI-web to alert about an event which corresponded to a verified outbreak (TP) from the Empres-i database (immediate notification or follow-up report) or an event that was judged as relevant by a veterinary epidemiologist (one of the authors of this work, EA). A False Positive (FP) was an event not found in the Empres-i database or an event which was annotated by the veterinary epidemiologist as irrelevant.

Sensitivity is the capacity of PADI-web to alert about an exceptional epidemiologic event (immediate notification) reported in the Empres-i database (TP). From January to June 2016 a total of 11 outbreaks for ASF, 15 for FMD, 8 for BTV, and 26 for HPAI, respectively, were reported in the Empres-i database.

False Negatives (FN) were all outbreaks from the immediate notifications that were not detected by PADI-web.

4   Results

4.1   Performance of the Classification

Relevance Classification. The *keyword-based classification* obtained imbalanced performances: the precision for the relevant class was 0.76 while its recall reached 0.97 (Table 1). The three classifiers included in this study outperformed *keyword-based classification*, increasing the precision of the relevant class up to 0.96 (linear SVM classifier). Among all the classifiers, the linear SVM classifier obtained the highest accuracy (0.94) and F-scores (0.95 and 0.92 for the relevant and irrelevant classes respectively).

Table 1. Comparison of the keyword-based method and three supervised classifiers (RF: Random Forest, MLP: Multilayer Perceptron, linear SVM: linear Support Vector Machine) for the classification of the relevance, in terms of precision, recall, F-score per class, and overall accuracy. For each class, the best performances are shown in bold.

Classification	Class	Precision	Recall	F-score	Accuracy
Keyword-based	Relevant	0.76	0.97	0.85	0.80
	Irrelevant	0.91	0.53	0.67	
RF	Relevant	0.86	**0.98**	0.92	0.86 ± 0.03
	Irrelevant	**0.96**	0.75	0.84	
MLP	Relevant	0.93	0.95	0.94	0.93 ± 0.03
	Irrelevant	0.92	0.90	0.91	
Linear SVM	Relevant	**0.96**	0.94	**0.95**	**0.94 ± 0.02**
	Irrelevant	0.91	**0.93**	**0.92**	

Topic Classification. The overall performances of the topic classification were lower than the relevance classification (Table 2). MLP and RF classifiers obtained comparatively equal performances, obtaining the highest accuracy (0.75) and F-score (0.73). The enriched representation R_2 outperformed the bag-of-words representation R_1 regarding all the scores evaluated.

The recall was heterogeneous between the different classes (Table 3), varying from 0.34 (*preparedness*) to 0.90 (*confirmed outbreak*). The best F-scores were obtained for *confirmed outbreak* (F-score = 0.82) and *impact* (F-score = 0.82).

News wrongly classified as *confirmed outbreak* represented respectively 56%, 31% and 18% of the categories *preparedness*, *suspected outbreak* and *impact* (Table 4). Only 6% of the *confirmed outbreak*, 1% *suspected outbreak* and none of the *unknown outbreak* news were classified as *preparedness* or *impact* categories.

PADI-web 97

Table 2. Comparison of three supervised classifiers (RF: Random Forest, MLP: Multilayer Perceptron, linear SVM: linear Support Vector Machine) and two textual representation (R_1: bag-of-words representation, R_2: R_1 enriched which term-count representation) for the topic classification, in terms of weighted mean precision, recall, F-score, and overall accuracy. The best performances are shown in bold.

Classifier	Representation	Precision	Recall	F-score	Accuracy
RF	R_1	0.64	0.62	0.57	0.62 ± 0.04
	R_2	0.66	0.65	0.60	0.65 ± 0.04
MLP	R_1	0.68	0.68	0.65	0.70 ± 0.05
	R_2	**0.76**	**0.75**	**0.73**	$\mathbf{0.75 \pm 0.05}$
Linear SVM	R_1	0.70	0.70	0.68	0.70 ± 0.07
	R_2	0.75	**0.75**	**0.73**	$\mathbf{0.75 \pm 0.04}$

Table 3. Comparison of performances scores in the different classes with linear SVM classifier.

Class	Precision	Recall	F-score
Confirmed outbreak	0.75	0.90	0.82
Suspected outbreak	0.69	0.51	0.59
Unknown outbreak	0.76	0.79	0.78
Preparedness	0.68	0.34	0.45
Impact	0.82	0.78	0.79

Table 4. Normalized confusion matrix obtained from the topic classification step with linear SVM classifier. The figures correspond to the percentage of news in each actual class (rows) classified in each predicted class (columns).

	Predicted				
Actual	Confirmed outbreak	Suspected outbreak	Unknown outbreak	Preparedness	Impact
Confirmed outbreak	90	3	1	4	2
Suspected outbreak	31	51	16	0	1
Unknown outbreak	10	10	79	0	0
Preparedness	56	0	1	34	9
Impact	18	13	1	1	78

4.2   Performance of the Information Extraction

The accuracy of the IE step was higher than 0.80, with the lowest accuracy occurring when detecting locations and the highest for names of diseases and host species (Table 5). Similarly, in terms of F-score, the IE obtained a score of 0.80 for locations, 0.85 for dates, 0.95 for diseases, 0.85 for numbers of cases, and 0.95 for hosts in the free text news articles.

Table 5. Performance of the information extraction step of the text-mining approach.

Entity	Precision	Recall	F-score	Accuracy
Location	0.81	0.80	0.80	0.80
Date	0.82	0.88	0.85	0.88
Disease	0.95	0.96	0.95	0.96
Number of cases	0.86	0.85	0.85	0.85
Host	0.94	0.96	0.95	0.96

4.3   Overall Performance of PADI-web

In the first six months of its implementation, from January to June 2016, PADI-web alerted on 123 outbreaks for ASF, 191 for FMD, 71 for BTV, and 632 for AI. The precision of PADI-web to adequately alert for true outbreaks was 30% (37/123 events) for ASF, 27% (51/191 events) for FMD, 45% (32/71) for BTV, and 54% (342/632) for AI. The sensitivity of PADI-web to alert for exceptional epidemiological events was 73% (8/11 events) for ASF, 20% (3/15 events) for FMD, 13% (1/8 events) for BTV, and 62% (16/26 events) for HPAI.

5   Conclusion

PADI-web, which is operational since January 2016, is a text mining tool specialized in monitoring the Web for the emergence of new animal infectious diseases.

The implementations of a supervised classifier to automatically filter out irrelevant news significantly improved the accuracy of the relevance classification in PADI-web, which is crucial to control the amount of daily news presented to the FEIS experts. The evaluation of the integration of a topic classification step for the relevant news obtained promising results. However, numbers of *preparedness* and *suspected outbreak* news were classified as *confirmed outbreak*. The alert status in a country is generally consecutive to an on-going outbreak in another area, thus the topics *preparedness* and *confirmed outbreak* can overlap in a same news. In addition, the distinction between a confirmed and a suspected outbreak often solely relies on the use of a different adjective (use of "detected" or "confirmed" instead of "suspected" or "investigated"). In practice, misclassification between *confirmed outbreak*, *suspected outbreak* and *unknown outbreak* is not a strong limitation, since all these categories have a high priority level. Increasing the learning dataset and enriching the feature representation may improve the accuracy of topic classification.

The results from the evaluation of IE show that this method is suitable for the extraction of epidemiological indicators. However, the evaluation of the IE step on the data stream from PADI-web showed a lower precision to correctly alert on events of epidemiological relevance. For example, a number of alerts produced by PADI-web were actually false positives as they corresponded to a location irrelevant to a disease outbreak. This suggests that further improvements, possibly training the SVM algorithm on a larger dataset, can improve the outcomes of the predicted model from the IE step. Our first evaluations showed, however, a higher sensitivity of PADI-web, especially for HPAI and ASF. These results suggest that new epidemiological events are reported in the media, particularly zoonotic events or outbreaks of significant economic impact.

In its first version, PADI-web used Google News in English as the main source of information. This might have influenced the results of the performance evaluation, especially the sensitivity for BTV and FMD which mostly occurred in Central Europe, Southern Africa and Asia. In order to overcome this limitation, the new version of PADI-web integrates a multilingual module to increase the local and regional coverage.

We believe these updates significantly improved the operational use of PADI-web by animal health authorities interested in integrating an event-based surveillance component to their epidemic intelligence activities. Furthermore, in December 2019, PADI-web detected COVID-19 related news through RSS feeds designed for animal health, highlighting its genericity and its ability to detect public health emergence signals.

Acknowledgements. We thank J. de Goër, B. Belot, C. Hemeury, M. Devaud, and T. Filiol for their contribution in the development of PADI-web. We also thank the members of the French Epidemic Intelligence Team in Animal Health for their constructive comments during the development of PADI-web. This work has been supported by the French General Directorate for Food (DGAL), the French Agricultural Research Centre for International Development (CIRAD), the SONGES Project (FEDER and Occitanie), and the French National Research Agency under the Investments for the Future Program, referred as ANR-16-CONV-0004 (#DigitAg). This work has also been funded by the "Monitoring outbreak events for disease surveillance in a data science context" (MOOD) project from the European Union's Horizon 2020 research and innovation program under grant agreement No. 874850 (https://mood-h2020.eu/).

References

1. Ahlers, D.: Assessment of the accuracy of GeoNames gazetteer data. In: Proceedings of the 7th Workshop on Geographic Information Retrieval, pp. 74–81. ACM, New York (2013)

2. Arsevska, E.: Identification of terms for detecting early signals of emerging infectious disease outbreaks on the web. Comput. Electron. Agric. **123**, 104–115 (2016). https://doi.org/10.1016/j.compag.2016.02.010

3. Arsevska, E., et al.: Web monitoring of emerging animal infectious diseases integrated in the French Animal Health Epidemic Intelligence System. PLoS ONE **13**(8), e0199960 (2018). https://doi.org/10.1371/journal.pone.0199960

4. Bodenreider, O.: The Unified Medical Language System (UMLS): integrating biomedical terminology. Nucleic Acids Res. **32**(Database issue), D267–D270 (2004). https://doi.org/10.1093/nar/gkh061

5. Breiman, L.: Random Forests. Mach. Learn. **45**(1), 5–32 (2001). https://doi.org/10.1023/A:1010933404324

6. Brownstein, J.S., Freifeld, C.C., Reis, B.Y., Mandl, K.D.: Surveillance Sans Frontiéres: Internet-based emerging infectious disease intelligence and the healthmap project. PLOS Med. **5**(7), 1–6 (2008). https://doi.org/10.1371/journal.pmed.0050151

7. Collier, N., Doan, S.: GENI-DB: a database of global events for epidemic intelligence. Bioinformatics **28**(8), 1186–1188 (2012). https://doi.org/10.1093/bioinformatics/bts099

8. Collier, N., et al.: BioCaster: detecting public health rumors with a Web-based text mining system. Bioinformatics **24**(24), 2940–2941 (2008). https://doi.org/10.1093/bioinformatics/btn534

9. Joachims, T.: Text categorization with Support Vector Machines: learning with many relevant features. In: Nédellec, NédellC, Rouveirol, C. (eds.) ECML 1998. LNCS, vol. 1398, pp. 137–142. Springer, Heidelberg (1998). https://doi.org/10.1007/BFb0026683

10. Lejeune, G., Brixtel, R., Doucet, A., Lucas, N.: Multilingual event extraction for epidemic detection. Artif. Intell. Med. **65**(2), 131–143 (2015)

11. Madoff, L.C.: ProMED-Mail: an early warning system for emerging diseases. Clin. Infect. Dis. **39**(2), 227–232 (2004). https://doi.org/10.1086/422003

12. Murtagh, F.: Multilayer perceptrons for classification and regression. Neurocomputing **2**(5), 183–197 (1991). https://doi.org/10.1016/0925-2312(91)90023-5

13. Nahm, U.Y., Mooney, R.J.: Using information extraction to aid the discovery of prediction rules from text. In: Proceedings of the International Conference on Knowledge Discovery and Data Mining, KDD-2000 Workshop on Text Mining, pp. 51–58 (2000)

14. Paquet, C., Coulombier, D., Kaiser, R., Ciotti, M.: Epidemic intelligence: a new framework for strengthening disease surveillance in Europe. Euro. Surveill. **11**(12), 212–214 (2006). 665 [pii]

15. Pedregosa, F., et al.: Scikit-learn: machine learning in Python. J. Mach. Learn. Res. **12**, 2825–2830 (2011)

16. Steinberger, R., Fuart, F., van der Goot, E., Best, C., von Etter, P., Yangarbe, R.: Text Mining from the Web for Medical Intelligence. NATO Science for Peace and Security Series, D: Information and Communication Security, pp. 295–310 (2008)

17. Richardson, L.: Beautiful soup documentation (April 2007)

18. Robertson, C., Yee, L.: Avian influenza risk surveillance in North America with online media. PLoS ONE **11**(11), 1–21 (2016). https://doi.org/10.1371/journal.pone.0165688

19. Salton, G., Buckley, C.: Term-weighting approaches in automatic text retrieval. Inf. Process. Manage. **24**(5), 513–523 (1988). https://doi.org/10.1016/0306-4573(88)90021-0

20. Strotgen, J., Gertz, M.: HeidelTime: high quality rule-based extraction and normalization of temporal expressions. In: Proceedings of the 5th International Workshop on Semantic Evaluation, pp. 321–324 (July 2010)
21. Uno, T., Asai, T., Uchida, Y., Arimura, H.: LCM: an efficient algorithm for enumerating frequent closed item sets. In: Proceedings of Workshop on Frequent Itemset Mining Implementations, FIMI 2003 (2003)
22. Valentin, S., et al.: PADI-web: a multilingual event-based surveillance system for monitoring animal infectious diseases. Comput. Electron. Agric. **169**, 105163 (2020). https://doi.org/10.1016/j.compag.2019.105163

KRNNT: Polish Recurrent Neural Network Tagger Extended

Krzysztof Wróbel[✉] (iD)

Jagiellonian University, Kraków, Poland
krzysztof@wrobel.pro

Abstract. The article presents a state-of-the-art complete part-of-speech tagger for Polish which uses recurrent neural networks. The networks allow accessing the full left and right context of a sentence in comparison to a context window. The tagger uses an external morphological analyzer. In comparison to the best Polish taggers, it does not use word form as a feature for the classifier, there is no separate classifier for unknown words, and predictions are not limited to tags provided by a morphological analyzer. The accuracy is higher—it achieves 28% error reduction and 7% points higher accuracy for unknown words. The tagger also might work faster than others by utilizing GPU. The tagger participated in PolEval 2017 POS Tagging competition and won task B and task C. Additionally, results for PolEval 2020 Morphosyntactic tagging of Middle, New and Modern Polish are reported. The paper is an extension of the Language & Technology Conference paper [25].

Keywords: Part-of-speech tagging · Polish · Neural networks

1 Introduction

Part-of-speech taggers assign part-of-speech tags to each word (token) in a sentence. For inflectional languages, like Polish or Czech tagger, has to recognize values of morphological categories such as gender, number, and case. The categories highly increase the number of tags up to a thousand.

Recurrent neural networks have not been applied in natural language processing for Polish yet, despite they are usually used in state-of-the-art systems for other languages [6,7,9,23].

Czech is also a Slavic language. The best Czech taggers achieved the accuracy above 95%. Prague Dependency Treebank (PDT) is used as a training and test data. The main difference between PDT and National Corpus of Polish (NCP), which is the primary source of tagging data for Polish, is that PDT is not well-balanced and contains only articles from newspapers and journals. On the contrary, transcriptions of spoken conversations and user generated content from web forums are included in NCP. Both corpora were manually annotated by 2 annotators and 1 person who was resolving disagreements. [11] obtained scores higher by 0.75% point by training and testing only on newspaper subcorpus.

© Springer Nature Switzerland AG 2020
Z. Vetulani et al. (Eds.): LTC 2017, LNAI 12598, pp. 102–116, 2020.
https://doi.org/10.1007/978-3-030-66527-2_8

This work presents a part-of-speech tagger for Polish using bidirectional recurrent networks (named KRNNT). Source code and trained models are available at: https://github.com/kwrobel-nlp/krnnt. KRNNT can be easily run using Docker container and integrated to other systems by API.

In the next section training dataset and tagset are described. Also, it summarizes the best publicly available Polish taggers. Section 3 formulates recurrent neural networks and its bidirectional extension. Main Sect. 4 describes all modules of the tagger and the training parameters. Evaluation is described in Sect. 5. Section 6 presents results of PolEval contest. Last Sect. 7 presents conclusions and future works.

2 Polish Tagging

2.1 Dataset

The largest publicly available dataset for Polish is a manually annotated subcorpus of the National Corpus of Polish (NCP) containing above 1 million tokens. The corpus is balanced with respect to genres and subjects—it includes newspaper articles, books, transcriptions of spoken conversations, and user generated content from web forums.

Table 1. National Corpus of Polish statistics.

Paragraphs	18484
Sentences	85663
Tokens	1215513
Unique tokens	143478
Average number of tokens in a sentence	14.19
Unique tags	926
Number of tokens with the same tag	
Average	1312.65
Standard deviation	8389.12
Minimal number	1
Maximal number	223499

Table 1 presents statistics of NCP. Figure 1 shows frequency of tags in logarithmic scale. Full tags, as labels in a multi-class classification problem, are very unbalanced.

The NCP tagset consists of 35 grammatical classes, each having a set of grammatical categories. The number of all possible tags (grammatical classes with unique values of grammatical categories) is about 4000, but texts in NCP represent 926 tag variants.

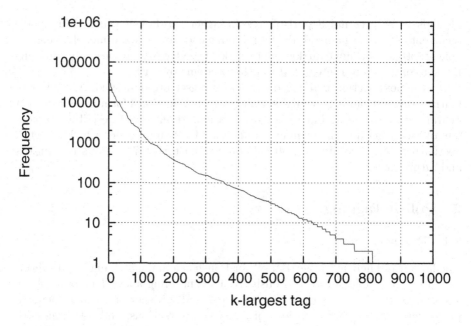

Fig. 1. Frequency of tags in logarithmic scale.

E.g. Polish adjectives have 2 numbers (singular, plural), 7 cases (nominative, genitive, dative, accusative, instrumental, locative, vocative), 5 genders (human masculine, animate masculine, inanimate masculine, feminine, neuter), and 3 degrees (positive, comparative, superlative)—210 variants in total.

Morphosyntactic tags are represented as a sequence of grammatical class and values of grammatical categories, e.g. `adj:sg:nom:m1:pos`, where `adj` is adjective, `sg` is singular number, `nom` is nominative case, `m1` is masculine gender and `pos` is positive degree.

2.2   Polish Taggers

The best Polish POS taggers that are publicly available include: Concraft [24], WCRFT [17], OpenNLP [11], WMBT [18,20], Pantera [1], TaKIPI [15]. Comparisons of Polish taggers were presented in [11,13,16].

So far Concraft has been the best Polilsh tagger. It uses an extended version of CRF algorithm [14] to tackle a high number of labels in an efficient way by restricting space of solutions to the set of tags defined in a morphosyntactic dictionary. For unknown words, morphosyntactic guessing is employed as a separate classifier.

WMBT is a tiered memory-based tagger. Each tier is assigned to a grammatical class or a category. A separate classifier is trained for known and unknown tokens. An algorithm used for each tier is kNN.

WCRFT is similar to WMBT, but kNN algorithm is replaced with CRF. Unknown words are pre-processed by appending potential tags based on analysis of a training data. Therefore, only one classifier is used while tagging.

Pantera uses rule induction algorithm driven by a modified version of Brill's transformation-based learning algorithm [2]. It uses two tiers in the process of tagging and operates on parts of labels, i.e. grammatical class and categories.

TaKIPI employs C4.5 decision tree algorithm. About 200 classes of ambiguity were defined and for each one the classifier was trained. The tagger uses also handwritten rules.

Apache OpenNLP library is a free implementation of NLP algorithms. Algorithm for POS tagging employs perceptron. No tiers were used and all tags are on the output of the neural network. The main difference to earlier mentioned taggers is that OpenNLP tagger does not use morphological analyzer having a token as the only input.

3 Recurrent Neural Network

A recurrent neural network is an extension to feedforward neural network, which is able to handle a variable-length sequence inputs. The RNN has a hidden state that is updated at each time step, therefore it has information about the left context of a sequence. The RNN shares parameters across all steps. The default behavior of RNN is to provide output for each step. However, RNNs are difficult to train to capture long-term dependencies, because the gradients tend to vanish or explode.

Long short-term memory (LSTM) [8] was developed to alleviate problems of raw RNNs. The LSTM introduces an additional memory cell and gates. Output, forget and input gates are responsible for modulating the amount of memory cell used to calculate the output of the LSTM and a new value of memory cell.

The output of the LSTM h_t at time t is:

$$h_t = o_t \tanh(c_t),$$

where o_t is output gate and c_t is the memory cell:

$$o_t = \sigma(W_o x_t + U_o h_{t-1} + V_o c_{t-1}),$$

$$c_t = f_t c_{t-1} + i_t \tanh(W_c x_t + U_c h_{t-1}).$$

σ is the logistic sigmoid function. Forget and input gates are computed as follows:

$$f_t = \sigma(W_f x_t + U_f h_{t-1} + V_f c_{t-1}),$$

$$i_t = \sigma(W_i x_t + U_i h_{t-1} + V_i c_{t-1}).$$

Gated Recurrent Unit (GRU) is similar to LSTM. It does not have the memory cell, but utilizes update and reset gates. It has fewer parameters than LSTM and therefore it's training is faster.

The output of the GRU is:

$$h_t = (1 - z_t)h_{t-1} + z_t \tanh(W x_t + U(r_t h_{t-1})),$$

where z_t is update gate and r_t is reset gate.

$$z_t = \sigma(W_z x_t + U_z h_{t-1})$$

$$r_t = \sigma(W_r x_t + U_r h_{t-1})$$

A simple extension to the RNN is a bidirectional recurrent neural network (BDRNN). It contains two RNNs, the first working forward, the second working backward. Outputs at each step are merged (i.e. concatenated). Forward RNN remembers the left context and backward RNN remembers the right context. This feature solves a problem of providing full context to a token in POS tagging. Many taggers are using context window to incorporate nearest neighbors of a token, but it is usually of limited size.

4   Polish RNN Tagger

4.1   Tokenization and Morphological Analysis

Text preprocessing is performed by external tools. Firstly, a text is segmented into sentences and into tokens using Toki [19]. Secondly, each token in a sentence is analyzed by morphological analyzer SGJP. Maca [19] integrates both tools and is used in this tagger (Concraft also uses Maca).

4.2   Morphological Guesser

A morphological guesser is a system that predicts potential tags for a word form. In practice, it is applied only for unknown words which are not present in the dictionary. In this work, a morphological guesser as a separate step is omitted— unknown words have no features associated with potential tags. To address this issue, features, based on the word form were added: first and last three letters of a word form. WCRFT and Concraft use a separate classifier that replaces morphological analysis for unknown words.

4.3   Morphological Disambiguator

Morphological disambiguator assigns one tag for each token. The decision is based on features (observations) of tokens. In this work, classification is performed by a recurrent neural network, so there is no need for creating context window which limits the length of a potential dependency. By using a bidirectional recurrent neural network, for each token, a classifier has information about full left and full right context. It creates an advantage in comparison to a window approach used in CRF. Both WCRFT and Concraft use the context of length 2.

4.4   Lemmatization

Lemmatization is a task closely related to morphological disambiguation. Concraft and WCRFT do not tackle this problem. From a dictionary, they choose all lemmas associated with the disambiguated tag. Unknown words are not lemmatized—for the higher scores, token form is set as the lemma.

In this work, a simple extension is provided. Training data is analyzed by counting lemmas for each pair of a token and disambiguated tag. During tagging, the tagger starts with prediction of a tag and then chooses the most frequent lemma for the pair: token and disambiguated tag.

4.5   Network Architecture

The network has two bidirectional GRUs. Dropout (fraction of units to drop during training) is applied to the linear transformation of the recurrent state. Dropout is also applied to the results from the second bidirectional GRU and processed by dense layer with the same shared weights for each step. The network is presented in Fig. 2. All features of words are encoded as one-hot vectors.

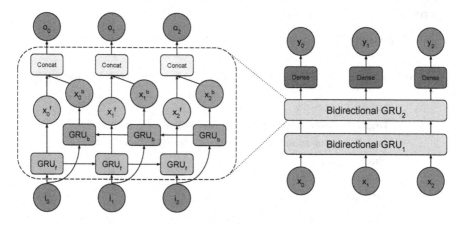

Fig. 2. A neural network used in this work. It allows variable-length sequences on input, but the figure shows the network for a sequence of length 3 $\{x_0, x_1, x_2\}$. Arrows represent directions of computation. A bidirectional GRU layer is presented on the left. Operation *Concat* concatenates outputs from two GRU layers into one matrix. Right diagram shows the full network with two bidirectional layers and a dense layer with shared weights on the output.

4.6   Features

The best Polish taggers use word form as a feature. In Polish, there are more than 3.8 million unique word forms. Treating them as one-hot vector would create too many inputs for a neural network. Generally, this problem of dimensionality reduction is being solved by using word embeddings. However, in this work word embeddings are not used—initial attempts have given worse results. Addition of word embeddings as an additional input to the model prevented learning from gaining above 90% of accuracy.

Eight sets of features were tested (the number of unique features is given in parentheses):

- **tags4** (388)—each tag is divided into two parts: grammatical class + case or person, and grammatical class + rest of grammatical categories [1],
- **tags5** (90)—case, and concatenation of number, case and gerund,
- **shape** (76)—collapsed shape of token - upper case letters are represented as u, lower case letters as l, digits as d, other characters as x (e.g. *Wrobel2017* gives *ulllldddd* and after collapsing *uld*),
- **cases** (5)—information whether word form is all lower cased, upper cased, capitalized, or a number,
- **interps** (55)—individual punctuation marks,
- **qubs** (226)—set of all particle adverbs,
- **3letters** (276)—first and last three letters of word form; this feature have source from morphological guesser in Concraft, in which prefixes and suffixes are generated,
- **separator** (2)—information about space before token.

Table 2 presents features generated for the word *obrazki*.

Table 2. Example of features generation for the word *obrazki* (images).

token	obrazki
tags4	1subst:nom, 2subst:pl:m3, 1subst:acc, 1subst:voc
tags5	pl:nom:m3, nom, pl:acc:m3, acc, pl:voc:m3, voc
shape	l
cases	islower
interps	
qubs	
3letters	P0o, P1b, P2r, S1i, S2k, S3z

4.7   Output Classes

To reduce the number of outputs WCRFT classifies each grammatical class and category separately while Concraft divides tags into two sets (the same as the feature **tags4**).

In comparison to other Polish taggers, KRNNT has undivided tags on output. A drawback of this approach is that tags not occurring in training data can not be predicted even if the morphological analyzer has information about possible correct tags.

4.8 Training

10% of training data is used as a validation set for early stopping. Early stopping criterion is checked after processing every 10,000 sentences with patience 10. The maximal number of epochs is 150. A loss (objective) function is categorical loss entropy. The last layer has softmax activation function.

Training is performed using Nadam optimizer, because it was proven to be effective for recurrent neural networks [4,21]. It is a combination of two algorithms: RMSProp and Nesterov momentum.

Training was performed on GPU NVIDIA Tesla K40 XL and took about 3 h (on NCP).

5 Evaluation

Many experiments testing different sets of features and neural network architectures were conducted (over 40,000 h on GPU). Taggers were assessed in terms of accuracy and speed of tagging. National Corpus of Polish with 10-fold cross-validation was chosen as a training corpus. Sentences from one paragraph are always in the same fold. Sentences incorrectly segmented by Maca (3.41% of NCP) are skipped during training (for simplicity).

Evaluation is performed with the whole pipeline including segmentation, morphological analysis and morphological disambiguation as proposed in [18].

The main metric is accuracy lower bound (Acc_{lower}). It penalizes all segmentation errors and is calculated as a percentage of all tokens that match tagger segmentation with correct tag. Additional metric accuracy upper bound (Acc_{upper}) treats segmentation errors as correctly tagged. It shows potential accuracy for a perfect tokenizer.

Two additional metrics are also provided: accuracy lower bound for known (Acc_{lower}^{K}) and unknown words (Acc_{lower}^{U}).

Table 3. Results of taggers in 10-fold cross-validation scheme using NCP.

Tagger	Acc_{lower}	Acc_{upper}	Acc_{lower}^{K}	Acc_{lower}^{U}
OpenNLP	87.24%		88.02%	62.05%
Pantera	88.95%		91.22%	15.19%
WMBT	90.33%		91.26%	60.25%
WCRFT	90.76%		91.92%	53.18%
Concraft	91.19%	91.53%	92.07%	60.64%
KRNNT	**93.72%**	94.05%	**94.43%**	**69.03%**

Comparison of results for each tagger is presented in Table 3. Scores for OpenNLP, Pantera, WMBT, and WCRFT originate from [11]. Evaluations for all taggers were performed on NCP and with 10-fold cross-validation scheme.

KRNNT significantly surpasses scores of other taggers, the error is reduced by 28% in comparison to Concraft. The accuracy of tagging unknown words is 7% points higher than in OpenNLP. Simple voting strategy over 10 models trained on the same data, but with different random initialization, increase accuracy lower bound up to 94.30% (tested without cross-validation).

For 0.55% tokens, the predicted tag was not in the set returned by the morphological analyzer. Despite that, KRNNT correctly assigns tags in 78.89%.

NCP has 2.81% unknown words according to Maca.

[11] developed ensemble of the first 5 taggers from Table 3. The best voting scheme achieves the accuracy lower bound slightly above 92%.

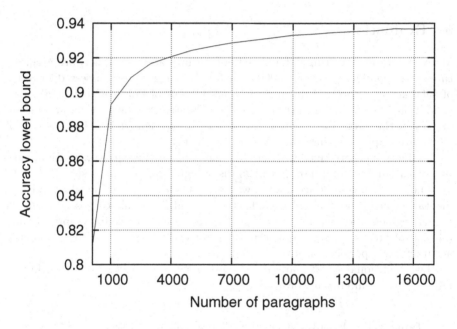

Fig. 3. Accuracy lower bound in function of a number of training paragraphs.

Figure 3 shows accuracy lower bound related to a number of paragraphs, that were used to train the tagger. More training data is needed to determine whether the classifier is saturated.

Table 4 shows tagging time in seconds of whole NCP including the start of a tagger. NCP was manually distributed for separate processes of Concraft because Concraft does not utilize more cores. KRNNT executes longer than Concraft. The analysis showed that the most time-consuming is to generate the features, execute Maca and parse its output. Distribution of these tasks to other cores decreases tagging time. Values in parentheses give the percentage of total time spent by GPU waiting for data. Implementation of the tagger in a statically typed language should improve performance.

Table 4. Time of tagging NCP measured in seconds. In the case of KRNNT, a percentage of tagging time when GPU is waiting for data is given in parentheses. KRNNT is also tested for corpus sorted by sentence (SS) length.

Tagger	1 core	2 cores	4 cores
Concraft	376	199	140
KRNNT GPU	556 (60%)	297 (50%)	223 (40%)
KRNNT GPU SS	423 (74%)	231 (64%)	141 (44%)

Processing sentences sorted by a number of words improves tagging time by 22%–37% because computations on GPU are performed in batches and all sentences need to be padded to the longest sentence in the batch. For sorted sentences, padding is minimal.

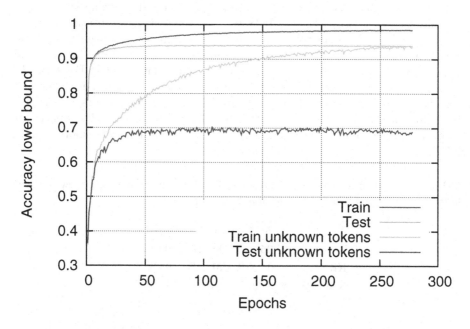

Fig. 4. Accuracy lower bound in function of a number of epochs.

Figure 4 shows accuracy lower bound for training and test data related to a number of epochs. The neural network can not memorize the training data (98.3%). Most likely this is caused by an insufficient representation of the input (word form is not a feature). The accuracy for test data is not raising after about 100 epochs and the model is not overfitting.

Manual analysis of 100 errors of the tagger showed that 6% relate to errors in manual annotation of NCP, e.g. *to były ostatnie słowa, jakie wypowiedział* (these were the last words that he had spoken) - the word *jakie* (that) is manually annotated as nominative, KRNNT tags it as accusative, which is correct. 9% of errors could only be avoided with the analysis of the whole paragraph, e.g. *Na baliku [...] bawiło się około 100 dzieci. Znaleźli się wśród nich* (About 100 children played at the ball. There were also among them) - the gender of word *nich* (them) is dependent on reference to *dzieci* (children). Dependencies longer than 5 words occur in 11% of errors. Including semantics and valency information could potentially reduce errors by 15%, e.g. *Władze miasta [...] szukają inwestora* (City authorities are looking for an investor) - the verb *szukać* (look for) takes objects with genitive case in this context (KRNNT assigns accusative case).

6   PolEval: POS Tagging

PolEval is a Polish version of SemEval—a contest for natural language processing tools. KRNNT participated in a task of morphosyntactic tagging [12]. It involves 3 subtasks: morphosyntactic disambiguation and guessing (subtask A), lemmatization (subtask B), and POS tagging (subtask C). Subtasks A and B are tested on gold segmented data, therefore systems do not need to perform tokenization and morphological analysis. Subtask C is tested on raw text and requires whole text processing pipeline. The training data is NCP. For subtasks A and B, the model was trained without morphological reanalysis. Therefore segmentation errors do not occur. Organizers prepared different testing corpus, they annotated over 1626 sentences for subtasks A and B and 1675 sentences for subtask C. Average number of tokens in sentences is around 16.5—more than in NCP.

Table 5. Results of best performing systems of PolEval subtask A. Accuracy is calculated separately and jointly for known and unknown words.

Tagger	Acc^K	Acc^U	Acc
Toygger	**95.24%**	**65.47%**	**94.63%**
KRNNT	94.49%	61.18%	93.80%
NeuroParser	94.21%	64.94%	93.61%

The best submission for task A (Table 5) was also prepared using bidirectional neural network. The main difference to KRNNT is a utilization of word embeddings and reduced output to separate grammatical classes and categories. KRNNT was placed second in the ranking.

Despite simple lemmatization module in KRNNT, it won subtask B (Table 6). NeuroParser has better accuracy in lemmatization of unknown words by 4% points, so there is a room for improvement.

Subtask C was also won by KRNNT (Table 7). However, lemmatization performed by NeuroParser was also better. Third place is taken by MorphoDiTaPL [22]—the framework achieving state-of-the-art results in Czech.

Table 6. Results of PolEval subtask B. Accuracy of lemmatization is calculated separately and jointly for known and unknown words.

Tagger	Acc^K	Acc^U	Acc
KRNNT	**98.19%**	80.86%	**97.84%**
NeuroParser	97.37%	**84.62%**	97.11%

Table 7. Results of PolEval subtask C. *POSAcc* is accuracy lower bound for morphosyntactic tagging, *LemAcc* is the accuracy of lemmatization and *OverallAcc* is the average of *POSAcc* and *LemAcc*.

Tagger	Tagging		
	POSAcc	*LemAcc*	*OverallAcc*
KRNNT	**92.98%**	96.91%	**94.95%**
NeuroParser	91.59%	**97.00%**	94.29%
MorphoDiTaPL	89.67%	95.78%	92.73%

6.1 PolEval 2020: Morphosyntactic Tagging of Middle, New and Modern Polish

PolEval 2020 hosts a shared task Morphosyntactic tagging of Middle, New and Modern Polish. The task organizers have shared training and development data analyzed by Morfeusz adapted to historical language. Additionally, each paragraph is annotated with a year of creation. In comparison to task A in PolEval 2017, the tokenization is ambiguous. A simple solution (used in Maca) is to calculate the shortest path in a directed acyclic graph representing possible tokenizations. Some difficulties may cause a lack of sentence boundaries.

The training and development data consist of 10755 and 244 paragraphs, respectively. 2% of tokens in development data are unknown to morphological analysis.

KRNNT was trained for 150 epochs using only training data, development data was not used for validation. The information about the year of text is omitted. As the data do not have information about sentence boundaries the training is performed on whole paragraphs. This approach should increase the accuracy, because some information from near sentences may be used to make better predictions. The model was tested on tokenized text using the shortest path method.

Evaluation results are presented in Table 8. We can expect higher results obtained by more recent model, e.g. contextual embeddings and by utilization of the year of analyzed text.

Table 8. Results of KRNNT on PolEval 2020 development data using only training data for training.

Tagger	Acc_{lower}	Acc_{lower}^{K}	Acc_{lower}^{U}
KRNNT	**92.07%**	**93.00%**	**48.63%**

7   Conclusion

This work presented Polish morphosyntactic tagger KRNNT. It achieves better accuracy than other publicly available taggers.

PolEval showed that better results could be achieved using bidirectional neural networks.

Lemmatization for unknown words may be improved by a separate neural network using a sequence to sequence architecture [3]. Bigger dictionary of named entities should boost the results.

Despite that tags have some structure, in this work they are treated separately. Therefore the system can not generalize well, e.g. "verb must be in every sentence" instead of "one of X tags describing verb must be in sentence". Tags should be partitioned on output, or more fine-grained outputs should be added.

Researchers should also focus on word embeddings for morphologically rich languages. Including them in a tagger should improve accuracy. What is more, representing tags as word embeddings might be beneficial because they can represent dependencies among them [5].

RNNs make local decisions for each token, incorporating CRF or hidden Markov models as the last layer will assign labels after seeing the word sequence [10].

Including information from the whole paragraph is essential for some ambiguities in tagging.

KRNNT is an open-source system and is available as a Docker container. The new model has been trained on NCP and PolEval 2017 data.

Acknowledgments. This research was supported in part by PLGrid Infrastructure.

References

1. Acedański, S.: A morphosyntactic Brill tagger for inflectional languages. In: Loftsson, H., Rögnvaldsson, E., Helgadóttir, S. (eds.) NLP 2010. LNCS (LNAI), vol. 6233, pp. 3–14. Springer, Heidelberg (2010). https://doi.org/10.1007/978-3-642-14770-8_3
2. Brill, E.: A simple rule-based part of speech tagger. In: Proceedings of the workshop on Speech and Natural Language, pp. 112–116. Association for Computational Linguistics (1992)
3. Cho, K., van Merrienboer, B., Gülçehre, Ç., Bougares, F., Schwenk, H., Bengio, Y.: Learning phrase representations using RNN encoder-decoder for statistical machine translation. CoRR abs/1406.1078 (2014). http://arxiv.org/abs/1406.1078

4. Dozat, T.: Incorporating Nesterov Momentum into Adam (2016)
5. Goldberg, Y.: A primer on neural network models for natural language processing. J. Artif. Intell. Res. (JAIR) **57**, 345–420 (2016)
6. Graves, A., Jaitly, N.: Towards end-to-end speech recognition with recurrent neural networks. In: Proceedings of the 31st International Conference on Machine Learning, ICML-14, pp. 1764–1772 (2014)
7. Graves, A., Mohamed, A., Hinton, G.: Speech recognition with deep recurrent neural networks. In: 2013 IEEE International Conference on Acoustics, Speech and Signal Processing (ICASSP), pp. 6645–6649. IEEE (2013)
8. Hochreiter, S., Schmidhuber, J.: Long short-term memory. Neural Comput. **9**(8), 1735–1780 (1997)
9. Huang, Z., Xu, W., Yu, K.: Bidirectional LSTM-CRF models for sequence tagging. CoRR abs/1508.01991 (2015). http://arxiv.org/abs/1508.01991
10. Huang, Z., Xu, W., Yu, K.: Bidirectional LSTM-CRF models for sequence tagging. arXiv preprint arXiv:1508.01991 (2015)
11. Kobyliński, Ł., Kieraś, W.: Part of speech tagging for Polish: state of the art and future perspectives. In: Proceedings of the 17th International Conference on Intelligent Text Processing and Computational Linguistics, CICLing 2016, Konya, Turkey (2016)
12. Kobyliński, Ł., Ogrodniczuk, M.: Results of the PolEval 2017 competition: part-of-speech tagging shared task. In: Proceedings of 8th Language & Technology Conference: Human Language Technologies as a Challenge for Computer Science and Linguistics. Wydawnictwo Poznańskie i Fundacja Uniwersytetu im. A. Mickiewicza, Poznań, Poland (2017)
13. Kuta, M., Chrzaszcz, P., Kitowski, J.: A case study of algorithms for morphosyntactic tagging of polish language. Comput. Inform. **26**(6), 627–647 (2012)
14. Lafferty, J., McCallum, A., Pereira, F.C.: Conditional random fields: Probabilistic models for segmenting and labeling sequence data (2001)
15. Piasecki, M.: Polish tagger TaKIPI: rule based construction and optimisation. Task Q. **11**(1–2), 151–167 (2007)
16. Pohl, A., Ziółko, B.: Using part of speech n-grams for improving automatic speech recognition of Polish. In: Perner, P. (ed.) MLDM 2013. LNCS (LNAI), vol. 7988, pp. 492–504. Springer, Heidelberg (2013). https://doi.org/10.1007/978-3-642-39712-7_38
17. Radziszewski, A.: A tiered CRF tagger for Polish. In: Bembenik, R., Skonieczny, L., Rybinski, H., Kryszkiewicz, M., Niezgodka, M. (eds.) Intelligent Tools for Building a Scientific Information Platform. Studies in Computational Intelligence, vol. 467. Springer, Heidelberg (2013). https://doi.org/10.1007/978-3-642-35647-6_16
18. Radziszewski, A., Acedański, S.: Taggers gonna tag: an argument against evaluating disambiguation capacities of morphosyntactic taggers. In: Sojka, P., Horák, A., Kopeček, I., Pala, K. (eds.) TSD 2012. LNCS (LNAI), vol. 7499, pp. 81–87. Springer, Heidelberg (2012). https://doi.org/10.1007/978-3-642-32790-2_9
19. Radziszewski, A., Śniatowski, T.: Maca – a configurable tool to integrate Polish morphological data. In: Proceedings of the 2nd International Workshop on Free/Open-Source Rule-Based Machine Translation (2011)
20. Radziszewski, A., Śniatowski, T.: A memory-based tagger for polish. In: Proceedings of the 5th Language & Technology Conference, Poznań, pp. 29–36 (2011)
21. Sutskever, I., Martens, J., Dahl, G., Hinton, G.: On the importance of initialization and momentum in deep learning. In: International Conference on Machine Learning, pp. 1139–1147 (2013)

22. Walentynowicz, W.: MorphoDiTa-based tagger for polish language (2017). http:// hdl.handle.net/11321/425. CLARIN-PL digital repository
23. Wang, D., Nyberg, E.: A long short-term memory model for answer sentence selection in question answering. In: ACL, vol. 2, pp. 707–712 (2015)
24. Waszczuk, J.: Harnessing the CRF complexity with domain-specific constraints. the case of morphosyntactic tagging of a highly inflected language. In: COLING, pp. 2789–2804 (2012)
25. Wróbel, K.: KRNNT: Polish recurrent neural network tagger. In: Vetulani, Z., Paroubek, P. (eds.) Proceedings of the 8th Language & Technology Conference: Human Language Technologies as a Challenge for Computer Science and Linguistics pp. 386–391. Fundacja Uniwersytetu im. Adama Mickiewicza w Poznaniu (2017)

Less-Resourced Languages

Experiments with Automatic and Semi-automatic Detection of Sparse Word Forms in Old Braj

Rafał Jaworski[(✉)] and Krzysztof Stroński

Adam Mickiewicz University in Poznań, Poznań, Poland
rjawor@amu.edu.pl

Abstract. This paper presents the work on automatic converb detection in Old Braj poetry from the 15–17 centuries. This is a part of research on non-finite verbal forms in early New Indo-Aryan (NIA) language corpora comprising data from Old Rajasthani, Awadhi, Braj, Dakkhini and Pahari [8]. The goal of the detection mechanism is to successfully identify a plaintext word as a converb or non-converb. Such mechanism facilitates further converb description and analysis, which is of great importance in research on historical syntax of NIA. In order to develop the automatic detector, a selection of state-of-art statistical classification mechanisms was used.

1 Introduction

Only in recent years there have been attempts to compile early NIA digitalized corpora, such as for example The Lausanne platform for early new Indo-Aryan digitized texts https://wp.unil.ch/eniat/. However, in comparison to Old Indo-Aryan or Middle Indo-Aryan, early NIA period which is crucial for understanding the main mechanisms of syntatic change in the history of IA lacks annotated corpora. An important desideratum is thus developing annotated corpora and tools detecting selected grammatical forms and facilitating their annotation. The present study which focuses only on one early NIA variety, namely Old Braj, is a modest attempt to fill this gap.

2 Converb in Indo-Aryan

The category of the converb has been recognized in the typological literature in early 90's and since then widely used as a crosslinguistically valid notion defined as "a nonfinite verb form whose main function is to mark adverbial subordination" [6]. In the context of IA languages this is one of the most central categories due to its role in clause linking as well as in defining of what is labelled 'the South Asian Linguistic Area', cf. [5,11]. Therefore the converb has received considerable attention in the IA linguistics. It has been analysed from various perspectives and at various stages of the development of IA (cf. [4,16,17] to name

© Springer Nature Switzerland AG 2020
Z. Vetulani et al. (Eds.): LTC 2017, LNAI 12598, pp. 119–130, 2020.
https://doi.org/10.1007/978-3-030-66527-2_9

just a few). So far, however, there has been no attempt to analyse non-finites using more advanced corpus oriented quantitative and qualitative methods. The first step to such an analysis is an attempt at converb detection which in turn results in creation of possibly large set of converbal forms. In the present paper we present results of converb detection in one particular variety of early NIA, namely Old Braj. The corpus developed so far for Old Braj consists of more than 10000 words of prose and verse composed by authors such as Indrajit of Orchā [12], Hita Harivaṃśa[1], Viṣṇnudās [3] and Bhūṣaṇ Tripāṭhī [14] between the 15–17 centuries. However, for the purpose of the present paper we have included only two authors, namely Hita Harivaṃśa and Bhūṣaṇ Tripāṭhī but as we will demonstrate in the next sections the tagged portions of the texts by the two authors were supplemented by the untagged ones which were the basis for the training of the detection mechanism. The selection of texts was not random. The texts belong to two genres (epic and lyrical poetry) and they seem to be quite representative for Old Braj.

3   Previous Research on Awadhi Converbs

Elswhere [8] we have demonstrated how converb detection can facilitate a multilayered analysis of non-emebeded structures along the lines presented in a multivariate analysis model [1]. The converb detector made it possible to extrapolate the results of the analysis performed on the mannually tagged corpus to an untagged one. We have chosen quite a homogenous corpus of early Awadhi (Jāyasī's Padmāvat - a verse text from the 16th century). We studied the main argument marking and subject identity constraint (SIC). Moreover, we nalysed the scope of the operators such as Tense (T), Illocutionary Force (IF) and Negation (NEG) operators which we have selected from the list of proposed variables in [1].

In our targeted research, we selected 236 converbal chain constructions (116 from the tagged part and 120 from the untagged one). We concluded that main argument marking follows the same rules as in the case of finite verb constructions, i.e. the A^2 marking depends exclusively on the transitivity of the main verb. What is more, unmarked A's outnumber marked ones which may result from the decay of the inflectional system leading to the disappearance of the ergative alignment in eastern varieties of IA. We have also noticed that Differential Object Marking (DOM) in early NIA is in the stage of development along the animacy and definiteness lines – in the 16th century we still find unmarked animate and definite O's (see example (1) with an animate and definite O *suā* 'parrot'). This conforms to the previous research on O marking in other branches of IA (see for example [18] for Nepali).

---

[1] http://wp.unil.ch/eniat/2015/05/hymns-by-hita-harivam.sa/.

[2] We use here consistently two of the three Dixonian primitive terms [2] i.e. A - subject of a transitive verb and O - object of a transitive verb.

(1) Old Awadhi

dhāi	suā	lai
wet-nurse.NOM.F.SG	parrot.**O**.NOM.M.SG	take.CVB

mārai	gaī
kill.INF.OBL	go.PPP.F.SG

'The wet-nurse having haken the parrot went to kill it'.
(JayP86.1)

SIC violation seems to have less constraints than in contemporary NIA. It is permitted not only when the subject of the converb is inanimate and the converb denotes a non-volitional act (see example (2) with an implicit animate subject and volitional action) unlike in modern NIA, cf. [16].

(2) Old Awadhi

sakami	hamkāri	phāmdi	giyam	melā
power	call.CVB	noose	neck.OBL.F.SG	put.PP.M.SG

'[birds] having called [one another] with power [loudly],
the noose was put on their neck.' (JayP72.3)

Preliminary analysis of the T-scope has brought interesting conclusions pertaining to the aspectual value of the IA converb. We assume that the fact that the IA converb is not congruent with the present tense reference gives direct evidence to its perfectivity (compare examples (3) and (4) in which T scope is conjunct or disjunct respectively). Interestinlgly we could find typological parallels to it as well.

(3) Old Awadhi

pamkhi-nha	dekhi	saba-nhi	dara
bird.OBL.M.PL	see.CVB	all.OBL.PL	fear.M.SG

khāvā
eat.PPP.M.SG

'... birds saw all of that and got scared.' (JayP69.2)

(4) Old Awadhi

sidha	darahim	nahim	apane	jīvam
holyman	fear.3PL.PRS	not	self	life.OBL.M.SG

kharaga	dekhi	kai	nāvahim	givām
sword	see.CVB	CVB	bow.3Pl.PRS	neck.OBL.F.SG

'Holly men do not fear for their own life but they <u>bow</u>
their necks <u>having seen</u> a sword.' (JayP240.3)

In Awadhi, IF and NEG scope result from the position of the markers. If the WH-word is in the right most position in the clause, then its scope is prefereably conjunct but when it preceeds the main verb and the converb is preposed, the scope remains local. If the NEG marker is in front of the postposed main verb or the converbal clause, the scope of negation is local. However, if the NEG marker occurs in front of the preposed converbal clause, the scope of negation is also local but it does not extend to the adjacent converbal clause. Similar conclusions were arrived at in the preliminary research on Rajasthani [7].

4   Automatic Converb Detection in Braj

The aim of an automatic converb detection mechanism is to annotate all converbs in a text. The input is a plain text written in Old Braj, which is transliterated according to ISO 15919 rules. The text is split into sentences and tokenized. The detector is run on each tokenized sentence. The input for the detection mechanism is therefore a series of plaintext, unannotated words, for which the detector is to provide yes/no answers, indicating whether they are converbs or not. This makes it a binary classification problem.

4.1   Experimental Data

In order to collect the data for the experiment, a portion of Braj documents were annotated manually on word level. This task was facilitated by the *IA tagger* system [7], available at

http://rjawor.vm.wmi.amu.edu.pl/tagging/

Login credentials can be obtained on request to the authors. The system enables the users to provide annotations for individual words in the sentence on several annotation levels. Among the levels there are: grammatical and morpho-syntactic information, POS-tags, semantics and pragmatic information (the list of levels and tags is configurable and typically tailored to a specific annotation task). The annotation is displayed as a table, where individual words are column headers and the annotation levels are represented in rows.

In our experiment, 10 001 words were tagged on the levels: grammar (according to the Leipzig Glossing Rules) and part-of-speech. The information about a word being a converb in the annotated texts is stored on the POS level in the appropriate tag: CVB. An example of an annotated Old Braj sentence is shown in Fig. 1.

Out of the total 10 001 words, 263 were annotated as converbs, which constitutes 2.6%.

4.2   Baseline System

The first step was to construct a baseline system, which works under the following principle: all the converbs encountered in the training set are added to the dictionary. Then, all the words in the test set which appear in the converb dictionary are annotated as converbs, while the remaining words are annotated as non-converbs. This step was taken in order to establish, whether converb detection in the collected data is a trivial or non-trivial problem.

All developed binary classifiers, including the baseline, were tested in the same scenario of 10-fold cross validation. The results obtained by the baseline system are shown in Table 1.

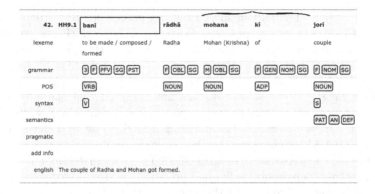

Fig. 1. Old Braj sentence annotated in the IA tagger system

Table 1. Results of the baseline converb detector.

Metric	Score
Accuracy	96.91%
Precision	56.8%
Recall	55.3%
F-score	56.0%

As the converbs are rare in the training and test sets, the accuracy measure is not reliable (most classifiers under such circumstances are heavily biased towards non-converb predictions, which are often correct). However, the precision and recall scores for the baseline system can provide useful information.

The baseline approach was supposed to yield a high precision but low recall score. The recall score was indeed low, which follows from the fact that converbs, as specific forms of verbs, are an open class and therefore there is a relatively high probability that a converb in the test set had not been detected during training.

However, the precision score achieved by the baseline system is also surprisingly low. The interpretation of this result is the following: some, not rare homonymous words can in some contexts be converbs, while in other contexts – not.

In conclusion, the converb detection problem in this scenario is definitely non-trivial.

4.3   VW and Converb Detection in Other Early NIA

Experiments in automatic processing of texts in early NIA had been done before [9]. The first phase of the research involved developing an automatic part-of-speech tagger for Rajasthani. All training data was acquired by the means of the IA tagger system. Each word was assigned exactly one POS tag, coming from a tagset of 22 tags. The part-of-speech tagging process for Rajasthani was seen by the authors as a multi-class classification problem which was approached with an algorithm based on the Maximum Entropy principle. As some of the tags in the tagset formed hierarchies (e.g.. there was a "NOUN" tag and its child, "NOUN-SINGULAR") the authors computed the accuracy of the automatic POS tagger in two scenarios - the first required that the automatically assigned tag matched exactly the expected tag (exact matching), while the second allowed partial matching, e.g. assigning a NOUN tag to a word tagged as NOUN-SINGULAR. Unfortunately, the accuracy results achieved in both scenarios fell below expectations, yielding 57.9% for exact matching and 64.1% for partial matching. Separately, individual precision and recall scores for detecting only converbs were computed for this system. The precision was as low as 33% with the recall not exceeding 7%.

This failure turned us towards developing systems focused on the one part-of-speech of particular interest – the converb. As a result, we developed specialized converb detectors for Rajasthani [7] and Awadhi [9]. The algorithms relied on statistical binary classifiers. The results achieved by the binary converb detectors are shown in Table 2.

Table 2. Results of the converb detectors for Rajasthani and Awadhi.

	Rajasthani	Awadhi
Precision	83%	80.2%
Recall	39%	64.4%
F-score	53%	71.4%

4.4   Vowpal Wabbit Classifier for Braj

In order to tackle the problem of automatic converb detection in Old Braj, the Vowpal Wabbit (VW) software [10] was used. VW is a well-established, robust statistical classification toolkit, which combines numerous classification and regression algorithms. VW is well optimized for the use of sparse features.

A useful functionality of VW is the analysis of the impact of individual features on the predictions (the *vw-info* program).

The design of the converb detector for Old Braj was inspired by previous research on similar problems for Old Rajasthani and Old Awadhi [9]. The feature engineering process resulted in finding the following most informative features:

– three letter suffix of the word
– two letter suffix of the word
– one letter suffix of the word (i.e. the last letter of the word)
– literal form of the previous word in the sentence
– distributional similarity class of the previous word in the sentence
– distributional similarity class form of the next word in the sentence
– (binary feature) is the word first or last in the sentence

Distributional similarity (often abbreviated *distsim*) is a method for categorizing words in a large corpus based on their contexts. Each word falls into a category with other words that appeared in similar contexts. The id of such a category can be used as a word feature.

In order to compute distributional similarity classes, an unannotated modern Rajasthani corpus of 81 843 words was used. It was processed with the help of word2vec software, described in [13]. The words were categorized into 209 classes, each containing between 1 and 66 words.

For example, one of the classes contained the following words: *te* 'this', *teha* 's/he', *bi* 'two', *bewai* 'both', which are all pronouns.

The final results of the developed VW converb detector are shown in Table 3.

Table 3. The final results of the developed VW converb detector for Braj enhanced by the Rajasthani training set.

Metric	Score
Accuracy	98.50%
Precision	81.8%
Recall	74.4%
F-score	77.9%

4.5   Interpretation of Achieved Results

The results of the VW converb detector show a considerable improvement in comparison with the baseline. They allow to use the developed converb detector in a scenario previously applied in the aforementioned research on Old Awadhi. It is possible to take a large unannotated collection of texts in Old Braj and run the converb detector on the text. Next, the sentences containing automatically detected converbs are presented to a team of linguists who manually verify the results of the detection. High precision results of the detector ensure that in most cases (4 out of 5) the linguists are presented with actual converbs, which saves vast amounts of their labour. On the other hand, the recall of nearly 75% indicates that only 1 in 4 de facto converbs in the unannotated text will not be presented to the linguists. This loss, however, is acceptable in the light of the following fact: as manual analysis of the whole large corpus would be unfeasible, such project would not start altogether, leading to a 100% converb loss. Note also that any recall score above the score achieved by dictionary detection implies that the detector is able to detect converbs unseen in the training data. Such new converb examples are of high interest to the linguists.

5   Experiments

This section presents two experiments carried out according to the scenario described in Sect. 4.5. In both cases the complete data set collected within the

IAtagger was used as the training set for the Braj converb detector. The detector was then run on two different larger texts written in Old Braj: 'Hita caurāsī' from the 16th c. by Hita Harivaṃśa and 'Śivrājbhūṣaṇa' from 1673 by Bhūṣaṇ Tripāṭhī. A short presentation of results of the linguistics analysis of converbal constructions based on both texts is provided in Sect. 5.3.

5.1 Experimental Scenario

In both cases the experimental scenario consisted in the following procedure:

1. Training the VW converb detector with the data from IAtagger
2. Running the detector on the longer text in order to obtain converb annotations
3. Upload of all the sentences with at least one converb annotated back to the IAtagger
4. Manual verification of annotations.

Based on the data provided by manual verification it is possible to compute the precision score of the VW converb detector in a real-life scenario. It is not possible, however, to assess the recall score based only on the experimental data due to the fact that the longer texts were not fully analyzed and the actual number of converbs in these texts remains unknown.

Nevertheless, the precision score gives an assessment of usability of the proposed research procedure. The higher the precision score, the less time the linguist spends on analyzing false positives.

5.2 Results

The VW converb detector executed on the fragment of 'Hita caurāsī' annotated **84** converbs out of which **75** were manually verified as actual converbs. This yields the precision score of **89.3%**. This result can be attributed to the fact that converb forms in this text exhibited high level of homogeneity.

The results on the 'Śivrājbhūṣaṇa' text were lower due to larger variety of converbal forms. The detector annotated **381** potential converbs out of which **270** were verified as such. This yields the precision score of **70.9%**.

Nevertheless, the precision achieved by the VW converb detector can still be considered satisfactory, as it allows for an efficient and effective linguistic analysis.

5.3 Old Braj Converbal Constructions - Linguistic Analysis

A part of each author was included in the main training set consisting of more than 10000 words. Based on selection of converbal forms both from the tagged set (69 and 115 for Hita Harivaṃśa and Bhūṣaṇ respectively) and the untagged one (78 and 286 for Hita Harivaṃśa and Bhūṣaṇ respectively) we were able to carry out a targeted search for converbal forms and formulate conclusions regarding morphosyntactic and scopal properties of converbs in Old Braj. Similarly to

Early Awadhi, firstly we dealt with main argument marking in converbal chain constructions. Interestingly, in the fully annotated part of the Braj corpus comprising both Hita Harivaṃśa's and Bhūṣan's works there were no instances of marked A but in the untagged part of Bhūṣan's work we have found single cases of marked A. As per general rule, main argument marking is triggered exclusively by the transitivity of the main verb (see example (5) with the main verb and converb being transitive and marked main argument 'you' and (6) with the main verb intransitive and a transitive converb and unmarked main argument 'Aurangzeb').

(5) Old Braj

lari	*lari*	*sarajā*	*soṁ*	*jaṁga*	*nipaṭa*
fight.CVB	fight.CVB	lion	with	fight.M.SG	excesively
abhaṁga	*gaḍha*	*koṭa*	*saba*	*hāre*	*taiṁ*
unbroken	fort	fortress	all	lose.PPP.M.PL	you.AG

'You fought excessively with the lion (i.e. Śivaji) and you lost all undefeated forts and fortresses.' (Ś.250.2)

(6) Old Braj

eka	*samai*	*saji-kai*	*saba*	*sena*
one	time.OBL	prepare-CVB	all	army
sikāra	*kauṁ*	*ālamagīra*	*sidhāe*	
hunting	DAT	Aurangzeb	proceed.PPP.M.PL	

'Once Aurangzeb prepared all soldiers and took off for hunting.' (Ś.90.1)

As regards O marking, we have a clear tendency of to develop gradually O marking along the lines of the animacy/defintness hierarchy. Therefore, we do not notice any O marking in Hita Harivaṃśa's work but they are present in Bhūṣan's work which is at least 100 years later. In Bhūṣan's work we find not only O marking with animate and definite O's such as 'king' in (7) but with inanimates as well, such as 'wealth' in (8).

(7) Old Braj

aiseṁ	*umarāva*	*lai*	*cale*	*manāya*
so	noble	take	move.PPP.M.PL	convince.CVB
sivarāja	*mahārāja*	*kauṁ*	*lakhi*	
Shivraj	king	DAT	see.CVB	

'...having seen the great king Shivaji the nobles proceeded to convince him...' (Ś.34.3)

(8) Old Braj

lakhi	*sampati*	*kauṁ*	*alakāpati*	*lājai*
see.CVB	wealth.F.SG	DAT	god of wealth	feel ashamed.3SG.PRS

'...having seen the wealth the god of wealth feels ashamed.' (Ś.16.2)

Unlike in modern NIA cf. [16] SIC violation is also permitted in the case where the subject of the converb is animate and the converb denotes a volitional act. Moreover, if there is a possessor/experiencer relation between the subject of the converb and the subject of the main clause, SIC rule is not obeyed. In example (9) the subject of the converb denoting a volitional act is animate and there is a possessor/experiencer relation between 'couple' and 'joy').

(9) Old Braj

āju	*prabhāta*		*latā*	*maṁdira*		*meṁ*
today	morning		creeper	palace		in
susa	*barasata*		*ati*	*haraṣi*		*jugala bara*
joy	rain down.PRS.SG	so		be happy.CVB		couple beautiful

'Today [at] dawn in the palace of creepers, the best couple is happy and (their) joy rains down.' (HH.5.1)

After our corpus analysis of seleced scopal properties of converbs in Old Braj, we conclude that the IF scope in commands is local when the main verb is preposed, as in (10). Another factor determining the IF-scope in questions may be the position of Wh-words, for instance if the Wh-word occurs in front of the main verb with the main clause postposed, the scope is preferably local (11).

(10) Old Braj

deṣi	*sambhāri*		*pīta*	*pata*	*ūpara*
see.2SG.IMP	put on.CVB		yellow	cloth	above
kahāṁ	*cūnarī*		*rātī*		
where	(partly dyed) cloth	red			

'See! [how] you have put on the yellow cloth, where [is] the red garment?' (HH.20.3)

(11) Old Braj

haraṣita	*iṁdu*		*tajata*		*jaisai*		*jaladhara*	*so*
delighted	moon		abandon.PRS.SG	like			cloud	so
bhrama	*dhūṁḍhi*		*kahāṁ*		*saba*		*hauṁ*	*pāūṁ*
error	search.CVB	where			all		I.NOM	get.1SG.SBJV

'As the delightful moon abandons the cloud, so having searched, where would I find out the error (delusion)?' (HH.14.4)

As for the T-scope, we could see in Old Awadhi a preference for conjunct scope with main verb in past tense forms whereas other forms usually implied local reading (compare (5) with conjunct and (12) with local scope).

(12) Old Braj

alikula-kalita		*kapola*	*dhyāya*		*lalita*	*anaṁdarūpa*
surrounded by bees		cheek	meditate.CVB		lovely	delightful beauty
sarita		*moṁ*	*bhūṣana*		*anhāiyai*	
stream		in	Bhushan		bathe.3SG.SBJV.PASS	

'May Bhushan take bath in the river of delightful beauty having meditated [with] cheeks surrounded by bees.' (Ś.1.3)

The last variable taken into account was the scope of the NEG-operator. In sequential constructions in early NIA, NEG-scope was predominantly conjunct but the conjunct reading results from the position of the main clause (or the subject) as well. Thus, with the main clause postposed (13) the NEG-scope is conjunct whereas with the main clause preposed (14) the NEG-scope is local. This is in line with findings in other dialectal groups such as Awadhi, Rajasthani and Dakkhini cf. [15].

(13) Old Braj

rājakāja	*dekhi*	*koū*	*pāvata*		*na bheu*	*hai*
royal act	see.CVB	who	get.PRS.SG		not secret	be.3SG.PRS

'The one who <u>has not</u> seen your royal act <u>does not</u> reach (your) secret.
lit. Having seen the royal act, no one reaches (your) secret.' (Ś.67.3)

(14) Old Braj

calahi	*na*	*capala*	*bāla*	*mṛga*	*nainī*
walk.IMP.2SG	not	restless	girl	deer	one with eyes
taji	*ba*	*mavana*			
abandon.CVB	now	silent			

'<u>Do not</u> walk, the restless deer-eyed girl, having abandoned now the silence?.' (HH39.5)

6   Conclusions

In this paper we have presented results of the research on one early New Indo-Aryan variety, namely Old Braj which involved linguistic analysis aided by computational methods. For the purpose of annotation of early NIA corpus the IA tagger system was designed. The system allows for manual word-level text annotation on several levels, which is performed by linguists. Collected data serves both for direct linguistic analysis and for training and testing automatic classifiers.

As the recent research on Old Braj is focused on specific verb forms – converbs – automatic converb detector for Braj was developed. The mechanism is a binary classifier, which relies on the Vowpal Wabbit machine learning toolkit. Features for the binary classifier are based on the literal form of a word, its context and the information on distributional similarity. The performance of the classifier allows for its use in a scenario, where converbs are automatically detected in a large unannotated text collection and then verified by linguists. This method has at least two significant advantages: firstly it enriches the lexicon of targeted non-finite forms and secondly it facilitates the process of a multilayered linguistic analysis of any type of converbal constructions.

Acknowledgements. This research was supported by Polish National Science Centre grant 2013/10/M/HS2/00553.

References

1. Bickel, B.: Capturing particulars and universals in clause linkage: a multivariate analysis. In: Bril, I. (ed.) Clause Linking and Clause Hierarchy : Syntax and Pragmatics, pp. 51–102. No. 121 in Studies in Language Companion Series, John Benjamins, Amsterdam (2010). http://dx.doi.org/10.5167/uzh-48989

2. Dixon, R.M.: Ergativity. Cambridge Studies in Linguistics, Cambridge University Press (1994). https://books.google.pl/books?id=fKfSAu6v5LYC
3. Dvivedī, L.: Viṣ?udās kavki?t Rāmāyana kathā. Sāhitya bhavan limited (1972)
4. Dwarikesh, D.P.S.: Historical syntax of the conjunctivc participle phrase in new indo-aryan dialects of madhyadesa (midland) of northern india. University of Chicago Ph.D. dissertation (1971)
5. Emenau, M.: The sanskrit gerund: a synchronic, diachronic and typological analysis. Language **32**, 3–16 (1956)
6. Haspelmath, M.: The converb as a cross-linguistically valid category. In: Haspelmath, M., König, E. (eds.) Converbs in Cross-Linguistic Perspective: Structure and Meaning of Adverbial Verb Forms - Adverbial Participles, Gerunds, pp. 1–55. No. 13 in Empirical approaches to language typology, Mouton de Gruyter, Berlin (1995)
7. Jaworski, R., Jassem, K., Stroński, K.: Manual and Automatic Tagging of Indo-Aryan Languages. Human Language Technologies as a Challenge for Computer Science and Linguistics, pp. 550–554 (2015)
8. Jaworski, R., Stroński, K.: New perspectives in annotating early new indo-aryan texts. In: Proceedings of the 32nd South Asian Languages Analysis Round Table SALA-32, Lisbon, Portugal, pp. 66–68 (2016)
9. Jaworski, R., Stroński, K.: Recognition and multi-layered analysis of converbs in early NIA. In: Proceedings of the 33rd South Asian Languages Analysis Round Table SALA-33, Poznań, Poland, pp. 55–56 (2017)
10. Langford, J., Li, L., Zhang, T.: Sparse online learning via truncated gradient. In: Advances in Neural Information Processing Systems, pp. 905–912 (2009)
11. Masica, C.P.: Defining a Linguistic Area: South Asia. Chicago University Press, Chicago (1976)
12. McGregor, R.: The Language of Indrajit of Orchā: A Study of Early Braj Bhāsā Prose. University of Cambridge Oriental Publications, Cambridge University Press (1968). https://books.google.pl/books?id=EjI3vgAACAAJ
13. Mikolov, T., Chen, K., Corrado, G., Dean, J.: Efficient estimation of word representations in vector space. CoRR abs/1301.3781 (2013). http://arxiv.org/abs/1301.3781
14. Misra, V.P.: Bhusana granthavali. Nai Dilli, Vani Prakasan (1994)
15. Stroński, K., Tokaj, J., Verbeke, S.: A diachronic account of converbal constructions in old rajasthani. In: Cennamo, M., Fabrizio, C. (eds.) Historical Linguistics 2015, Selected papers from the 22nd International Conference on Historical Linguistics, Naples, 27–31 July 2015, pp. 424–441. No. 348 in Current Issues in Linguistic Theory, John Benjamins, Amsterdam/Philadephia (2019)
16. Subbārāo, K.: South Asian Languages: A Syntactic Typology. South Asian Languages: A Syntactic Typology, Cambridge University Press (2012). https://books.google.pl/books?id=ZCfiGYvpLOQC
17. Tikkanen, B.: The Sanskrit gerund: a synchronic, diachronic, and typological analysis. Studia Orientalia, Finnish Oriental Society (1987). https://books.google.pl/books?id=XTkqAQAAIAAJ
18. Wallace, W.D.: Object-marking in the history of nepali: a case of syntactic diffusion. Stud. Linguist. Sci. **11**(2), 107–128 (1981)

Towards Better Text Processing Tools
for the Ainu Language

Karol Nowakowski$^{(\boxtimes)}$ ⓘ, Michal Ptaszynski ⓘ, and Fumito Masui ⓘ

Department of Computer Science, Kitami Institute of Technology, 165 Koen-cho,
Kitami, Hokkaido 090-8507, Japan
nowakowski.karol.p@gmail.com, ptaszynski@cs.kitami-it.ac.jp,
f-masui@mail.kitami-it.ac.jp

Abstract. In this paper we present our research devoted to the development of Natural Language Processing technologies for the Ainu language, a critically endangered language isolate spoken by the Ainu people, the native inhabitants of northern parts of the Japanese archipelago. In particular, we focused on improving the existing tools for transcription normalization, word segmentation (tokenization) and part-of-speech tagging. In the experiments we applied two Ainu language dictionaries from different domains (literary and colloquial) and created a new data set by combining them. The experiments confirmed the positive effect of these modifications on the overall performance of the tools, especially with objective samples unrelated to the training data. We also discuss further improvements obtained by applying corpus-driven language models to the problem of word segmentation and using a state-of-the-art tool for training part-of-speech taggers.

Keywords: Ainu language · Endangered languages · Under-resourced languages · Transcription normalization · Word segmentation · Tokenization · Part-of-speech tagging

1   Introduction

The Ainu language is a critically endangered language spoken by the Ainu people, the native inhabitants of northern Japan. We believe that technologies being developed within the field of Natural Language Processing (NLP) have a great potential to support the urgent tasks of documenting, analysing and revitalizing endangered languages, such as Ainu. This paper presents our research aimed at the development of text processing tools for the Ainu language. In particular, it is an extended version of the paper presented at the 8th Language & Technology Conference (LTC 2017) in Poznan, Poland [10], explaining the improvements introduced in POST-AL, a tool for automatic linguistic analysis of the Ainu language, initially developed by Ptaszynski and Momouchi [17]. Furthermore, we include a short description of recently obtained results being a continuation of the research presented at the conference.

© Springer Nature Switzerland AG 2020
Z. Vetulani et al. (Eds.): LTC 2017, LNAI 12598, pp. 131–145, 2020.
https://doi.org/10.1007/978-3-030-66527-2_10

The remainder of this paper is organized as follows. In Sect. 2 we briefly describe the characteristics and the current situation of the Ainu language. In Sect. 3 we provide an overview of related research. Section 4 presents the proposed tools for transcription normalization, word segmentation and part-of-speech tagging. In Sects. 5 and 6 we introduce the training data and test data used in the experiments. Section 7 summarizes the evaluation methods we applied. In Sect. 8 we present the results of the evaluation experiments. Finally, Sect. 9 contains conclusions.

2 The Ainu Language

In terms of typology, Ainu is an agglutinating, SOV (subject-object-verb) language, with elements of polysynthesis, such as noun incorporation [20]. Although there have been many attempts to relate Ainu to individual languages (e.g., Japanese) or to groups of languages, such as Paleo-Asiatic, Ural–Altaic, or Malayo–Polynesian languages (see [8]), none of these hypotheses gained wider acceptance. Thus, Ainu is regarded as a language isolate.

For the most part of its history, the Ainu language did not have a written form, but instead had a rich tradition of oral literature. One of the best-known examples of the Ainu literary forms are the *yukar*, narrative epics about deities and heroes.

2.1 Current Situation

While the exact number of Ainu language speakers is difficult to determine, in a survey conducted in 2017 by the Hokkaidō regional government [5], only 4.1% out of 671 respondents answered that they were able to communicate using the Ainu language. This situation is a consequence of the language shift from Ainu to Japanese that started in the 19th century and resulted in the mother tongue of the Ainu people no longer being transmitted to next generations [19].

That being said, in the last few decades Ainu people started to regain pride in their culture, which resulted in an increase of interest in the Ainu language, too. Currently, a number of Ainu language courses are offered throughout Hokkaidō. The Foundation for the Research and Promotion of Ainu Culture (FRPAC) holds an annual Ainu language speech contest ("Itak an ro" [Let's speak Ainu!]) and collaborates with the STV Radio in Sapporo in broadcasting a series of Ainu language courses ("Ainugo Rajio Kōza"[1]). A magazine in the Ainu language, "Ainu Taimuzu" [Ainu Times][2], is being published since 1997. There are also musicians performing in the Ainu language, such as "Oki Dub Ainu Band" and "Marewrew".

---

[1] https://www.stv.jp/radio/ainugo/index.html.
[2] https://otarunay.at-ninja.jp/taimuzu.html.

3   Related Research

One of the early attempts at creating Natural Language Processing tools for the Ainu language was the POST-AL system, developed by Ptaszynski and Momouchi [17]. It performed five main tasks: transcription normalization (modification of parts of text that do not conform to modern rules of transcription – e.g., *kamui→kamuy*), tokenization (this step also encompasses the process of word segmentation, needed for the analysis of texts where space-delimited segments are too coarse-grained – which is often the case in older documents), part-of-speech (POS) tagging, morphological analysis and word-level translation into Japanese. It was trained using a dictionary compiled by Kirikae [7] and performed morphosyntactic disambiguation based on lexical n-grams obtained from sample sentences included in the dictionary. The tokenizer was based on the maximum matching algorithm. The system also included several heuristic rules for normalization of transcription in older texts.

It two later studies, Ptaszynski et al. [16,18] investigated the possibility of improving the system's performance in part-of-speech tagging and word segmentation, by testing it with four different dictionaries. In 2017, Nowakowski, Ptaszynski and Masui [10] achieved further improvements by using a combination of two dictionaries instead of one and applying a hybrid method of part-of-speech disambiguation, based on n-grams and Term Frequency, as well as a segmentation algorithm maximizing mean token length. They also enhanced the list of transcription normalization rules and made their execution conditional on the output of the tokenizer, in order to correctly handle ambiguous cases. Nowakowski et al. [14] applied the SVMTool [4] – a state-of-the-art generator of sequential taggers based on Support Vector Machines – to the task of part-of-speech tagging, with good results.

Apart from the development of tools for automatic processing of the Ainu language, recently Nowakowski, Ptaszynski and Masui [11] launched a project to create a large-scale annotated corpus of Ainu. Utility of the corpus for NLP applications was positively verified by Nowakowski, Ptaszynski and Masui [12, 13], who used it to train an n-gram language model-based word segmentation system.

In the remaining part of this paper, we will discuss the results originally presented by by Nowakowski, Ptaszynski and Masui [10] at the LTC'17 conference, along with further improvements obtained by Nowakowski et al. [14] and Nowakowski, Ptaszynski and Masui [13].

4   System Description

In this section we present the technical details of the proposed algorithms for transcription normalization, word segmentation, and part-of-speech tagging. Unless otherwise specified, the description pertains to the system developed by Nowakowski, Ptaszynski and Masui [10].

4.1 Transcription Normalization

The role of this preprocessing step is to detect all substrings of a given input string that are equal to the upper part of any of the transcription change rules shown in Table 1, and generate a list of all possible transcriptions, where each of such substrings is either substituted with its modern equivalent (the lower part of the corresponding rule) or retained without modification. Given an input string with n substrings to be potentially modified, a list of 2^n strings will be generated. An example is shown in Table 2. Unlike in the original system [17], transcription change rules are optional – the decision as to which of them should be applied (which of the possible permutations of the input string to select) is made in the next step by the word segmentation algorithm.

Table 1. Transcription change rules.

Original transcription													
ch	sh(i)	ai	ui	ei	oi	au	iu	eu	ou	mb	b	g	d
c	s	ay	uy	ey	oy	aw	iw	ew	ow	np	p	k	t
Modern transcription													

Table 2. A fragment of text before and after processing with the normalization algorithm.

Input string	List of output strings	Meaning
chepshuttuye	cepsuttuye	"to exterminate fish"
	cepshuttuye	
	chepsuttuye	
	chepshuttuye	

4.2 Tokenization Algorithm

The tokenizer proposed by Nowakowski, Ptaszynski and Masui [10] takes as input a list of all possible permutations of the input string – which has been generated in the previous stage – and performs a dictionary lookup in order to find a single token or the shortest possible sequence of tokens from the lexicon, such that after concatenation is equal to any string from the input list (i.e., it iterates through all variants of the input string and selects the one that allows for a complete match with the shortest sequence of lexicon items – an example is shown in Table 3). If more than one variant can be matched with a sequence containing a certain number of tokens, priority is given to the variants following the modern transcription rules listed in the lower part of Table 1 (the matching algorithm iterates through the strings with modernized transcription, before proceeding to the ones where normalization rules were not applied).

Table 3. A fragment of text before and after processing with the tokenization algorithm.

Input	Shortest sequence of tokens to match
cepsuttuye	cep sut tuye
cepshuttuye	
chepsuttuye	
chepshuttuye	

There are two reasons for delegating the decision as to which variant of the input string should be selected, to the tokenizer: firstly, in the two dictionaries applied as the system's training data (see Sect. 5) there are more than one hundred items containing character sequences included in Table 1 as obsolete transcription rules, which means that there are exceptions to those rules. Secondly, older texts often include long space-delimited segments which need to be split into multiple tokens by the word segmentation algorithm. As a result, character combinations corresponding to one of the transcription change rules often occur at token boundaries (an example is shown in Table 4), in which case no modification should be applied. That, however, becomes clear only after word segmentation has been performed. This means that the dictionary lookup algorithm described here not only detects word boundaries but also performs disambiguation of transcription change rules.

Table 4. Example of a situation where the transcription change rules should not be applied.

Input segment	Strings generated by the transcription normalization algorithm	Gold standard transcription and word segmentation	Meaning
setautar	setawtar	seta utar	"dogs"
	setautar		

4.3 Part-of-Speech Tagger

Originally, POST-AL [17] performed part-of-speech disambiguation based on usage examples included in the dictionary. Namely, for each ambiguous token found in the input text it extracted lexical n-grams (2- and 3-grams) containing that token, searched for them in the dictionary and returned the part-of-speech tag of the candidate entry with the highest number of matches. However, the training data extracted from lexicons is very sparse, which means that for many n-grams there exist few or no relevant samples. To compensate for that, Nowakowski, Ptaszynski and Masui [10] created a modified tagging algorithm, which in such cases also takes into account the Term Frequency (number of occurrences in the training set) of each candidate term and returns the tag assigned to the item with the highest value. For instance, the form *sak* used as transitive verb (meaning 'to lack; not to have') appears 14 times in our dictionary, whereas the noun *sak* ('summer') has three occurrences, which means that according to the proposed disambiguation method "transitive verb" should be selected as the POS tag for the token *sak*.

In order to verify the performance of different part-of-speech disambiguation methods, three variants of the tagging algorithm were tested:

- With n-gram based disambiguation (as in the original system [17]);
- With TF (Term Frequency) based disambiguation;
- N-grams + TF (TF based disambiguation is only applied to cases where n-gram based disambiguation is insufficient).

4.4   Further Improvements

As an extension of the research carried out by Nowakowski, Ptaszynski and Masui [10], two main elements of the NLP pipeline for the Ainu language were further improved: word segmentation and part-of-speech tagging. In particular, Nowakowski, Ptaszynski and Masui [13] developed MiNgMatch Segmenter – a corpus-based word segmentation algorithm minimizing the number of n-grams needed to match the input text. They compared their system with segmenters utilizing state-of-the-art n-gram language modelling techniques and a neural model performing word segmentation as character sequence labelling. The experiments indicated that it is capable of achieving overall results comparable with the other best-performing models. Furthermore, Nowakowski et al. [14] used the SVMTool [4] to train a part-of-speech tagging model and compared its performance to that of POST-AL.

In this paper, we present a subset of the experimental results obtained in the two studies described above. For further details, however, please refer to the respective papers.

5   Training Data

5.1   Ainu Shin-Yōshū Jiten

Initially, Ptaszynski and Momouchi [17] trained their system using the information extracted from a single dictionary, namely the *Ainu shin-yōshū jiten* by Kirikae [7]. It is a lexicon to Yukie Chiri's *Ainu shin-yōshū* ("Ainu Songs of Gods", a collection of 13 *yukar* stories first published in 1923). The dictionary contains 2,019 entries. Each headword is assigned with a part-of-speech tag. Additionally, a subset of the entries include usage examples. Later we will refer to this dictionary as "KK".

5.2   Ainu Conversational Dictionary

Ainugo kaiwa jiten ("Ainu conversational dictionary") was one of the first dictionaries of the Ainu language, published in 1898 [6]. The original dictionary contains 3,847 entries.

In 2010, Bugaeva et al. [1] released the A Talking Dictionary of Ainu: A New Version of Kanazawa's Ainu Conversational Dictionary, which was an online dictionary based on the *Ainugo kaiwa jiten*. Apart from the original content (Ainu words and phrases and their Japanese translations), the dictionary was augmented with additional information, including modernized Latin transcription, modern Japanese translations, part-of-speech classification and English translations.

Nowakowski, Ptaszynski and Masui [10] converted the A Talking Dictionary of Ainu to a format compatible with POST-AL and used it as an alternative training set. Apart from isolated headwords, the original dictionary included 2,459 multi-word items (phrases and sentences). Such entries were divided into separate single-word entries. The original multi-word entries were added to the modified dictionary as usage examples, which can be used by the part-of-speech tagging system for disambiguation. Finally, unification of duplicate entries (entries containing words appearing in multiple entries of the original dictionary) was performed, using the translations provided Bugaeva et al. to determine which headwords to merge, and which should be retained as separate homonymous entries. These modifications resulted in a dictionary containing 2,555 single-word entries, of which 1,496 contain usage examples. The total number of usage examples in the dictionary is 12,513 (including duplicates). In the following sections we will refer to this dictionary as "JK".

Furthermore, leveraging the fact that each word in the aforementioned multi-word entries included in the A Talking Dictionary of Ainu was assigned a part-of-speech label – in a fashion similar to part-of-speech annotated text – Nowakowski et al. [14] converted it into a small corpus of 12,952 token-tag pairs (excluding punctuation). Apart from a subset of 1,701 tokens held out for testing, they used that data to train a part-of-speech tagging model based on Support Vector Machines.

5.3   Combined Dictionary

In addition to using each of the two dictionaries independently, they have been combined into a single data set. This was achieved by extracting entries containing words listed in both dictionaries and automatically unifying them, based on their Japanese translations (namely, homonymous entries with at least one *kanji* character in common in their translations were merged). That resulted in a dictionary containing 4,161 entries. In the following sections we will refer to this data set as "JK+KK".

5.4   Training Corpus for Word Segmentation

In 2018, Nowakowski, Ptaszynski and Masui [11] started a project to build a large-scale corpus of the Ainu language. At present, the corpus comprises a total of 410 thousand segments of source text, in a uniform format allowing for its use in Natural Language Processing applications. Recently, Nowakowski, Ptaszynski and Masui [13] applied it as the training data in their experiments with the MiNgMatch Segmenter and other word segmentation systems.

6   Test Data and Gold Standard

In evaluation experiments, four different data sets were used:

- **Y9–13**: Five out of thirteen *yukar* epics (no. 9–13) from the *Ainu shin-yōshū* [3]. Apart from the original text by Chiri, we also used the version with transcription and word segmentation modernized by Kirikae [7]. The modernized version comprises a total of 1608 tokens. In [13], only the last two stories were used (hereinafter referred to as "Y12–13").
- **JK samples**: A held out subset of the A Talking Dictionary of Ainu. Apart from the original text by Jinbō and Kanazawa [6], we also used the modernized version by Bugaeva et al. [1]. In [10], a sample of 62 sentences (428 tokens) was used. [14] and [13] performed their experiments on larger subsets (1701 and 1259 tokens, respectively).
- **Sunasawa**: Both datasets mentioned above are either obtained directly from one of the dictionaries applied as the training data for our system (JK samples) or from the collection of *yukar* epics on which one of these dictionaries was based (Y9–13). To investigate the performance with texts unrelated to the system's training data, we decided to apply other datasets as well. As the first one we used a colloquial text sample included in *The Languages of Japan* [20], namely a fragment (154 tokens) of Kura Sunasawa's memoirs written in the Ainu language, *Ku sukup oruspe* ("My life story") [21].
- **Mukawa**: We also used a sample (11 sentences, 87 tokens) from the Japanese–Ainu Dictionary for the Mukawa Dialect of Ainu [2].

Below we describe the different variants of the test data applied in testing each element of our system.

6.1   Test Data for Transcription Normalization

In experiments with transcription normalization, two data sets were used: Y9–13 and JK samples. Each of them was prepared in two versions: (i) original texts by Chiri or Jinbō and Kanazawa, without any modifications, and (ii) original texts preprocessed by removing word delimiters (whitespaces) from each line. As the gold standard data, the modernized versions of both texts [1, 7] were used.

6.2 Test Data For Tokenization

In [10], the tokenizer was evaluated on three different variants of the test data: (i) with modern transcription and spaces removed (all four data sets), (ii) original texts by Chiri (Y9–13) and Jinbō and Kanazawa (JK samples), and (iii) original texts with spaces removed (Y9–13 and JK samples).

In [13], where transcription normalization was not included in the processing pipeline, an additional variant was used: the original word segmentation was retained, but in order to prevent differences between transcription rules applied in original and modernized texts from affecting word segmentation experiment results, the texts were preprocessed by unifying their transcription with the modern (gold standard) versions.

Below is a fragment of text from Y9–13 in original transcription by Chiri [3], modernized transcription by Kirikae [7], two versions without spaces used in [10] and the variant used in [13]:

- **Original transcription**: unnukar awa kor wenpuri enantui ka;
- **Modern transcription**: un nukar a wa kor wen puri enan tuyka;
- **Modern transcr., spaces removed**: unnukarawakorwenpurienantuyka;
- **Original transcr., spaces removed**: unnukarawakorwenpurienantuika;
- **Modern transcr., orig. spaces**: unnukar awa kor wenpuri enantuy ka;
- **Meaning**: "When she found me, her face [took] the color of anger."

Modern transcriptions of all four texts [1, 2, 7, 20] were used as the gold standard.

6.3 Test Data for Part-of-Speech Tagging

Nowakowski, Ptaszynski and Masui [10] used two texts to evaluate part-of-speech tagging: a single *yukar* from the *Ainu shin-yōshū* (no. 10 – hereinafter "Y10") annotated by Yoshio Momouchi (see [9]) and 428 tokens from the A Talking Dictionary of Ainu (JK samples), with annotations by Bugaeva et al. [1]. In [14], taggers were tested on five epics (Y9–13) and 1701 tokens from the A Talking Dictionary of Ainu.

7 Evaluation Methods

The results of normalization and tokenization experiments were calculated by the means of Precision (P), Recall (R), and balanced F-score (F). In the case of transcription normalization, Precision is calculated as the percentage of correct single-character edits (deletions, insertions or substitutions) within all edits performed by the normalization algorithm, and Recall as the percentage of correct edits performed by the system within all edits needed to normalize the transcription of a given text.

$$P = \frac{correct\ edits}{all\ returned\ edits} \tag{1}$$

$$R = \frac{correct\ edits}{all\ gold\ standard\ edits} \qquad (2)$$

In the case of tokenization, Precision is calculated as the proportion of correct separations (spaces) within all separations returned by the system, whereas Recall is the number of correct spaces the system returned divided by the number of spaces in the gold standard.

$$P = \frac{correctly\ predicted\ spaces}{all\ returned\ spaces} \qquad (3)$$

$$R = \frac{correctly\ predicted\ spaces}{all\ gold\ standard\ spaces} \qquad (4)$$

Balanced F-score is the harmonic mean of Precision and Recall:

$$F = 2\frac{PR}{(P+R)} \qquad (5)$$

Part-of-speech tagging experiments were evaluated by the means of Accuracy, which is defined as the proportion of tokens correctly tagged by the system within all tokens of the test sample.

8   Results And Discussion

8.1   Transcription Normalization Results

Table 5 shows the results of transcription normalization experiments originally reported in [10]. System trained on Kirikae's lexicon achieved the highest scores for the Y9–13 dataset, which is not surprising, since the dictionary is based on *yukar* epics. In the case of JK samples, however, performance with the combined dictionary (JK+KK) was as good as with the JK dictionary only. Furthermore, the combined dictionary yielded the best results on average. The results for texts with original word segmentation retained were slightly better in all test configurations. Relatively low values of Recall in experiments on JK samples, observed across all combinations of dictionaries and input text versions, can be explained by a high occurrence of forms transcribed according to non-standard rules modified by Bugaeva et al. in the modernized version of the dictionary, but not included in the list of transcription change rules applied in this research, such as 'ra'→'r' (e.g., *arapa→arpa*), 'ri'→'r' (e.g., *pirika→pirka*), 'ru'→'r' (e.g., *kuru→kur*), 'ro'→'r' (e.g., *koro→kor*) or 'ei'→'e' (e.g., *reihei→rehe*). This has two reasons: firstly, these rules are not universal (they are observed only in a portion of texts requiring normalization). Secondly, preliminary experiments revealed that employing these rules can negatively affect the performance of the system when processing texts to which they don't apply (e.g., the *Ainu shin-yōshū*).

Table 5. Transcription normalization experiment results (best results in bold).

Dictionary			Y9–13	JK Samples	Avg.	Input Text Version:
Dictionary	JK	Precision	0.968	**0.985**	0.976	Original transcription
		Recall	0.857	**0.655**	0.756	
		F-score	0.909	**0.787**	0.848	
		Precision	0.971	0.926	0.948	Original transcription, no spaces
		Recall	0.825	0.630	0.727	
		F-score	0.892	0.750	0.821	
	KK	Precision	**0.990**	0.947	0.968	Original transcription
		Recall	0.925	0.620	0.772	
		F-score	**0.957**	0.749	0.853	
		Precision	**0.990**	0.929	0.959	Original transcription, no spaces
		Recall	0.914	0.585	0.750	
		F-score	0.951	0.718	0.834	
	JK+KK	Precision	0.983	**0.985**	0.984	Original transcription
		Recall	**0.928**	**0.655**	**0.792**	
		F-score	0.955	**0.787**	**0.871**	
		Precision	0.986	**0.985**	**0.986**	Original transcription, no spaces
		Recall	0.912	0.645	0.779	
		F-score	0.948	0.779	0.864	

8.2   Tokenization Results

The results of tokenization experiments using the algorithm proposed in [10] are shown in Table 6. Table 7 shows a fragment from Y9–13 before and after segmentation (and transcription normalization). Similarly to transcription normalization, the F-score on Y9–13 was the highest when the *Ainu shin-yōshū jiten* (KK) was used to train the system. In the case of JK samples, however, the performance (F-score-wise) with the combined dictionary (JK+KK) was as good as when using the JK dictionary only, and Precision was higher. Furthermore, the combined dictionary provided notably better performance on test data unrelated to the training data (Sunasawa and Mukawa) and the highest average Precision and F-score. On the other hand, average Recall was higher with the KK dictionary. To some extent it can be explained by the differences in word segmentation between the two dictionaries: many expressions (e.g., *oro wa*, 'from' or *pet turasi*, 'to go upstream') written as two separate segments by Kirikae (both in the lexicon part of the *Ainu shin-yōshū jiten* and in his modernized transcriptions of the *yukar* epics), are transcribed as a single unit (*orowa*, *petturasi*) by Bugaeva et al. [1]. Once these forms are added to the lexicon, the word segmentation algorithm, which prefers long tokens over shorter ones, stops applying segmentation to the tokens *orowa* and *petturasi*, resulting in lower Recall. This phenomenon occurs in the opposite direction as well.

Table 6. POST-AL tokenizer evaluation results (best results in bold).

			Y9–13	JK Samples	Sunasawa + Mukawa	Avg.	Input Text Version:
Dictionary	JK	Precision	0.536	0.922	0.653	0.704	Modern transcription, no spaces
		Recall	0.811	0.904	0.836	0.851	
		F-score	0.646	**0.913**	0.734	0.764	
		Precision	0.591	0.830	n/a	0.710	Original transcription
		Recall	0.871	**0.959**	n/a	0.915	
		F-score	0.704	0.890	n/a	0.797	
		Precision	0.529	0.772	n/a	0.651	Original transcription, no spaces
		Recall	0.799	0.850	n/a	0.824	
		F-score	0.637	0.809	n/a	0.723	
	KK	Precision	0.913	0.667	0.604	0.728	Modern transcription, no spaces
		Recall	0.887	0.869	**0.869**	0.875	
		F-score	0.900	0.755	0.713	0.789	
		Precision	**0.947**	0.692	n/a	0.819	Original transcription
		Recall	**0.917**	0.951	n/a	**0.934**	
		F-score	**0.932**	0.801	n/a	0.866	
		Precision	0.892	0.587	n/a	0.740	Original transcription, no spaces
		Recall	0.878	0.831	n/a	0.855	
		F-score	0.885	0.688	n/a	0.787	
	JK+KK	Precision	0.899	**0.937**	**0.745**	0.860	Modern transcription, no spaces
		Recall	0.852	0.891	0.860	0.868	
		F-score	0.875	**0.913**	**0.798**	0.862	
		Precision	**0.947**	0.835	n/a	**0.891**	Original transcription
		Recall	0.894	0.951	n/a	0.922	
		F-score	0.919	0.889	n/a	**0.904**	
		Precision	0.880	0.784	n/a	0.832	Original transcription, no spaces
		Recall	0.840	0.840	n/a	0.840	
		F-score	0.860	0.811	n/a	0.835	

Authors of older documents in the Ainu language tended to use spaces more sparingly than contemporary experts (recall the example in Sect. 6.2). The opposite (i.e., insertion of a whitespace where it would not be inserted in a modern transcription), however, is not frequent. In line with that observation, in experiments on texts (Y9–13 and JK samples) with original transcription, the tokenizer yielded higher scores for test data with original word boundaries retained than for the variants where spaces were removed. This means that the original word segmentation contributes positively to the tokenization process.

In addition to the results previously presented by Nowakowski, Ptaszynski and Masui [10], in Table 8 we report the results of an experiment comparing their system with the MiNgMatch Segmenter proposed in [13]. The performance of both systems was measured on the last two stories from the *Ainu shin-yōshū*, preprocessed as in [13] (see Sect. 6.2). The n-gram based word segmentation algorithm trained on a relatively (in comparison to dictionaries) large amount of textual data, is better in handling ambiguous cases, which results in higher overall scores.

Table 7. A fragment from Y9–13 before and after normalization and tokenization.

Input:	kekehetakchepshuttuyechikikushnena
Tokenizer output:	keke hetak cep sut tuye ciki kusne na
Gold standard:	keke hetak cep sut tuye ci ki kusne na
Meaning:	"Now I'm going to show you how to make fish extinct" [15]

Table 8. Comparison of POST-AL tokenizer and MiNgMatch Segmenter on Y12–13.

System:	POST-AL [10]		MiNgMatch [13]
Dictionary:	KK	JK+KK	n/a
Precision	0.940	0.937	0.957
Recall	0.940	0.923	0.933
F-score	0.940	0.930	0.945

Table 9. Results (Accuracy) of the part-of-speech tagging experiments with different variants of POST-AL tagger and each of the dictionaries (best results in bold).

Dictionary		Y10	JK Samples	Average	Disambiguation method:
Dictionary	JK	0.540	0.965	0.753	TF
		0.492	0.967	0.730	N-grams
		0.556	**0.977**	0.767	N-grams & TF
	KK	0.807	0.563	0.685	TF
		0.840	0.526	0.683	N-grams
		0.845	0.575	0.710	N-grams & TF
	JK+KK	0.850	0.960	0.905	TF
		0.861	0.942	0.902	N-grams
		0.877	**0.977**	**0.927**	N-grams & TF

8.3 Part-of-Speech Tagging Results

The results of part-of-speech tagging experiments performed by Nowakowski, Ptaszynski and Masui [10] are presented in Table 9. The results indicate that as long as the tagger is trained with language data belonging to the same type as the target data (i.e., classical Ainu of the *yukar* epics, also covered in the KK dictionary, and colloquial language of the JK dictionary), disambiguation based on lexical n-grams is more accurate than the method using Term Frequency. But they also show that combining both approaches (with priority given to n-grams) provides the best performance in each case.

Although the two dictionaries belong to different domains (colloquial and classical language), combining them into a single data set had positive effect on

Table 10. Comparison of tagging accuracy between POST-AL and SVMTool [14]

	Y9–13	JK Samples
POST-AL	0.670	0.959
SVMTool	0.783	0.978

overall tagging performance (and in the case of Y10, yielded the best results of all combinations).

In Table 10, we present a subset of the results obtained by Nowakowski et al. [14], who compared the performance of the part-of-speech tagger proposed in [10] with a tagging model trained on the same data (A Talking Dictionary of Ainu) using the SVMTool [4]. Tagging models generated by the SVMTool learn to generalize to words unseen in the training data, based on a variety of features, such as word and tag n-grams and morphological features (prefixes and suffixes). As a result, the SVM-based tagger outperformed POST-AL (which has no mechanism for handling Out-of-Vocabulary words), especially when applied to out-of-domain data (Y9–13).

9   Conclusions

We have reported the results of our research (originally described in [10]) focused on improving POST-AL, a tool for computer-aided processing of the critically endangered Ainu language. In addition to developing better algorithms for transcription normalization, word segmentation and part-of-speech tagging, we expanded the system's training data by combining two comprehensive Ainu language dictionaries. Experiments confirmed the positive effect of these modifications on the overall performance, especially with objective samples unrelated to the training data.

Moreover, we shortly described recent work extending the aforementioned project: a word segmentation system based on an n-gram model learned from a corpus (MiNgMatch Segmenter), and experiments with a state-of-the-art part-of-speech tagging tool based on the Support Vector Machines.

References

1. Bugaeva, A., Endō, S., Kurokawa, S., Nathan, D.: A Talking Dictionary of Ainu: A New Version of Kanazawa's Ainu Conversational Dictionary (2010). http://lah.soas.ac.uk/projects/ainu/
2. Chiba University Graduate School of Humanities and Social Sciences: Ainugo Mukawa Hōgen Nihongo - Ainugo Jiten [Japanese - Ainu Dictionary for the Mukawa Dialect of Ainu] (2014). http://cas-chiba.net/Ainu-archives/index.html
3. Chiri, Y.: Ainu shin-yōshū [Collection of Ainu mythic epics]. Kyōdo Kenkyūsha, Tōkyō (1923)

4. Giménez, J., Márquez, L.: SVMTool: a general POS tagger generator based on Support Vector Machines. In: Proceedings of the 4th International Conference on Language Resources and Evaluation (LREC 2004) (2004)
5. Hokkaidō Government, Environment and Lifestyle Section: Heisei nijū-ku-nen Hokkaidō Ainu seikatsu jittai chōsa hōkokusho [Report of the Survey on the Hokkaidō Ainu actual living conditions in 2017] (2017). http://www.pref.hokkaido. lg.jp/ks/ass/H29_ainu_living_conditions_survey_.pdf
6. Jinbō, K., Kanazawa, S.: Ainugo kaiwa jiten [Ainu conversational dictionary]. Kinkōdō Shoseki, Tōkyō (1898)
7. Kirikae, H.: Ainu shin-yōshū jiten: tekisuto, bumpō kaisetsu tsuki [Lexicon to Yukie Chiri's Ainu Shin-yōshū with text and grammatical notes]. Daigaku Shorin, Tōkyō (2003)
8. Majewicz, A.: Ajnu. Lud, jego język i tradycja ustna [Ainu. The people, its language and oral tradition]. Wydawnictwo Naukowe UAM, Poznań (1984)
9. Momouchi, Y., Azumi, Y., Kadoya, Y.: Research note: construction and utilization of electronic data for "Ainu shin-yōsyū". Bulletin of the Faculty of Engineering at Hokkai-Gakuen University **35**, 159–171 (2008)
10. Nowakowski, K., Ptaszynski, M., Masui, F.: Improving tokenization, transcription normalization and part-of-speech tagging of ainu language through merging multiple dictionaries. In: Proceedings of the 8th Language & Technology Conference (LTC 2017), pp. 317–321 (2017)
11. Nowakowski, K., Ptaszynski, M., Masui, F.: A proposal for a unified corpus of the Ainu language. IPSJ SIG Tech. Rep. **237**, 1–6 (2018)
12. Nowakowski, K., Ptaszynski, M., Masui, F.: Word n-gram based tokenization for the Ainu language. In: Proceedings of International Workshop on Modern Science and Technology (IWMST 2018), pp. 58–69 (2018)
13. Nowakowski, K., Ptaszynski, M., Masui, F.: MiNgMatch - a fast N-gram model for word segmentation of the Ainu language. Information **10**, 317 (2019). https://doi. org/10.3390/info10100317
14. Nowakowski, K., Ptaszynski, M., Masui, F., Momouchi, Y.: Applying support vector machines to POS tagging of the Ainu language. Proc. Workshop Comput. Methods Endangered Lang. **2**, 17–23 (2019)
15. Peterson, B.: Project Okikirmui. The complete Ainu legends of Chiri Yukie, in English (2013). http://www.okikirmui.com/
16. Ptaszynski, M., Ito, Y., Nowakowski, K., Honma, H., Nakajima, Y., Masui, F.: Combining multiple dictionaries to improve tokenization of Ainu language. In: Proceedings of The 31st Annual Conference of the Japanese Society for Artificial Intelligence (2017)
17. Ptaszynski, M., Momouchi, Y.: Part-of-speech tagger for Ainu language based on higher order Hidden Markov Model. Expert Syst. Appl. **39**, 11576–11582 (2012)
18. Ptaszynski, M., Nowakowski, K., Momouchi, Y., Masui, F.: Comparing multiple dictionaries to improve Part-of-speech tagging of Ainu language. In: Proceedings of The 22nd Annual Meeting of The Association for Natural Language Processing, pp. 973–976 (2016)
19. Refsing, K.: The Ainu Language. The Morphology and Syntax of the Shizunai Dialect. Aarhus University Press, Aarhus (1986)
20. Shibatani, M.: The Languages of Japan. Cambridge University Press, London (1990)
21. Sunasawa, K.: Ku sukup oruspe [My life story]. Miyama Shobō, Sapporo (1983)

Speech Processing

The Harmonia Corpus – A Dialogue Corpus for Automatic Analysis of Phonetic Convergence

Jolanta Bachan[1]([⊠]) [iD], Mariusz Owsianny[2] [iD], and Grażyna Demenko[1] [iD]

[1] Faculty of Modern Languages and Literatures, Adam Mickiewicz University in Poznań, al. Niepodległości 4, 61-874 Poznań, Poland
{jbachan,lin}@amu.edu.pl
[2] Poznan Supercomputing and Networking Center, ul. Jana Pawła II 10, 61-139 Poznań, Poland
mowsianny@man.poznan.pl

Abstract. The work presents the creation of a dialogue corpus for analysis and formal evaluation of phonetic convergence in spoken dialogues in human-human and human-machine communication, with the goal of comparing dialogue features at all levels of language use. The Harmonia corpus was created within a project which aims at (1) extracting phonetic features which can be mapped on a synthetic signal, (2) creating dialogue models applicable in a human-machine interaction and (3) practical evaluation of the convergence. For the corpus the following language groups were recorded: 16 pairs of Polish speakers speaking Polish (native speech), 10 pairs of German speakers speaking German (native speech), 12 pairs of German and Polish speakers speaking Polish (non-native speech), and 10 pairs of Polish and German speakers speaking German (non-native speech). The speakers could hear each other, but could not see each other. The recording scenarios consisted of controlled, neutral and expressive tasks and provided over 27 h of speech. This scenario combination is novel and promises to provide an empirical foundation for both linguistic and computational dialogue modelling of both face-to-face and man-machine dialogue.

Keywords: Dialogue corpus · Phonetic convergence · Recording scenarios · Human-computer interaction

1 Introduction

Phonetic convergence in a dialogue is a natural phenomenon. Phonetic convergence involves shifts of segmental as well as suprasegmental features in pronunciation towards those of a communicative partner [22]. The research on this phenomenon has its origin in the Communication Accommodation Theory (CAT) that has been established in the 1970s [14, 15]. The main assumption of this theory is that interpersonal conversation is a dynamic adaptive exchange involving both linguistic and nonverbal behaviour between two human interlocutors. This theory started as a model of interpersonal communication and has since been developed to encompass insights from a number of disciplines, including linguistics, sociology and psychology. One central ingredient of CAT is the

© Springer Nature Switzerland AG 2020
Z. Vetulani et al. (Eds.): LTC 2017, LNAI 12598, pp. 149–163, 2020.
https://doi.org/10.1007/978-3-030-66527-2_11

attention that speakers and listeners direct at the speech of their interlocutors. Individual adjustments to speech are assumed to subserve the function of controlling (maintaining, reducing or increasing) social distance. The speaking style of conversational partners thus converges, diverges or remains unchanged, depending on the strategies applied by the interlocutors. Most studies in the CAT framework aim at finding social motivations for accommodation behaviour and share the assumption that the processes underlying the manipulation of speech behaviour are – at least partially – under the speaker's conscious control.

Speakers accommodate their behaviour on semantic, lexical, syntactic, prosodic, gestural, postural and turn-taking levels [24]. The function of inter-speaker accommodation is to support predictability, intelligibility and efficiency of communication, to achieve solidarity with, or dissociation from, a partner and to control social impressions. The significant role of such adaptive behaviour in spoken dialogues in human-to-human communication has important implications for human-computer interaction. In the context of speech technology applications, communication accommodation is important for a variety of reasons: models of convergence can be used to improve the naturalness of synthesised speech (e.g. in the context of spoken dialogue systems, SDS), accounting for accommodation can improve the prediction of user expectations and user satisfaction/frustration in real time (in on-line monitoring) and is essential in establishing a more sophisticated interaction management strategy in SDS applications to improve the efficiency of human-machine interaction.

Studies on phonetic convergence rest on the assumption that the incoming speech signal undergoes an early, front-end analysis, which decomposes the speech signal into a set of features. In principle, each feature can be the target of convergence processes in production. Acoustic features investigated include (e.g. [3, 9, 12, 16, 31]): voice-onset time (VOT), formants, voicing, F0 range and register, pitch accents, intensity, duration, pausing, and speaking rate, as well as the long-term average spectrum (LTAS). Such acoustic measures can be complemented by perceptual judgements of the presence or degree of convergence.

Communicative adaptation has been viewed as a potential functionality in human-machine interaction to improve system performance [1, 6, 10, 11, 26, 27]. It can be assumed that a responsive human-computer interface that accommodates some features of the human interlocutor may be perceived as more user-friendly and may even lead to enhanced learning. The phenomenon of phonetic convergence that occurs naturally and partly automatically in human-human communication has not yet been exploited sufficiently in human-machine communication systems and the manipulation of the phonetic structure of speech generated in SDS environment with the aim of converging to the human speech pattern has been hardly investigated so far (cf. [2, 4, 18, 19, 21, 28]).

Apart from being an information exchange, it is widely recognised that human conversation also is a social activity that is inherently reinforcing. As such, new conversational interfaces are considered social interfaces, and when we participate in them we respond to the computer linguistically and behaviourally as a social partner. Human-computer interfaces that mimic human communication (and thus account for accommodation/convergence phenomena) will constitute next-generation conversational interfaces for speech technology applications. The benefits of using speech as an interface are multiple: simplicity (speech is the basic means of communication), quickness, robustness, pleasantness (related to social aspects of spoken communication, building relations), convenience (it can be used in hands-free and eyes-free situations or when other interfaces are inconvenient), it can be used as an alternative interface for the disabled and has some technical benefits (readily available hardware such as a telephone is sufficient).

Although the literature on communication accommodation in spoken dialogues in human interaction is fairly extensive, research on human-computer interaction has yet to face the challenge of investigating whether users of a conversational interface likewise adapt their speech systematically to converge with a computer software interlocutor. At this moment, the application of phonetic convergence in speech technology applications is not feasible for two reasons. The first one is related to the lack of an efficient quantitative description of this complex behavioural phenomenon as it occurs in spoken language. Past research on interpersonal accommodation has focused on qualitative descriptions of the social dynamics and context involved in linguistic accommodation. It also has relied on global correlation measures to demonstrate linguistic accommodation between two interlocutors. Only quantitative predictive models that account for the magnitude and rate of adaptation of different features, the factors that drive dynamic adaptation and re-adaptation, and other key issues will be valuable in guiding the design of future conversational interfaces and their adaptive processing capabilities. The second reason is that current SDS architectures are not designed to accommodate natural dialogue with human users, therefore a platform for testing quantitative models of inter-speaker accommodation does not yet exist.

The present paper describes creation of the Harmonia spoken dialogue corpus for analysis and objective evaluation of phonetic convergence in human-human communication. In Sect. 2 the corpus design is presented: the information about the subjects, the reading and repetition tasks, the scenarios of the dialogues, and the recording setup in a professional studio. Section 3 outlines the annotation specifications of the corpus. The last section concludes the paper and presents works carried out on the Harmonia corpus.

2   Corpus Design

The dialogue corpus Harmonia was created within a project which aimed at (1) extracting phonetic features which could be mapped on a synthetic signal, (2) creating dialogue models applicable in human-machine interaction and (3) practical evaluation of the types and degree of phonetic convergence. The Harmonia dialogue corpus contains dialogues with different configuration of speakers' L1/L2:

- subcorpus Harmonia_PL1_PL1: Polish L1 speaker with Polish L1 speaker
- subcorpus Harmonia_PL1_DE1_PL2: Polish L1 speaker with German L1/Polish L2 speaker
- subcorpus Harmonia_DE1_DE1: German L1 speaker with German L1 speaker

- subcorpus Harmonia_DE1_PL1_DE2: German L1 speaker with Polish L1/German L2 speaker

The subcorpus of dialogues between Poles is the biggest and the richest, containing a wider range of dialogue scenarios and complex annotation. The subcorpora of German dialogues of native and non-native speech, and the Polish dialogues between a German and a Pole are much smaller and the annotation is simpler.

2.1 Subjects

Polish L1 Speaker with Polish L1 Speaker
For the native Polish subcorpus, 16 pairs of Polish speakers were recorded: 8 male-male pairs and 8 female-female pairs who knew each other and/or were close friends. From all the subjects the following metadata was collected: name, sex, age, height, weight, education, profession, information on languages spoken and proficiency levels.

The youngest subject was 19 years old and the oldest was 58 years old (recorded in pair with a 50-year-old), the biggest age difference was 12 years and the average age difference was 3 years. Only 3 pairs of female speakers were above 30 years old, all the other subjects were younger than 29 years. The average age of the subjects was 27 years. Additionally, in each session a 33-year-old female teacher/phonetician carried out 3 dialogues with each of the subjects as a confederate.

Polish L1 Speaker with German L1/Polish L2 Speaker
For the non-native Polish subcorpus, 12 pairs of native Polish speakers with native German speaking Polish were recorded: 6 male-male pairs and 6 female-female pairs. All the speakers spoke Polish fluently, but their command differed a lot: one speaker had lived in Poland only for 5 months, 4 speakers lived in Poland for 19–30 years, some speakers were born in Germany in Polish families. Because of the General Data Protection Regulation (GDPR) no additional information about the subjects was collected. The confederate in the non-native Polish subcorpus was the same teacher as in the native Polish dialogues.

German L1 Speaker with German L1 Speaker[1]
For the native German subcorpus, 10 pairs of German speakers were recorded: 3 male-male pairs and 6 female-female pairs. The metadata collected from the subjects included information about: age, gender, height, weight, mother tongue (for all subjects it was German), highest school-leaving qualification or university degree, job or field of study, languages and language level, region and city of childhood/youth, a note if the speakers knew each other or were strangers and information if they knew the confederate or not, and additional annotation about the course of the recording.

The youngest participant was 19 years old and the oldest was 55. The biggest age differences were 21 and 36 for two pairs, the other pairs belonged to the same age group

---

[1] The German dialogues between the German native speakers and Polish L1/German L2 speakers were recorded at Saarland University by the Phonetic group led by Prof. Dr. Bernd Möbius who was the partner of the Harmonia project.

around 22 years old. Each subject talked also to a confederate – a 21-year-old German student trained to carry out the task.

German L1 Speaker with Polish L1/German L2 Speaker

For the non-native German subcorpus, 10 pairs of native German speakers were recorded with Poles speaking German: 1 male-male pair and 9 female-female pairs. The metadata collected about the subjects was the same as for the native German group. The youngest participant was 19 years old and the oldest was 57. It was hard to find Polish students speaking German in Germany where the recordings took place, so the age of subjects varied a lot. The mean age of the participants was 27 years old, while the mean age difference between the speakers was 12 years. All the subjects were also recorded in a dialogue carried out with a confederate – the same German student as in the native German subcorpus.

2.2   Scenarios for Pairs: Native Speakers of Polish Speaking Polish

The recording session was composed of a few short tasks. The first tasks were controlled reading and repetition. These tasks were introduced to assess the speakers's talent to adapt their speech to the model voice and their expressiveness while reading an enthusiastic interview with a music star. The next set of dialogues were task-oriented (neutral): either the dialogues were cooperative with no leader or in the dialogue the leader was specified and it was expected that the interlocutor would adapt to the leader's voice. Additionally, a set of expressive scenarios was recorded. These dialogues were also cooperative with no leader in the dialogue when recorded in pairs of common speaker, but when each of the speakers was to talk with the teacher/phonetician, it was expected that the speaker would adapt to the teacher in their expressiveness, liveliness and language.

Such a choice of scenarios was made to apply the developed convergence models to speech technology scenarios at different kinds of call centers, automatic information services or computer games.

Controlled Scenarios

There were 3 tasks in the controlled scenarios. In the first task, the subject heard a recording of a short sentence over the headphones by a male or a female speaker and the subject's task was to repeat the sentence in a way to best imitate the melody of the original. The sentence "Jola lubi lody" (Eng. "Jola likes ice-creams") was played 6 times with a stress on different syllables: "Jola **lu**bi lody" or "**Jo**la lubi lody" or "Jola lubi **lo**dy". Figure 1 shows the short sentence uttered by the male or female speaker, with the stress marked by "+" on different syllables. The blue line on the spectrogram is the fundamental frequency (F0) of the speakers – the lower for the male and the higher of the female.

The second task was to read a dialogue. It was an interview by a reporter and a singer. The dialogue was constructed in such a way to contain neutral and expressive phrases with exclamations.

In the third task the subject was asked to read/repeat the phrases of the same dialogue as in the previous task, but imitating the melody of phrases of the pre-recorded speech

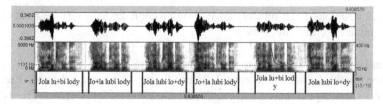

Fig. 1. The sentence "Jola lubi lody" with stress marked by "+" on different syllables.

(a similar task as in the first one, but this time the sentences were longer and their expressiveness differed).

These controlled recordings were carried out to evaluate general speakers possibilities to produce segmental and suprasegmental structures (accent type and placement, consonant cluster production) and to assess whether the speakers had talent to imitate other's speech and whether they could be expected to phonetically converge with the other speaker to a great extent. While recording the corpus, two phoneticians carrying out the recordings assessed perceptually that one speaker had little tendency to adjust his speech to the speech recordings.

Task-Oriented Scenarios

The task-oriented (neutral) scenarios consisted of 4 dialogues. The first was a decision-making dialogue in which the interlocutors were to decide together what to take to a desert island to survive. They could choose 5 items from the following list: TV set, binoculars, matches, nails, soap, favourite teddy bear, mattress, knife, petrol, tent, pen, bowl, book, hammer, kite. This was a cooperative dialogue, there was to be no role asymmetry and the maximum convergence was expected.

The second dialogue was based on a diapix task [30] where in a cooperative dialogue the subjects were to find 3 differences between two pictures. There was no role asymmetry and the subjects had to describe their pictures in order to find the differences. The diapixes are presented in Fig. 2. There are 10 differences between the pictures, but preliminary recordings revealed that finding all differences was taking too long and the task was simplified to finding only 3 differences.

Fig. 2. Diapixes for neutral scenario: describe and find 3 differences [30].

The last two dialogues from the task-oriented scenarios were map-tasks. One of the speakers was asked to play a tourist in a foreign city who just arrived at the main station and the other was to pretend to be a receptionist in a hotel. The tourist was calling the hotel at which he booked a room to ask how to get there from the main station. The subjects had the map of the city to be used in the dialogue (Fig. 3). There was asymmetry in the dialogue and it was expected that the tourist would converge to the receptionist, i.e. the leader of the dialogue. The map-task was recorded twice with the speakers exchanging their roles and with different maps.

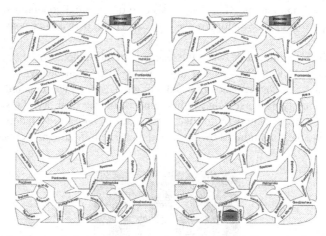

Fig. 3. Maps for the map-task: tourist's map on the left, receptionist's map on the right; "Dworzec Główny" means "Main Station."

Expressive Scenarios

The set of expressive dialogues was divided into 4 groups: a) asymmetry: power – dominant vs. submissive (entertainment scenario), b) asymmetry: emotionally coloured speech – valence: positive vs. negative (fun vs. sadness/fear, terrorist attack scenario), c) no role asymmetry: both speakers in agreement vs. both speakers in disagreement (provocation in art) and d) dialogues with the teacher (also agreement and disagreement).

In the first scenario one of the speakers played the role of a tourist information centre assistant of a big city and his task was to provide information about events and interesting places in the city and to convince the caller to choose at least one of his offers. If he had convinced the caller, the assistant would have received an award from his boss. The other person was a party-goer who wanted to find out what attractions the city offered at night. The dialogue was asymmetric, designed to boost a strong convergence to the tourist information assistant, the leader of the dialogue, who showed great enthusiasm. The same scenario was used again, but with the exchanged speakers roles.

In the second scenario, the tourist information assistant was informed about terrorist attacks in the city and was unwilling to provide any information about the entertaining events in the city. Despite the threat of another attack, the assistant had to inform the caller about the interesting places in the city, but the best procedure was to suggest only the safest options or to convince the caller to stay at home. The other speaker was again the party-goer who despite the threat of terrorist attacks wanted to go out to have some fun. The dialogue was to show a strong asymmetry and convergence to the assistant, the leader, who showed no enthusiasm to provide any information and even scared the caller that going out might put his life in danger. After the dialogue was finished, the subjects changed their roles and carried out a similar dialogue again.

Dialogues on provocation in art were designed to elicit mutual convergence as there was to be no role asymmetry. The subjects saw pictures of a very provocative content and their tasks were to discuss them and approve this form of art in the first scenario, in the following dialogue they both were asked to oppose and condemn such art. The same set of approve/oppose dialogues was also carried out between each subject and the teacher.

Finally, the last dialogue between the teacher and the subject was about Madonna's provocative performance. Both parties strongly supported their own views: the teacher – the opponent – was very conservative and thought Madonna was evil and condemned Madonna for crucifying herself during her concert, on the contrary, the subject – the supporter – was a fan of modern art, liked provocations and loved Madonna. The task was to exchange their opinions of the presented photo from Madonna's concert (Fig. 4).

Fig. 4. Picture for the expressive scenario: Madonna on the cross [23].

The dialogues with the teacher allowed to control the course of the dialogue, boost more expressiveness in subjects if needed, add more fun or show extreme indignation. The teacher could also control the length of the dialogues and make it longer if she thought the given subject did not speak long enough in the previous tasks.

2.3 Scenarios for Pairs: Native Speaker of Polish with Native Speaker of German Speaking Polish

Controlled Scenario: Reading
The first task was for the German to read 4 sentences presented below. In the sentences were there typical for Polish fricatives, africates, nasal sounds and consonant clusters.

Kasia zanosi koszyk z kaszą do kasy.
Na wczasach często czytuję kiczowate czasopisma.
Zaczął oglądać film, ale zaraz zasnął.
Potrzeba matką wynalazku.

Expressive Scenarios
There were 2 dialogues recorded between the German and the Pole. In the first dialogue, the speakers were to agree on the presented picture (see Fig. 5). In the second dialogue, the speakers were to oppose the show by Madonna who crucified herself on the cross during her concert (see Fig. 4).

Fig. 5. Provocative art [17].

The last dialogue was recorded between the German and the Polish teacher/confederate. The teacher conducted the dialogue according to the following scenario:

- While entering a "virtual" exhibition, the subject had to repeat the 4 selected Polish sentences after the teacher (the sentences from the first task).

- Room 1: the subject talked with the teacher about provocative art, picture 1 (Fig. 5). After ca. 3-min discussion, the persons "changed the room" and the subject had to say the 4 sentences after the teacher (the sentences from the first task).
- Room 2: the subject talked with the teacher about provocative art, picture 2 (Fig. 4). After ca. 3-min discussion, on leaving the exhibition, the subject had to repeat the 4 sentences after the teacher (the sentences from the first task).

Such scenario was a source of spontaneous dialogues, but also provided three times sentences repeated by the teacher. These repeated sentences, together with the read sentences from the first task, constitute a clear material for analysis of segmental (sound) alignment of a non-native Polish speaker to the teacher in the course of discourse.

2.4 Scenarios for German Dialogues

The scenarios for German dialogues, groups (1) a pair of German native speakers and (2) a German native speaker talking to a Pole speaking German were the same and were composed of a reading task, a neutral dialogue, two expressive dialogues and a dialogue with a confederate.

Controlled Scenario: Reading
The first task was to read the following sentences including sounds which do not exist in the Polish language.

Das Leben ist eben angenehm.
In der Nacht entfachten sie ein Feuer, es war prachtvoll.
Ich wurde beim Essen von Nina angesprochen.
Hörst du die Schönheit der Wörter?

Neutral Scenario
The first dialogue recorded was a decision-making task in which the interlocutors were to decide together what to take to a desert island to survive. The list of items was the same as in the Polish desert island task, but translated into German. The dialogue was cooperative and there was no role asymmetry.

Expressive Scenarios
The expressive scenarios were the same as in the Polish-German pairs speaking Polish. There were 2 dialogues recorded: in the first dialogue, the speakers were to agree on the presented picture (see Fig. 5), in the second dialogue, the speakers were to oppose the show by Madonna (see Fig. 4).

The last dialogue conducted with each of the subject was led by the confederate. The confederate was the moderator and the scenario was the same as in the non-native Polish dialogues. The subjects visited a "virtual" exhibition with two pictures: Fig. 5 and then Fig. 4. To enter the exhibition, move to another room and to leave the place the subjects had to repeat the 4 sentences in German – the same sentences as in the reading task.

2.5   Recording Session

The recordings of Polish dialogues were carried out in a professional recording studio at the Faculty of Modern Languages and Literatures, Adam Mickiewicz University in Poznań, Poland. The German dialogues were recorded at the Department of Language Science and Technology, Saarland University in Saarbrücken, Germany.

Each recording session started by signing a consent in Polish or German by the subjects to the recording their voices for the academic project purposes. Speakers answered also the questions concerning basic personal information described in Sect. 2.1 Subjects.

For the dialogue recording, the studios were specially prepared according to the highest standards [13]. In Poznań participants of the experiment felt free and could hear each other over the headphones, but could not see one another. One of the recorded persons was closed in the insulated reverberation cabin while the second speaker sat in the corner of the studio which was separated by sound absorbing panels. In Germany the recording setup was a bit different. The two speakers were sitting in a big soundproof booth separated by a thin partition wall which made it impossible to see the interlocutor, but they could hear each other.

The speech prompts were on a piece of paper, but during the recording the speakers were asked to put the paper on a small table nearby. Holding a paper is a classic source of noise and for the future recordings a music stand will be used during the recording sessions.

In Poland four professional microphones were used for recordings: 2 overhead microphones (DPA 4066 omnidirectional headset microphone) and 2 stationary microphones (condenser, large diaphragm studio microphone with cardioid characteristic – Neumann TLM 103). Microphones were plugged into the high performance audio interface Roland Studio Capture USB 2.0 equipped with 12 microphone preamps. The recordings were carried out using Cakewalk Sonar X1 LE software [29]. This setup provided 4 mono channels of recordings, 2 for each speaker, at 44.1 kHz sampling frequency and 16 bit depth. Exemplary screenshot showing the process of recording a dialogue is presented in Fig. 6. First speaker's voice was recorded in sound insulation cabin (anechoic chamber): first sound track is recorded using studio stationary microphone and the third sound track was recorded with the headset microphone. Second and fourth sound tracks concern respectively studio and headset microphones used by second speaker in the acoustically separated by sound absorbing panels corner of the studio. One recording session of Polish native dialogues lasted approximately 2 h and provided about 1 h of speech. The recordings between the Poles and the Germans lasted about 30 min. During the recordings, the speakers were asked to drink mineral water to refresh their throats. Short breaks were also taken if needed.

In Germany only 2 microphones were used – one for each speaker. This gave 2 sound tracks of recordings. The voice of the other speaker was heard in the background of the main recorded speaker, but was silent enough not to cause trouble in speech analysis.

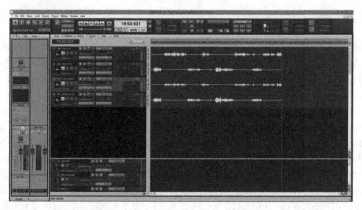

Fig. 6. Screenshot of the Sonar X1 LE recording software. 1st and 3rd sound tracks – speaker A, 2nd and 4th sound tracks – speaker B [29].

The set of scenarios for each of the subcorpus provided altogether 27.5 h of speech recordings. In the corpus 96 different speakers and two confederates were recorded. The summary of the corpus is presented below:

- Polish L1 speaker with Polish L1 speaker: 13 h, 16 pairs of speakers
- Polish L1 speaker with German L1/Polish L2 speaker: 3 h 46 min, 12 pairs of speakers
- German L1 speaker with German L1 speaker: 5 h 20 min, 10 pairs of speakers
- German L1 speaker with Polish L1/German L2 speaker: 5 h 24 min, 10 pairs of speakers

3   Annotation Specifications of the Dialogue Corpus

The first annotation specification was designed to be carried out on 7 tiers in Praat [5]:

1. ort_A – orthographic and prosodic annotation, speaker A
2. DA_A – dialogue acts, speaker A
3. info_A – metadata: information about speaker, e.g. excited, information about relation between speakers, e.g. dominant, any additional information, speaker A
4. ort_B – orthographic and prosodic annotation, speaker B
5. DA_B – dialogue acts, speaker B
6. info_B – metadata, speaker B
7. agree – parts of dialogues where both speakers agree or not, information about convergence in dialogue.

The annotation tiers were described in detail in [8] and the annotation work continued for a few weeks on the native Polish subcorpus (see Fig. 7). However, the process was very time consuming and the annotation was reduced to only two tiers 1 and 4, i.e. orthographic and prosodic annotation of speaker A and speaker B, respectively (see Fig. 8).

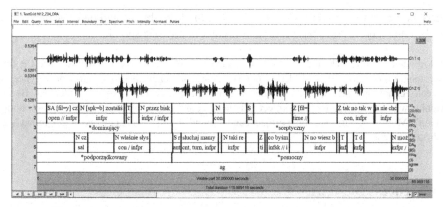

Fig. 7. Sample dialogue annotation on 7 tiers in Praat, a Polish dialogue.

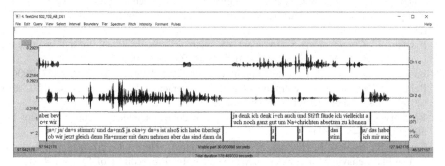

Fig. 8. Sample dialogue annotation on 2 tiers in Praat, a German dialogue.

4   Discussion and Conclusion

In the present paper, the creation of the Harmonia dialogue corpus for phonetic convergence analysis and modelling was presented. The dialogue scenarios and controlled speech prompts were shown in detail and the recording method and equipment setup in two professional studios were presented. Finally, the annotation specifications of spontaneous speech were outlined. The scenario combination and annotation specifications are novel and promise to provide an empirical foundation for both linguistic and computational dialogue modelling of both face-to-face and man-machine dialogue by providing systematic quantitative data on convergence in a set of plausible scenarios.

The analysis of Polish vowels on the Harmonia corpus was presented in [20] and the preliminary analysis of segmental and lexical convergence in German dialogues between native speakers of German and native speakers of Polish was presented in [25]. Additionally, three perception tests were carried out to see if people could sense differences in speaker voices even in short fragments of recorded speech. Three factors were evaluated in those tests: pitch, tempo and meaning of speech. The tests showed that people could sense the changes in all three investigated factors. More about the analyses carried out in the Harmonia project can be found at [7]. The analysis of the corpus will serve in the future for creation of convergence models which could be implemented in spoken dialogue systems based on spontaneous, expressive speech.

Acknowledgements. The present study was supported by the Polish National Science Centre, Harmonia project no.: 2014/14/M/HS2/00631, "Automatic analysis of phonetic convergence in speech technology systems" and was conducted in cooperation with the project partner Prof. Bernd Möbius who was the leader of a project "Phonetic convergence in Human-Machine Communication". More about the Harmonia project can be found at: http://wczt.pl/technologia_mowy/spe ech_convergence.html

References

1. Bachan, J.: Modelling semantic alignment in emergency dialogue. In: Proceedings of 5th Language & Technology Conference: Human Language Technologies as a Challenge for Computer Science and Linguistics, Poznań, Poland, 25–27 November 2011, pp. 324–328 (2011)
2. Bachan, J.: Communicative alignment of synthetic speech. Ph.D. thesis. Institute of Linguistics, Adam Mickiewicz University, Poznań, Poland (2011)
3. Baumann, S., Grice, M.: The intonation of accessibility. J. Pragmat. **38**, 1636–1657 (2006)
4. Beňuš, Š.: Social aspects of entrainment in spoken interaction. Cogn. Comput. **6**(4), 802–813 (2014). https://doi.org/10.1007/s12559-014-9261-4
5. Boersma, P., Weenink, D.: PRAAT, a system for doing phonetics by computer. Glot Int. **5**(9/10), 341–345 (2001)
6. Carlson, R., Edlund, J., Heldner, M., Hjalmarsson, A., House, D., Skantze, G.: Towards human-like behaviour in spoken dialog systems. In: Proceedings of Swedish Language Technology Conference (SLTC), Gothenburg, Sweden (2006)
7. Demenko, G. (ed.): Phonetic Convergence in Spoken Dialogues in View of Speech Technology Applications, Akademicka Oficyna Wydawnicza EXIT, Warszawa (in press)
8. Demenko, G., Bachan, J.: Annotation specifications of a dialogue corpus for modelling phonetic convergence in technical systems. In: Studientexte zur Sprachkommunikation - Proceedings of 28th Conference on Electronic Speech Signal Processing (ESSV), Saarbrücken, Germany, 15–17 March 2017 (2017)
9. Duran, D., Lewandowski, N.: Cognitive factors in speech production and perception: a socio-cognitive model of phonetic convergence. In: Matešić, M., Memišević, A. (eds.) Language and Mind: Proceedings from the 32nd International Conference of the Croatian Applied Linguistics Society, pp. 15–31. Peter Lang, Berlin (2020)
10. Edlund, J., Gustafson, J., Heldnera, M., Hjalmarssona, A.: Towards human-like spoken dialogue systems. Speech Commun. **50**(8–9), 630–645 (2008)
11. Edlund, J., Heldner, M., Gustafson, J.: Two faces of spoken dialogue systems. In: Inter-speech 2006. Pittsburgh, PA, USA (2006)
12. Gessinger, I., Raveh, E., Le Maguer, S., Möbius, B., Steiner, I.: Shadowing synthesized speech – segmental analysis of phonetic convergence. In: ISCA, pp. 3797–3801 (2017)
13. Gibbon, D., Moore, R., Winski, R.: Handbook of Standards and Resources for Spoken Language Systems. Mouton de Gruyter, Berlin (1997)
14. Giles, H.: Accent mobility: a model and some data. Anthropol. Linguist. **15**, 87–105 (1973)
15. Giles, H., Coupland, N., Coupland, J.: Accommodation theory: communication, context, and consequence. In: Giles, H., Coupland, N., Coupland, J. (eds.) Contexts of Accommodation: Developments in Applied Sociolinguistics, pp. 1–68. Cambridge University Press (1991)

16. Gorisch, J., Wells, B., Brown, G.: Pitch contour matching and interactional alignment across turns: an acoustic investigation. Lang. Speech **55**, 57–76 (2012)
17. Hanuka, A.: Underworld. http://www.asafhanuka.com/underground. Accessed 02 Nov 2020
18. Jankowska, K., Kuczmarski, T., Demenko, G.: Human converging responses to natural speech and synthesized speech. Lingua Posnaniensis (in press)
19. Lelong, A., Bailly, G.: Study of the phenomenon of phonetic convergence thanks to speech dominoes. In: Esposito, A., Vinciarelli, A., Vicsi, K., Pelachaud, C., Nijholt, A. (eds.) Analysis of Verbal and Nonverbal Communication and Enactment. The Processing Issues. LNCS, vol. 6800, pp. 273–286. Springer, Heidelberg (2011). https://doi.org/10.1007/978-3-642-25775-9_26
20. Maleszewski, P.: Analiza iloczasu polskich samogłosek w dialogach (Eng. Analysis of the Polish vowel length in dialogues). MA thesis. Institute of Ethnolinguistics, Adam Mickiewicz University, Poznań, Poland (2020)
21. Oertel, C., Gustafson, J., Black, A.: On data driven parametric backchannel synthesis for expressing attentiveness in conversational agents. In: Proceedings of Multimodal Analyses enabling Artificial Agents in Human-Machine Interaction (MA3HMI), Satellite Workshop of ICMI 2016 (2016)
22. Pardo, J.S.: On phonetic convergence during conversational interaction. J. Acoust. Soc. Am. **119**, 2382–2393 (2006)
23. Patoleta, R.: Penis na krzyżu – gdzie przebiegają granice prowokacji? http://robertpatoleta. bloog.pl/id,5640692,title,penis-na-krzyzu-gdzie-przebiegaja-graniceprowokacji,index.html. Accessed 15 Jan 2016
24. Pickering, M.J., Garrod, G.: Toward a mechanistic psychology of dialogue. Behav. Brain Sci. **27**, 169–225 (2004)
25. Pikus, S.: An analysis of speech alignment in German dialogues between native speakers of German and Polish. MA thesis. Institute of Ethnolinguistics, Adam Mickiewicz University, Poznań, Poland (2020)
26. Porzel, R., Baudis, M.: The Tao of CHI: towards effective human-computer interaction. In: Dumais, S., Roukos, S. (eds.) HLT-NAACL 2004: Main Proceedings (Boston, Massachusetts, USA, 2–7 May 2004), pp. 209–216. Association for Computational Linguistics (2004)
27. Porzel, R., Scheffler, A., Malaka, R.: How entrainment increases dialogical efficiency. In: Proceedings of Workshop on Effective Multimodal Dialogue Interfaces, Sydney (2006)
28. Savino, M., Lapertosa, L., Caffò, A., Refice, M.: Measuring prosodic entrainment in Italian collaborative game-based dialogues. In: Ronzhin, A., Potapova, R., Németh, G. (eds.) SPECOM 2016. LNCS (LNAI), vol. 9811, pp. 476–483. Springer, Cham (2016). https://doi. org/10.1007/978-3-319-43958-7_57
29. Sonar X1 LE. https://www.roland.fi/products/sonar_x1_le/. Accessed 11 Mar 2017
30. van Engen, K.J., Baese-Berk, M., Baker, R.E., Choi, A., Kim, M., Bradlow, A.R.: The wildcat corpus of native-and foreign-accented English: communicative efficiency across conversational dyads with varying language alignment profiles. Lang. Speech **53**(4), 510–540 (2010)
31. Ward, A., Litman, D.: Automatically measuring lexical and acoustic/prosodic convergence in tutorial dialog corpora. In: Proceedings of the SLaTE Workshop on Speech and Language Technology in Education (2007)

Resources and Tools for Automated Speech Segmentation of the African Language Naija (Nigerian Pidgin)

Brigitte Bigi[1]($\boxtimes$), Oyelere S. Abiola[2], and Bernard Caron[3]

[1] LPL, CNRS, Aix-Marseille Univ.,
5, Avenue Pasteur, 13100 Aix-en-Provence, France
`brigitte.bigi@lpl-aix.fr`
[2] UMR 7114, MODYCO-CNRS, Université Paris Nanterre,
200 Avenue de la Republique, 92000 Nanterre, France
`biolaoye2@gmail.com, oyelere_sa@parisnanterre.fr`
[3] Llacan, CNRS, Inalco, Paris, France
`bernard.caron@cnrs.fr`
`http://www.lpl-aix.fr/~bigi`

Abstract. The development of HLT tools inevitably involves the need for language resources. However, only a handful number of languages possesses such resources. This paper presents the development of HLT tools for the African language Naija (Nigerian Pidgin), spoken in Nigeria. Particularly, this paper is focusing on developing language resources for a tokenizer, an automatic speech system for predicting the pronunciation of the words and their segmentation.

The newly created resources are integrated into SPPAS software tool and distributed under the terms of public licenses.

Keywords: Speech · Segmentation · Naija · HLT

1 Introduction

The development of Human Language Technologies (HLT) tools is a way to break down language barriers. There are approximately 6000 languages in the world but unfortunately, only a handful possess the linguistic resources required for implementing HLT technologies [1]. Large corpora datasets for most of the under-resourced languages are created by HLT researchers for Natural Language Processing (NLP) or for Speech Technologies. African languages form about 30% of the world languages and their native speakers form 13% of the world population [9]. However, "NLP in Africa is still in its infancy; of about 2000 languages, a very few have featured in NLP research and resources, which are not easily found online." [10]. With a 182M population, Nigeria counts about 92M Internet users for June, 2015, i.e. 51.1% of its population[1] and it's constantly

---

[1] Source: http://www.internetworldstats.com/af/ng.htm - 2017-10.

© Springer Nature Switzerland AG 2020
Z. Vetulani et al. (Eds.): LTC 2017, LNAI 12598, pp. 164–173, 2020.
https://doi.org/10.1007/978-3-030-66527-2_12

growing. The official language is English but over 527 individual languages are spoken in Nigeria[2]. Recently, the Igbo language was one of the Nigerian languages investigated for NLP [10].

Among these Nigerian languages, Nigerian Pidgin English (NPE) is a post-creole continuum that is spoken as a first language (L1) by 5 million people, while over 70 million people use it as a second language (L2) or as an inter-ethnic means of communication in Nigeria and in Nigerian Diaspora communities. The origin of NPE is generally described as a development out of an English-lexified jargon attested in the 18th Century in the coastal area of the Niger delta (River State), with some lexical influence from Krio through the activities of missionaries from Sierra Leone [6]. The heartland of NPE is the Niger Delta, with Lagos and Calabar as secondary extensions. NPE is identified by "pcm" in the iso-639-3 language codes.

Since the independence of Nigeria in 1960, NPE has been rapidly expanding from its original niche in the Niger delta area, to cover two-thirds of the country, up to Kaduna and Jos, and is now deeply rooted in the vast Lagos conurbation of over 20 million people. It has become, over the last 30 years the most important, most widely spread, and perhaps the most ethnically neutral lingua franca used in the country today. In this geographical expansion, and as it conquers new functions (e.g. in business, on higher education campuses, in the media and in popular arts), NPE is subject to extensive contact and influence from its original lexifier, i.e. English and from the multitude of vernacular Nigerian languages. A mixed language has emerged that is fast expanding (both in geography and function) and rapidly changing. The name **Naija** (meaning 'Nigeria' in NPE) is used to describe this language learnt and used as an L2 in most of Nigeria, and differenciate it from the creolised variety (NPE) spoken as an L1 in the Niger delta (see [5] for a short characterization of Naija). Naija is the object of this paper on the development of HLT as part of the NaijaSynCor project[3]. It aims at describing the language in its geographical and sociological variations, based on a 500k word corpus annotated and analyzed with cutting edge HLP tools developed for corpus analysis.

Then, the development of speech technologies for Naija faces the following problems: (1) lack of language resources (lexicon, corpora, ...), not to mention digital resources; and (2) acoustic and phonological characteristics that still need to be properly investigated. These issues are shared with most under-resourced languages, and linguists are currently looking for solutions to solve or to avoid them. Nevertheless, language data collection is still a challenging and fastidious task.

This paper describes the development of a corpus and some language resources for Naija as part of a corpus-based project, which aims at evaluating the nativization of the language. Such newly created linguistic resources were integrated into SPPAS software tool [2] for a tokenizer, an automatic speech system for predicting the pronunciation of words and their segmentation. This

---

[2] Source: https://www.ethnologue.com/country/NG - 2017-06.

[3] http://naijasyncor.huma-num.fr/.

paper describes such resources at two stages of the process: at the beginning of
the project and at its end.

2 Corpus Creation

For the NaijaSynCor project, a total of 384 samples of oral corpus (monologues,
dialogues), an 6 min each, is to be collected from 380 speakers so as to represent
the widest scope of functions and locations of Naija in the country. The speech
recordings are done using professional digital recorders and wireless microphones
- one per speaker.

At the initial stage of the project, 8 of these recordings were partially manually transcribed and time-aligned at the phonetic level (Table 1). The transcriptions use the Extended Speech Assessment Methods Phonetic Alphabet
(X-SAMPA) code, a machine-readable phonetic alphabet that was originally
developed by [12]. The recordings are totalling 3 min 29 s in length, 4 men (M)
and 4 women (W). Only these files were available to construct our first HLT
tools; recordings were being collected and their orthographic transcription was
done gradually during the project.

Table 1. Description of the transcribed corpus, manually time-aligned at the phoneme
level.

File (wav)	Recording duration (in sec.)	Speech duration (in sec.)	Nb of phon-	Speech rate (phon/sec)
M_1	32.578	20.817	254	12.20
M_2	35.155	23.509	281	11.95
M_3	48.431	35.960	403	11.20
M_4	40.243	20.213	233	11.53
W_1	33.698	28.708	360	12.54
W_2	35.926	28.174	258	9.16
W_3	37.311	27.790	284	10.22
W_4	34.239	24.087	263	10.92

At the end of the project, 80 files representing about 8 h of recordings were
manually annotated:

1. orthographic transcription time-aligned into Inter-Pausal Units;
2. time-aligned tokens;
3. X-SAMPA transcription time-aligned at the syllables level.

Orthographic transcription is often the minimum requirement for a speech
corpus, as it is the entry point for most HLT tools. Corpora are using the orthography developed by [4] in her work on Lagos Nigerian Pidgin. This etymological

orthography - adapted from the lexifier language orthography, i.e. English - has been chosen preferably to the phonological script used by linguists as it is spontaneously used by educated Nigerians, and thus easier to teach to transcribers. Code-switched to English sections are identified by dedicated boundaries.

The following are examples of transcribed speech:

(W_2) '*so, all Edo people wey don travel go different-different-different places, everybody go come travel come back.*' So, all the Edo people who have travelled far and wide, everybody will return home.

(M_2) '*So, we don carry di matter come again, as we dey always carry am come.*' So, we have brought the topic again, as we always bring it.

3   Phonetic Description of Naija

At the beginning of the project, the phonetic transcription of a few minutes of speech enabled us to establish the list of the phonemes that are mostly used by the speakers. While the list of consonants is pretty close to the English one, the list of vowels used in Naija language is very different.

As shown in Table 2, only the sounds /dZ/ and /tS/ were observed infrequently, and the English /D/ and /T/ were not observed in the initial manual phonetic transcription but in the whole corpus. The other consonants of English are shared by both languages except /Z/.

But as also shown in the Table 2, the set of vowels Naija and English are sharing is only: /E/, /i/, /u/ and diphtongs. Six different nasalized vowels were observed in the corpus, but with a small number of occurrences.

Table 2. Phonemes inventory and occurrences in the manually time-aligned files and in the whole corpus

b	d	g	k	p	t	tS	dZ		m	n	N	j	w	
37	163	52	90	54	119	3	3		98	140	4	26	89	
5248	16549	5715	8361	5069	12062	970	1483		8173	11361	122	5539	7790	

l	r\	S	f	s	z	v	h	T	D		OI	aI	aU	eI
57	58	23	48	141	12	24	12	0	0		0	37	8	0
5647	5643	1581	6210	13991	1780	2454	1105	32	113		352	5421	1100	151

a	e	E	i	o	O	u	a~	e~	E~	i~	O~	u~	
203	123	111	221	93	126	74	21	1	18	18	20	3	
17141	12348	9601	22368	9110	14088	8195	2486	431	1983	2964	3901	385	

4   Creating Resources for HLT Tools

4.1   Vocabulary

A lexicon was created with both all English words and specific words observed in the corpus. In a first version of the lexicon, established in 2017, we added about 700 words. Gradually, as more corpora were transcribed, new words with their orthographic variants were added.

At the end of the project, the orthographic transcriptions 8 h recordings of the corpus contains 90k words. They represent a vocabulary of 4,600 different words; and among them 1,540 (33%) are specific to Naija language: there are not in the English vocabulary. Among the 4,600 words of the vocabulary, 2,040 (44%) are occurring only once. The 10 most frequent words are covering 24% of the pronounced words:

1. dey: 3658 occurrences
2. go: 2794 occurrences
3. i: 2697 occurrences
4. di: 2554 occurrences
5. you: 2168 occurrences
6. na: 1745 occurrences
7. for: 1634 occurrences
8. sey: 1597 occurrences
9. e: 1497 occurrences
10. no: 1477 occurrences

4.2 Pronunciation Dictionary

A pronunciation dictionary was manually created including observed words only. The dictionary was originally created by extracting the lexicon of the corpus published in annex of [4]. The observed pronunciations of the corpus of Table 1 were added to the dictionary. Created en 2017, the first dictionary was made of 4.7k pronunciations of 3.7k words.

At the end of the project, the pronunciations observed in the whole corpus were automatically extracted and added to the dictionary. The final dictionary contains 10.6k pronunciations of 5.7k words. Here is an extract of its content:

```
above [above] a b O f
above(2) [above] a b o f
above(3) [above] a b o v
abroad [abroad] a b r\ O
abroad(2) [abroad] a b r\ O d
figures [figures] f i g O
file [file] f aI l
fill [fill] f i
fill(2) [fill] f i l
```

4.3 Acoustic Model

Acoustic models were created using the HTK Toolkit [13], version 3.4. The models are Hidden Markov models (HMMs). Typically, HMM states are modeled by Gaussian mixture densities whose parameters are estimated using an expectation maximization procedure. Acoustic models were trained from 16 bits, 16,000

HZ wav files. The Mel-frequency cepstrum coefficients (MFCC) along with their first and second derivatives were extracted from the speech in the standard way (MFCC_D_N_Z_0). The training procedure is based on the VoxForge tutorial. The outcome of this training procedure is dependent on both: 1/ the availability of accurately annotated data; and 2/ on good initialization.

Of course, such requirements are difficult to fit in, particularly for under-resourced languages. The initialization of the models creates a prototype for each phoneme using time-aligned data. In the specific context of this study particularly at the beginning of the project with a lack of training data, this training stage has been switched off. It has been replaced by the use of phoneme prototypes already available in some other languages. The articulatory representations of phonemes are so similar across languages that phonemes can be considered as units which are independent from the underlying language [11]. In SPPAS package, nine acoustic models of the same type - i.e. same HMMs definition and same MFCC parameters, were freely distributed with a public license so that the phoneme prototypes can be extracted and reused: English, French, Italian, Spanish, Catalan, Polish, Mandarin Chinese, Southern Min.

To create an initial model for Naija language, most of the prototypes of English language were used. The others were extracted from French language in majority then Southern Min (3 of the nasals), Italian (3 sounds) Polish (/O/) and Spanish (/T/).

The following fillers were also added to the model in order to be automatically time-aligned too: silence, noise, laughter.

This approach enables the acoustic model to be trained by a small amount of target language speech data [7]. The initial Naija model was created by using the 8 files described in Table 1 only. At the end of the project, the whole corpus was introduced in the training procedure and an updated model was created.

5   HLT Tools

In recent years, the SPPAS software tool has been developed to automatically produce annotations, including the alignment of recorded speech sounds with its phonetic annotation. The multi-lingual approaches that are proposed enabled us to adapt some of the automatic annotations of SPPAS to Naija language. An example of Text Normalization, Phonetization and Alignment of a Naija speech segment is proposed in Fig. 1.

5.1   Automatic Tokenization and Phonetization

Tokenization of the Naija language is very similar to the English one. For the purpose of our multimodal studies, we slightly adapted the Text Normalization [1] and Phonetization systems [3]. For the text normalization, we had only to add the list of words of the Naija language into the "resources" folder of SPPAS. From the orthographic transcription, the text normalization system produces tokens (first line of annotations in Fig. 1). These can then be used by the automatic

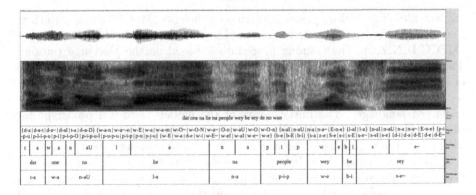

Fig. 1. Example of result of the automatic annotations of Naija

phonetization system (second line of Fig. 1). For that purpose, we simply copied the pronunciation dictionary of Naija into the "resource" folder of SPPAS.

5.2 Automatic Alignment

Forced-alignment is the task of automatically positioning a sequence of phonemes in relation to a corresponding continuous speech signal. Given a speech utterance along with its phonetic representation, the goal is to generate a time-alignment between the speech signal and the phonetic representation. Experiments in this paper were carried out by adding the Naija acoustic model into the "resources" folder of SPPAS. The automatic alignment can be carried out either using HTK (HVite) or Julius decoder engines [8]. Julius is the default aligner used in SPPAS. It produced the alignment of phonemes and tokens as shown in the 3rd and 4th lines of Fig. 1.

5.3 Experiments

Some experiments were conducted to evaluate the accuracy of the phoneme alignments. It was evaluated using the Unit Boundary Positioning Accuracy - UBPA that consists in the evaluation of the delta-times (in percentage) comparing manual phonemes boundaries with the automatically aligned ones. Obviously, the main acoustic model can't be evaluated because all the available data was used to train the model. However, we performed some experiments to have a quick glance at the accuracy of the alignments.

An initialization model was created only from the prototypes already available in the other languages, i.e. without using any Naija data nor training procedure. UBPA of such model is 88.57% in a delta-time of 40 ms. This first result confirms the suitability of a cross-lingual approach to create an acoustic model for the speech segmentation task, at least to create an initial one.

The other experiments were performed using the leave-one-out algorithm: 8 models were created. Each model was trained on 7 of our files, and the model

Table 3. UBPA of Naija automatic alignment

Delta T(automatic)-T(manual)	Initial model Count	Initial model Percent	Final model Count	Final model Percent
$-0.030 \leq$ Delta < -0.040	39	1.67%	56	2.40%
$-0.020 \leq$ Delta < -0.030	91	3.90%	97	4.15%
$-0.010 \leq$ Delta < -0.020	222	9.50%	207	8.86%
$0 \leq$ Delta < -0.010	616	26.37%	622	26.63%
$0 <$ Delta $< +0.010$	550	23.55%	579	24.79%
$+0.010 \leq$ Delta $< +0.020$	401	17.17%	441	18.88%
$+0.020 \leq$ Delta $< +0.030$	169	7.24%	144	6.16%
$+0.030 \leq$ Delta $< +0.040$	55	2.35%	52	2.23%
UBPA	91.35%		94.09%	

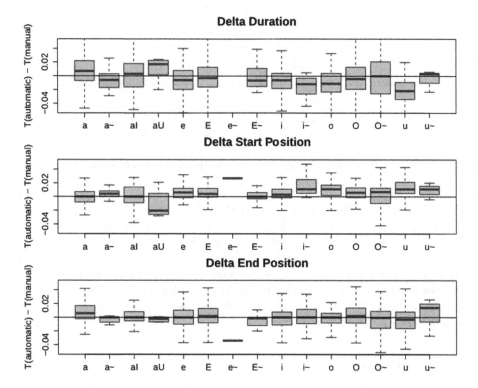

Fig. 2. Detailed results of Naija automatic alignment of vowels

was used to time-align the remaining file. The resulting UBPA is then 91.35%, with a detailed result in Table 3. Of course, introducing Naija manually created data into the training procedure increased significantly the accuracy, even if such

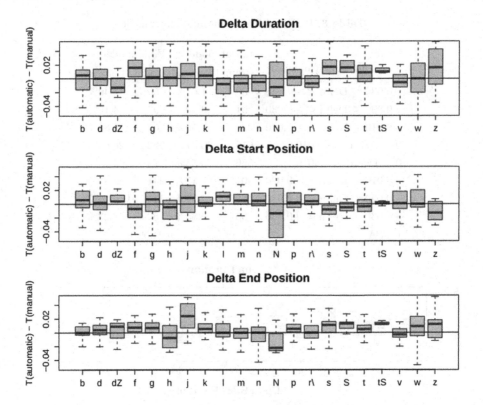

Fig. 3. Detailed results of Naija automatic alignment of consonants

data are only 3 min 29 s long. Finally, at the end of the project, accuracy of the model was enhanced with the whole data corpus as shown in Table 3.

UBPA is a unique measurement suitable to get a quick idea of the accuracy of a model or to compare the quality of several models. However, phoneticians often prefer a qualitative evaluation, as we propose in Fig. 3 and 2 for the final model. We can observe that the automatic system is slightly reducing the duration of the vowels except for /a/, and /aU/, mainly because the beginning of the vowels occurs later that the expected one.

6   Conclusion

This paper presented the first linguistic resources for the Naija language. It is shown that they are useful for HLT tools: it made Text Normalization (including a tokenizer), Phonetization and Alignment automatic annotations available for Naija. These resources were gradually improved and updated as the project progresses. The lexicon, the pronunciation dictionary and the acoustic model are all freely distributed into SPPAS since version 1.9 for the initial model and version 3.0 for the final model.

Acknowledgements. This work was financed by the French "Agence Nationale pour la Recherche" (ANR-16-CE27-0007).

References

1. Bigi, B.: A multilingual text normalization approach. In: Vetulani, Z., Mariani, J. (eds.) LTC 2011. LNCS (LNAI), vol. 8387, pp. 515–526. Springer, Cham (2014). https://doi.org/10.1007/978-3-319-08958-4_42
2. Bigi, B.: SPPAS - multi-lingual approaches to the automatic annotation of speech. Phonetician **111–112**, 54–69 (2015). http://www.isphs.org/Phonetician/Phonetician_111-112.pdf#page=54
3. Bigi, B.: A phonetization approach for the forced-alignment task in SPPAS. In: Vetulani, Z., Uszkoreit, H., Kubis, M. (eds.) LTC 2013. LNCS (LNAI), vol. 9561, pp. 397–410. Springer, Cham (2016). https://doi.org/10.1007/978-3-319-43808-5_30
4. Deuber, D.: Nigerian Pidgin in Lagos: Language Contact, Variation and Change in an African Urban Setting. Battlebridge Publications (2005)
5. Esizimetor, D., Egbokhare, F.: Naija. Hawai University Web Site. Language Varieties, 4 July 2014. http://www.hawaii.edu/satocenter/langnet/definitions/naija.html
6. Faraclas, N.: A Grammar of Nigerian Pidgin. Ph.D. thesis, Berkeley University of California (1989)
7. Le, V., Besacier, L., Seng, S., Bigi, B., Do, T.: Recent advances in automatic speech recognition for Vietnamese. In: International Workshop on Spoken Languages Technologies for Under-resourced languages, pp. 47–52. Hanoi, Vietnam (2008). http://www.lpl-aix.fr/~bigi/Doc/le2008sltu.pdf
8. Lee, A., Kawahara, T.: Recent development of open-source speech recognition engine Julius. In: Asia-Pacific Signal and Information Processing Association, pp. 131–137. Annual Summit and Conference, International Organizing Committee (2009)
9. Lewis, M., Gary, F., Charles, D.: Ethnologue: Languages of the World, 18th edn. Dallas, Texas (2015)
10. Onyenwe, I.: Developing Methods and Resources for Automated Processing of the African Language Igbo. Ph.D. thesis, University of Sheffield (2017)
11. Schultz, T., Waibel, A.: Language-independent and language-adaptive acoustic modeling for speech recognition. Speech Commun. **35**(1), 31–51 (2001)
12. Wells, J.: SAMPA computer readable phonetic alphabet. In: Handbook of Standards and Resources for Spoken Language Systems, vol. 4 (1997)
13. Young, S.J., Young, S.: The HTK hidden Markov model toolkit: design and philosophy. University of Cambridge, Department of Engineering (1993)

Speaker Variability for Emotions Classification in African Tone Languages

Moses Ekpenyong[1]([✉]) [iD], Udoinyang Inyang[1] [iD], Nnamso Umoh[1],
Temitope Fakiyesi[1], Okokon Akpan[1], and Nseobong Uto[2] [iD]

[1] University of Uyo, P.M.B. 1017, Uyo 520003, Nigeria
{mosesekpenyong,udoinyanginyang}@uniuyo.edu.ng,
nnamso_obong@yahoo.com, okokonakpan@uniuyo.edu.ng
[2] University of Saint Andrews, St Andrews KY 169SS, UK
npu@st-andrews.ac.uk

Abstract. In this paper, we investigate the effect of speaker variability on emotions and languages, and propose a classification system. To achieve these, speech features such as the fundamental frequency (F0) and intensity of two languages (Ibibio, New Benue Congo and Yoruba, Niger Congo) were exploited. A total of 20 speakers (10 males and 10 females) were recorded and speech features extracted for analysis. A methodological framework consisting of 4 main components: speech recording, knowledge base, preprocessing and analytic. ANOVA was used to test the intra- and inter-variability among speakers of various languages on emotions and languages, while the predictive analysis was carried out using support vector machine (SVM). We observed that language and speech features are dependent on speakers' characteristics. Furthermore, there exists a highly significant variability in the effect of emotions and languages on speech features. Results of SVM classification yielded 66.04% accuracy for emotions classification and 79.40% accuracy for language classification. Hence, classification performance favored the language classifier compared to emotion classifier, as the former produced low root mean squared error (RMSE) when compared with the later.

1 Introduction

Speech is a natural way of communication between humans. It represents a reliable footprint that embeds phonetic and emotional characteristics of any speaker. A speech signal therefore provides cues for expressing emotions – as it represents time-varying indicator that conveys multiple layers of information (words, syllables, languages, etc.). This phenomenon does not only convey linguistic content of a message but also the expression of attitude and speakers' emotion (Mozziconacci and Hermes 1999). The role of speech features on classification performance is vital in speaker and emotion recognition, as most of the existing recognition systems are executed under certain acoustic conditions. Features commonly utilized in speech based emotion classification systems should capture both emotion-specific information and speaker-specific information (Sethu et al. 2013), and the absolute fidelity of content that defines the classification performance

© Springer Nature Switzerland AG 2020
Z. Vetulani et al. (Eds.): LTC 2017, LNAI 12598, pp. 174–185, 2020.
https://doi.org/10.1007/978-3-030-66527-2_13

rests on how the speech is produced by the speaker ('acted' vs. 'spontaneous') and/or the environment in which the speech is produced ('optimal' vs. 'suboptimal'). Batliner and Huber (2007) addressed these interrelated issues of speaker characteristics (personalization) and suboptimal performance of emotion classification, with the argument that:

- inherent multi-functionality and speaker-dependency of speakers makes its use as a feature in emotion classification less promising, and
- constraints on time and budget often prevent the implementation of an optimal emotion recognition module.

Ideally, the only source of variability in extracted speech features arise from differences in the emotions being expressed. Speech features variability may also come from other reasons as well, including linguistic content (differences between what is being said) and speaker identity (differences between who said it), and these additional sources of variability are known to degrade classification performance (Cao et al. 2015; Chakraborty et al. 2017). While previous studies discriminate speakers using static fundamental frequency (F0) parameters, recent works focus on the dynamic and linguistically structured aspects of F0 – owing to the dynamic nature of lexical tones. Chan (2016) explored the speaker-discriminatory power of individual lexical tones and of the height relationship of level tone pairs in Cantonese, and the effects of voice level and linguistic condition on their realization. Results showed that F0 height and F0 dynamics are separate dimensions of a tone and are affected by voice level and linguistic condition in different ways. Moreover, discriminant analyses reveal that the contours of individual tones and the height differences of level tone pairs are useful parameters for characterizing speakers.

It is assumed that the emotion system is governed by the central nervous system and it is fast to react, able to switch quickly from one state to another, and produces only one emotion instance at a time. However, the intensity of emotion is a non-monotonic function of deterrence to the goal of emotion. Several experiments using supporting data as well as selected theoretical problems have been carried out to support these assumptions (c.f. Brehm 1999). Each emotion induces physiological changes which directly affect speech (Kassam and Mendes 2013). Physiological changes include affects in measures of speech features such as pitch, intensity and speech rate or duration (Kim 2007). High arousal emotions trigger higher values of speech features. For instance, anger and joy emotions have same arousal state, but differ in affect (positive and negative valence), and consequently raising serious concerns on how to accurately discriminate emotions that are at the same level of arousal, and those that have lower values, and in same arousal space. Sethu et al. (2008) investigated the effect of speaker- and phoneme-specific information on speech-based automatic emotion classification. They compared the performance of the classification system using established acoustic and prosodic features (pitch, energy, zero crossing rate and energy slope) for different phonemes, in both speaker-dependent and speaker-independent modes, using the linguistic data consortium (LDC) emotional prosody speech corpus comprising of speech from professional actors trying to express emotions while reading short phrases consisting of dates and numbers in order to ensure no semantic or contextual information is available. Their results indicate that speaker variability is more significant than phonetic variations; and features commonly used in emotion classification systems do not completely disassociate emotion-specific information from speaker-specific information (Batliner and Huber 2007).

This paper investigates the intra- and inter-variability of speech features on speakers, emotions and languages. The speech features considered include the fundamental frequency: F0 (the acoustic correlate of speech) and intensity (a measure of loudness). A support vector machine (SVM) classification system is then developed to predict emotions and languages. The remainder of this paper is outlined as follows: Sect. 2 discusses the methods and includes the proposed system framework and their respective components. Section 3 presents the results, discussing the intra- and inter-variability analysis and the classification results. Section 4 concludes on study and offers future research perspective.

2   Methods

2.1   Proposed System Framework

The framework defining our methodology is presented in Fig. 1, and consists of four main components: speech recording, knowledge base, preprocessing and analytic modeling phases. The speech recording component captures the various speech emotions from speakers of various languages. In this paper, speech recordings were obtained from native speakers of Ibibio (New Benue Congo, Nigeria) and Yoruba (Niger Congo, Nigeria). The speech sources might emanate from multiple locations/sources/speakers—homogenous or heterogeneous, therefore may be having significantly varying and inconsistent data formats and types, ambiguous, poor quality and may pose some challenges during analysis. Pre-processing is an essential task adopted to adapt heterogeneous speech corpora into a homogenous corpus. This work reduces pre-processing into five stages as follows 1) data cleaning 2) transformation 3) Integration 4) feature extraction and 5) feature selection. Data cleaning and transformation detect and remove outliers, insert missing entries as well as other operations required to standardize the data-points into a format that is computationally less expensive to model. Both stages produce a reconciled version of the hitherto heterogeneous speech corpora, and make it suitable for automatic speech corpora integration. During integration, the speech corpora was fused into a uniform and consistent version through schema integration approaches, object matching and redundancy removal, and pushes them into the speech feature database. Syllable units of two speech features, F0 and intensity were extracted using a Praat script. These features are selected in addition to each speaker identity and emotion for the analytic phase.

The analytic engine performs two main tasks: speech feature variability analysis, and emotion and language predictive analytics. The speech feature variability component performs intra-language and inter-language variability assessments with analysis of variance (ANOVA) test, while the predictive analytics component builds and executes the support vector machine (SVM) model. The results are finally produced to a decision support engine for appropriate evaluation.

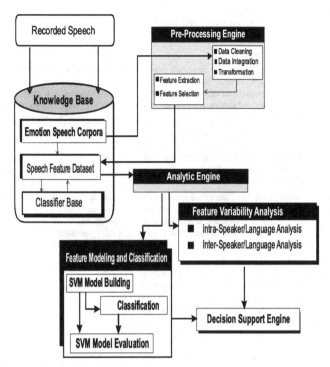

Fig. 1. Proposed system framework.

Emotion Speech Corpora. A total of 20 native speakers, 10 from each language (Ibibio and Yoruba), were selected for this study. The speakers were presented with two speech corpora that embed two negative emotions (anger and sadness), and told to act naturally, the respective emotions. Each speaker was recorded twice and the best speaking style selected. Table 1 documents the emotion corpora (column 4) use in this study with translations into Ibibio and Yoruba (column 3). Figure 2 shows a Praat speech analysis window showing the waveforms, spectrogram, point and syllable TextGrid annotations. In this paper, we are interested in the overall effect of the syllable units rather than specific units. Hence the blind labeling of the syllable units with the repeating label 'syl'. A future study is expected to address the effect of specific units with the right labels and compare their performance with other units such as phonemes and words.

Table 1. Recorded emotions speech corpora

Language	Emotion	Recorded speech	English gloss
Ibibio	Anger	sọp idem nyaak afọn. nsinam anduok etab? nsinam asueeñ eka mmi? eka mmi ado ñka mfo? yak akuppọ utebe inua mfo asọñ anye anyen. usukponoke owo? akpe maana asio uyo; nya ubeek edet ado	Come on! leave my shirt/dress. Why spit on me? Why insult my mother? Is my mother your mate? That you open your dirty mouth to insult her? Don't you have respect? If you speak again; I'll destroy your dentition
Yoruba	Anger	jowo fi aso mi sile. kilode ti ofi tuto simi lara? kilode ti ofi dojuti iya mi? se egbere ni iya mi je? ti ofi ya enu buruku re lati soro si won. se iwo koni aponle ni? to ba tun soro; mo ma ba eyin reje	
Ibibio	Sadness	hmmm! mbre mbre, udọñọ ami aya awod owo ama. nso ke adodo? afịd awo edifefeeñe korona, awo ikọọmọ owo ubọk aba. abasi mmi! ubọk mfo-o! hmmm! ñkitaña abiọọn. idaha ami, ndiweek iñweek ke anyen. abiọọn aya awot awo ama	Hmm! Bit by bit, this sickness will kill everyone. What is it? Everyone is afraid of coronavirus, no one greets with hand again! My God! I your hands I rest. Hmmm! Not to mention hunger. Now I breathe through the eyes. Hunger will kill us all
Yoruba	Sadness	hmm! die die, aisan yi yoo pa gbogbo eniyan. kini gan? gbogbo eniyan lohun beru korona, kosi eniti ofe bo eniyan lowo mo. oluwa, saanu! Hmm, ka ma tiso tebi. nisinyi, oju ni mofi hun mi. ebi yoo paniyan	

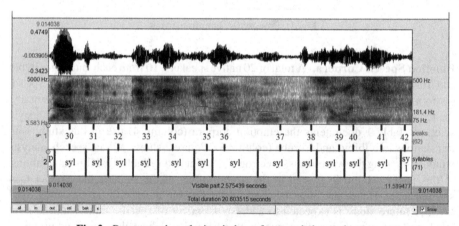

Fig. 2. Praat speech analysis window of a recorded speech corpus

Feature Extraction and Selection. A Praat script was then written to extract the syllable units of the pitch (F0) and intensity features. Syllable units were used because they are the closest and most stable features for detecting speech fluency and pronunciation, and reveals clearly, the syllable nucleus (most often a vowel) and an optional initial and final margins (typically, consonants). The extracted features were then labeled to form the speech feature datasets to which the classifier base connects with. A snippet of the labeled datasets is given in Table 2, with the parameters coded to reflect the source of

Table 2. Labeled datasets

Speaker	Emotion	Language	F0	Intensity
IM8	AN	IB	96.89	68.78
IM8	AN	IB	121.31	72.73
IM8	AN	IB	122.25	57.32
IM8	AN	IB	123.15	66.83
IM8	AN	IB	125.43	78.21
IM8	AN	IB	128.66	74.75
IM8	AN	IB	131.01	76.41
IM8	AN	IB	132.97	77.80
IM8	AN	IB	133.52	69.38
IM8	AN	IB	136.55	70.39
IM8	AN	IB	138.40	77.51
IM1	SAD	IB	112.49	72.61
IM1	SAD	IB	114.23	70.84
IM1	SAD	IB	116.95	64.87
IM1	SAD	IB	117.22	71.86
IM1	SAD	IB	118.38	74.74
IM1	SAD	IB	120.36	65.38
IM1	SAD	IB	120.56	74.50
IM1	SAD	IB	120.74	67.36
IM1	SAD	IB	123.00	71.33
IM1	SAD	IB	126.30	74.10
IM1	SAD	IB	126.67	72.81
IM1	SAD	IB	127.98	70.85
IM1	SAD	IB	128.06	77.61
YM1	SAD	YU	197.99	41.29
YM1	SAD	YU	199.13	53.44
YM1	SAD	YU	199.54	57.60
YM1	SAD	YU	200.03	62.04
YM1	SAD	YU	200.18	60.20
YM1	SAD	YU	201.50	60.70
YM6	AN	YU	188.93	61.01
YM6	AN	YU	191.18	58.54

(continued)

Table 2. (*continued*)

Speaker	Emotion	Language	F0	Intensity
YM6	AN	YU	191.84	57.56
YM6	AN	YU	192.02	59.06
YM6	AN	YU	194.95	57.04
YM2	AN	YU	195.32	59.04
YM2	AN	YU	195.80	47.12
YM2	AN	YU	199.13	50.13
YM2	AN	YU	202.12	53.76
YM2	AN	YU	205.21	63.15
YM2	AN	YU	206.54	61.68
YM2	AN	YU	207.57	57.26
YM2	AN	YU	207.83	66.04

the input, e.g. IM8 codes the eighth speaker of the Ibibio language, AN codes the anger emotion, and IB codes the Ibibio speaker.

3   Results

3.1   Intra-variability Analysis

A two-way analysis of variance (ANOVA) was performed on 10 speakers each of Ibibio and Yoruba languages, to examine the effect of speaker and emotion on F0 and intensity, respectively (see Table 3). The results in Table 4 show that there was significant difference in the mean values of F0 across speakers ($F = 2.38$, $p = 0.011$) and emotions ($F = 597.8$, $p = 0.00$) of Ibibio language. A similar result is also observed for mean values of

Table 3. ANOVA test for intra-variability analysis of emotions and languages on speech features

Response	Factors	Ibibio		Yoruba	
		F	p-value	F	p-value
F0	Emotion	116.55	0.00	597.84	0.00
	Speaker	2.38	0.011	74.36	0.00
	Interaction	8.81	0.00	8.48	0.00
Intensity	Emotion	17.68	0.00	34.97	0.00
	Speaker	6.43	0.00	13.28	0.00
	Interaction	17.17	0.00	11.07	0.00

intensity; speakers ($F = 6.43, p = 0.00$) and emotions ($F = 17.68, p = 0.00$). The mean differences of speakers and emotions are statistically significant in Yoruba language (p < 0.01, p $= 0.00$). However, the impact on F0 by emotion (F = 597.84) is the highest. Also noticed is the significant interaction between the intra-language effects of speakers and emotion on F0 ($F = 8.81, p = 0.00; F = 8.48; p = 0.00$) as well as intensity ($F = 11.07, p = .000; F = 8.48; p = .000$) for Ibibio and Yoruba languages respectively at 95% confidence level. This implies that F0 and intensity of speech in a given language are significantly dependent on the speaker as well as the speaker's emotion.

Table 4. Mean values of intra-lingual variability

Factors	Level	Ibibio		Yoruba	
		F0	Intensity	F0	Intensity
Speaker	S1	163.59	67.46	196.945	56.08
	S2	172.52	67.59	187.77	57.25
	S3	151.14	68.04	231.361	58.60
	S4	172.39	71.71	169.85	60.07
	S5	159.35	67.28	171.20	60.12
	S6	164.72	68.14	162.14	58.47
	S7	170.63	67.22	144.96	60.40
	S8	155.69	68.15	190.67	55.38
	S9	157.46	69.96	197.29	59.02
	S10	157.41	67.85	225.34	62.31
Emotion	Anger	179.10	69.08	211.998	59.85
	Sadness	145.88	67.60	163.507	57.69

In Table 4, a Turkey multi-comparison test shows that the mean F0 and mean intensity values of male speakers of Ibibio vary significantly for anger emotions (mean F0 = 179.10, mean intensity = 69.08) than sadness emotion (mean F0 = 145.88, mean intensity = 67.60). The same inference is drawn for Yoruba language.

3.2 Inter-variability Analysis

Table 5 is the results of a 2-way ANOVA for inter-variability analysis. For F0 response, we found that among speakers of various languages, there exist a highly significant variability in the effect of emotions on F0 (p < 0.01, p $= 0.00$). Similarly, for speakers expressing various emotions, there is significant variability in the effect of spoken languages on F0 (p < 0.01, p $= 0.00$). Moreover, there exists a highly significant variability in the effect of interaction between emotions and languages on F0. This implies that the variability in the effect of emotions on F0 changes for different languages. Also, variability in the effect of languages changes for different emotions (p < 0.00, p $= 0.00$).

Moreover, there exists a highly significant variability in the effect of interaction between emotions and languages on F0, this implies that the variability in the effect of emotions on F0 changes for various languages. Also, variability in the effect of languages changes for various emotions ($p < 0.01$, $p = 0.00$). For intensity response, we found that among speakers of various languages, there exist a highly significant variability in the effect of emotions on intensity ($p < 0.01$, $p = 0.00$). Also, for speakers expressing various emotions, there exist a highly significant variability in the effect of languages on intensity ($p < 0.01$, $p = 0.00$). However, there exist no significant variability in the effect of interaction between emotions and languages. This implies that variability in the effect of emotions on intensity does not change across various languages. Similarly, variability in the effect of languages on intensity does not change significantly for various emotions ($p > 0.05$, $p = 0.214$).

Table 5. ANOVA test for inter-variability analysis of emotions and languages on speech features

Response	Factor	F	p-value
F0	Emotion	417.10	0.00
	Language	159.51	0.00
	Interaction	14.58	0.00
Intensity	Emotion	45.08	0.00
	Language	1241.98	0.00
	Interaction	1.54	0.214

The results from Tukey's simultaneous tests (see Table 6) indicate that the mean level for intensity in Ibibio language (mean $= 68.34$) is significantly higher than that of Yoruba language (mean $= 58.78$) while the mean level of F0 for Yoruba language (mean $= 187.75$) is significantly higher than that of Ibibio language (mean $= 162.49$). However, for the various languages, anger emotions produce higher mean values of F0 (mean $= 195.55$) and intensity (mean $= 64.47$), compared to sadness emotions, which had mean F0 and intensity values of 154.60 and 62.64, respectively.

Table 6. Mean values of inter-lingual variability

Factors	Level	F0	Intensity
Emotion	Anger	195.55	64.47
	Sadness	154.60	62.64
Language	Yoruba	187.75	58.78
	Ibibio	162.49	68.34

3.3 SVM Classification

The objective of the SVM algorithm is to find a hyperplane in an N-dimensional space that distinctly classifies the data points, where N is the number of features. A binary SVM classifier was used in the classification and prediction of emotions and languages. Binary sequential minimal optimization (SMO) was adopted for the training of the SVM with linear kernel: $k(x, y) = <x, y>$ using F0 and intensity as input features. Results of emotion and language classification are discussed in this section. The emotion SVM model used 0.04 s for model building and execution with 62,600 kernel evaluations (60.695% cached), while the language SVM model evaluated and utilized a total of 48,475 kernels out of which 60.197% were cached in 0.07 s. The model coefficients of the input parameter as shown in Table 7 indicate that for emotions model, intensity has a higher coefficient value than F0, compared to the language model, which yielded a higher coefficient value for F0 than intensity.

Table 7. SVM model coefficients

Input parameter	Emotion SVM model	Language SVM model
F0	−6.1328	3.8187
Intensity	−1.7383	−7.7709
Error term	2.7082	3.7399

The classification performances are evaluated with derivatives of confusion matrix and receivers operating characteristics (ROC) curve including, true positive rate (TPR), Recall, precision and area under the curve (AUC) in addition to kappa statistics, root mean squared error (RMSE) and coverage. The classification confusion matrix of emotion and language is given in Table 8 and Table 9, respectively. The emotion classifier has an overall accuracy of 66.04% (1,849 instances), while the language classifier has an overall accuracy of 79.4% (2,224 instances). In terms of classification errors, the language SVM classifier performed better than emotion classifier (i.e., language RMSE = 0.4536; emotion RMSE = 0.58), despite a higher relative absolute error of 0.68.

Table 8. Confusion matrix for emotion classification

		Predicted	
		Anger	Sadness
Actual emotion	Anger	837	563
	Sadness	388	1012

The model performance evaluation for emotions and languages is documented in Table 10. As shown in Table 10, sensitivities (TPRs) of 60% and 72% are recorded for

Table 9. Confusion matrix for language classification

		Predicted	
		Ibibio	Yoruba
Actual emotion	Ibibio	1121	279
	Yoruba	297	1103

anger emotions and Sadness emotions, respectively, while false positive rates (FPRs) of 28% and 40% are produced by the classifier. The language SVM classifier outperforms the emotion SVM classifier in terms of sensitivity to instances in the respective language classes. Also, 78.80% of instances belonging to the Yoruba class were correctly predicted while the true positive rate (TPR) of instances in Ibibio class is 80.10%, i.e., the highest performance of the two classifiers. Furthermore, the overall weighted AUC for language and emotion prediction performance are 79.40% and 66.00%, respectively.

Table 10. SVM model performance evaluation for emotions and languages

Class label	Sensitivity	FPR	F-Measure	AUC
Anger	0.6000	0.2800	0.6380	0.6600
Sadness	0.7200	0.4000	0.6800	0.6600
Ibibio	0.8010	0.2120	0.7960	0.7330
Yoruba	0.7880	0.1990	0.7930	0.7940
Weighted average (emotion)	0.6600	0.3400	0.6590	0.6600
Weighted average (language)	0.7940	0.2060	0.7940	0.7940

4   Conclusion and Future Works

Classifying emotions using speech features is a relatively new area of research, and has many potential applications. But there exists considerable uncertainty as regards the best algorithm for classifying emotions. This uncertainty however deepens for tone languages – as there are no sufficient resources to empower rigorous research in this area. This paper proposed a generic framework using a state-of-the-art classify – the SVM, to further emotion research for African languages by exploiting basic speech features such as F0 and intensity, and concentrated on the overall effect of syllable units. Two negative emotions (anger and sadness) were used to investigate the intra- and inter-variability of the speech features on emotions and languages, using Ibibio and Yoruba as case study. Results obtained show valid implications useful for advancing speech processing of tone languages. Future directions of this paper include: (i) a study of other emotional types, as well as the creation of large corpus datasets –to enable efficient learning of speech features and prediction; (ii) adoption of a hybrid learning methodology – to improve the robustness of the classifier; (iii) elimination of insignificant feature(s) contribution, hence, reducing computational time; and, (iv) inclusion of more speakers and languages – to improve diversity of the classifier.

References

Batliner, A., Huber, R.: Speaker characteristics and emotion classification. In: Müller, C. (ed.) Speaker Classification I. LNCS (LNAI), vol. 4343, pp. 138–151. Springer, Heidelberg (2007). https://doi.org/10.1007/978-3-540-74200-5_7

Brehm, J.W.: The intensity of emotion. Pers. Soc. Psychol. Rev. 3(1), 2–22 (1999)

Cao, H., Verma, R., Nenkova, A.: Speaker-sensitive emotion recognition via ranking: studies on acted and spontaneous speech. Comput. Speech Lang. 29(1), 186–202 (2015)

Chakraborty, R., Pandharipande, M., Kopparapu, S.K.: Analyzing Emotion in Spontaneous Speech. Springer, Singapore (2017). https://doi.org/10.1007/978-981-10-7674-9_6

Chan, K.W.: Speaker variability in the realization of lexical tones. Int. J. Speech Lang. Law 23(2), 195–214 (2016)

Kassam, K.S., Mendes, W.B.: The effects of measuring emotion: physiological reactions to emotional situations depend on whether someone is asking. PLoS ONE 8(6), 1–8 (2013)

Kim, J.: Bimodal emotion recognition using speech and physiological changes. In: Robust Speech Recognition and Understanding, pp. 1–18. InTech (2007)

Sethu, V., Epps, J., Ambikairajah, E.: Speaker variability in speech based emotion models-analysis and normalisation. In: 2013 IEEE International Conference on Acoustics, Speech and Signal Processing, pp. 7522–7526. IEEE (2013)

Analysis of Polish Nasalized Vowels Based on Spatial Energy Distribution and Formant Frequency Measurement

Anita Lorenc[1], Katarzyna Klessa[2]([⊠]), Daniel Król[3], and Łukasz Mik[3]

[1] Institute of Applied Polish Studies, University of Warsaw,
Krakowskie Przedmieście 26/28, 00-927 Warszawa, Poland
anita.lorenc@uw.edu.pl
[2] Institute of Applied Linguistics, Adam Mickiewicz University in Poznań,
Niepodleg łości 4, 61-874 Poznań, Poland
klessa@amu.edu.pl
[3] University of Applied Sciences in Tarnów, Mickiewicza 8, 33-100 Tarnów, Poland
dankrol@gmail.com, l_mik@pwsztar.edu.pl

Abstract. In this paper, we discuss the results of the analysis of F1 and F2 frequency measurements in Polish nasalized vowels represented in writing by the graphemes ę and ą (realized before voiceless fricatives). The speech material included recordings of isolated word items provided by 20 adult native speakers of Polish (10 females and 10 males). According to the claims often presented in phonetic studies, the two vowels are phonetically realized as diphthongs composed of two subsequent stages of realization: an oral and a nasal stage. In our investigation, we refer to the results obtained by Lorenc et al. (cf. [13] and [14]) based on the analyses of spatial distribution of the acoustic field which indicate that the structure might be even more complex in certain cases and include three or even more stages. We measure formant frequencies within these stages using the stage timestamps obtained with a novel infrastructure composed of a multi-channel recorder with a circular microphone array. Among others, the results indicate that the two vowels differ significantly with regard to their internal structures as expressed by the number and types of the stages as well as frequency formant characteristics of those stages.

Keywords: Speech analysis · Acoustic camera · Formant frequency · Acoustic field energy distribution · Nasalized vowels

1   Introduction

The Polish vowel system is often referred to as relatively simple in terms of description and usage, as it consists of (only) six oral vowels [a, e, i, o, u, ɨ]

Research described in this paper was supported by grant no. 2012/05/E/HS2/03770 financed by The Polish National Science Centre (decision no. DEC-2012/05/E/HS2/ 03770).

Z. Vetulani et al. (Eds.): LTC 2017, LNAI 12598, pp. 186–196, 2020.
https://doi.org/10.1007/978-3-030-66527-2_14

[7]. However, some difficulties remain, and one of them is certainly the widely discussed problem of the status of two so-called nasalized (or nasal) vowels, denoted in writing by the graphemes ę and ą and phonotactically constrained to positions before fricative consonants (e.g., [15]). The questions are posed not only from the point of view of fundamental research, but also in the context of speech and language technology. One of the questions concerns the internal structure of the sounds resulting from their specific manner of articulation. According to many empirical studies, the nasalized vowels are produced asynchronously with (at least) two subsequent stages of realization (e.g., [3,16,20]). The first stage is often referred to as (prevalently) an oral one, while for the second one, an important influence of a nasal resonance is observed. Although a possibility of 3-stage realizations was mentioned earlier (e.g., by Wierzchowska [20]), they were usually described as oral or oral-nasal ones. The oral-nasal realizations were described as ones where the presence of nasality was increasing throughout the vowel, however, it was always preceded by the initial oral resonance.

The two-segment approach has been reflected in the publications dedicated to the description of the Polish phonemic inventory, as well as some of the works on automatizing experimental procedures in phonetics and technical solutions. For example, Steffen-Batogowa [18] supported the idea of transcribing the sounds with [eũ] and [oũ] respectively in her work on automation of grapheme-to-phoneme conversion rules for Polish. In the *Illustration of the IPA: Polish* by W. Jassem [7], the realizations of the sounds represented by the ę and ą graphemes are treated as sequences of two distinct phones, i.e. an oral vowel [e] or [o] followed by a nasalized component (an approximant or a nasal consonant depending on the context within the utterance).

On the other hand, the standard version of SAMPA (computer-readable alphabet, [19]) for Polish includes [e~], [o~] transcription labels for the two sounds, representing the the graphemes ę and ą, respectively. Therefore, in SAMPA the Polish vowel inventory includes eight vowels as not only does it include the oral vowels [a, e, i, o, u, ɨ] mentioned above but also the two nasalized sounds.

Following the postulates by Steffen-Batogowa [18], extended variants of the SAMPA alphabet were proposed and tested in studies in the context of speech technology or in corpus annotation (e.g., [10]). They included e.g., the two-symbol [ew~], [ow~] representations instead of [e~], [o~], as well as separate labels for the nasalized approximants [w~] [j~]. The realizations of ę and ą were thus treated as sequences of subsequent phones. From the practical point of view, however, it appears (e.g., to be very difficu)lt to define the position of boundaries between the oral and nasal segment (the "boundary" is actually a continuous transient and any segmentation applied for the needs of unit selection resulted in glitches and disfluencies in synthesized speech). Consequently, better speech synthesis results could be obtained by avoiding the segmentation into stages and treating the oral and nasalized sounds as the components of one inherently diversified vowel segment (corresponding to the graphemic representation). Similar, single-segment approach was assumed for speech recognition tasks (e.g., [21]).

The normative approaches to the problem indicate a lack of consensus, even in terms of defining the basic pronunciation standards or guidelines. For example, regardless of the results reported in the domains of acoustic or instrumental phonetic research, the dictionary of standard Polish pronunciation [8], recommends a synchronic and monophthong pronunciation of the vowels in front of fricatives and in the prepausal position. Two-segment, diphthong-like realizations are referred to as surprising and not common enough to be treated as normative.

In the present work, the realizations of the nasalized vowels are investigated with the use of data obtained from an innovative acoustic camera infrastructure developed by Król et al. [11] and by means of F1-F2 formant frequency analysis. While the results confirm that the two-stage realization is the most typical (although not exclusive) for the front vowel [ew̃], the back vowel [ow̃] appears to be more diversified in terms of both articulatory features and formant frequency values. As far as the notation is concerned, we follow the two-segment approach and transcribe the front vowel ę with [ew̃] and the back vowel ą with [ow̃]. It should be noted, however, that in the light of our findings these transcription labels might also become a matter of discussion especially with regard to the back vowel.

The structure of the paper is as follows: Sect. 2 of the paper provides information about the study material as well as the applied methods and tools. In Sect. 3, the results of the F1, F2 measurements are discussed. Section 4 includes conclusions and a concise outline of the future work.

2   Data, Methods and Tools

2.1   Speakers and Speech Material

Speech material was provided by 20 adult native speakers of Polish (10 females and 10 males) aged from 22 to 46. The speakers were selected from a larger group of candidates and at the preliminary stage, their pronunciation was carefully evaluated by a team of experts (phoneticians and speech therapists). All the speakers used contemporary standard Polish, declared having university education, and represented nine (out of sixteen) Polish voivodeships. The recordings are one of the outcomes of a larger project described by Lorenc [13].

The material selected for the present study consists of 161 wave files (16 bit PCM, 96 kHz) including recordings of isolated two-syllable words, treated as containers for the target nasalized vowels (76 realizations of [ew] and 85 realizations of [ow]). The target vowels were always located in the initial stressed syllables of the container words, before a voiceless fricative consonant [s]. The preceding context was either the voiceless plosive [p] or voiced fricative [v] consonant. The words were meaningful Polish words, such as for instance: węzeł [vew̃zɛw] (En. 'knot') or pąsy [pow̃sɨ] (En. 'blushes'). The word recordings were manually segmented into phones using Praat [1].

2.2   Vowel Stages and Stage Boundary Positions

Vowel stage boundary positions were established based on spatial acoustic field distribution in the function of time obtained from acoustic camera (for details see: [11,14]). The method of generating the information is illustrated in Fig. 1.

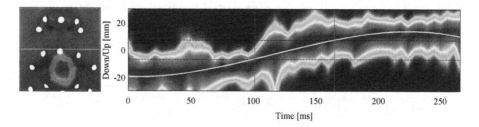

Fig. 1. The method of generating time-aligned spatial distribution of acoustic field.

A 60×60 mm fragment of the acoustic camera image was chosen for the needs of the present experiment. The fragment shows the area of the speaker's mouth and nose. The division point was set to a sensor located directly above the upper lip (point 0 at the vertical Down/Up axis in Fig. 1). The information about the sensor position enables constant adjustments to the movements of the speaker's head and consequently, stabilization of the selected area. We applied a 3rd polynomial approximation of maximum pressure, which made it possible to eliminate minor signal fluctuations (specific mainly to the stages where both oral and nasal resonances were active), and this way to obtain a clear image of the pressure changes over time (Fig. 2).

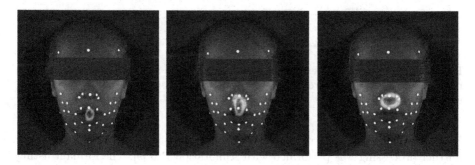

Fig. 2. Generating time-aligned spatial distribution of acoustic field. Example illustrations of three kinds of resonances: oral resonance (left), oronasal (middle), and nasal resonance (right).

For the purpose of the source of emission, a 3 dB acoustic pressure drop threshold was assumed. Three areas of acoustic field distribution were defined:

oral (range: -30 mm to -15 mm), oral-nasal (15 mm to $+15$ mm) and nasal ($+15$ mm do $+30$ mm). The areas correspond to three possible stages of vowel realization with either oral, both oral and nasal (oronasal), or nasal resonance, respectively. In Fig. 1, the areas are indicated by the black dotted lines. The information about the timestamps and durations of particular stages (red vertical markers in Fig. 1) was generated based on the cross points of the approximation polynomial line with the lines denoting spatial boundaries of particular phases.

2.3   Formant Measurement and Data Processing

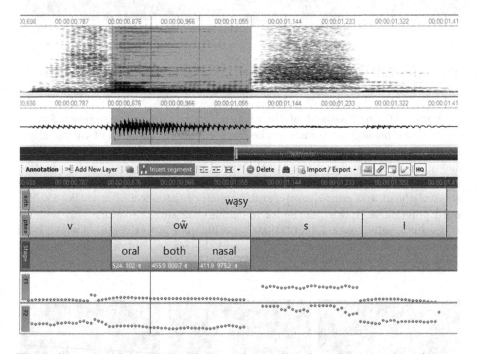

Fig. 3. Annotation Pro: an example view of time-aligned segmentation data, formant measurement results and three realization stages (oral, both, nasal) for the vowel [ow̃] in the word *wąsy*.

Formant frequencies were measured in Praat using the Formant listing option. Further processing was performed with Annotation Pro [9]. The formant listing files were automatically imported to annotation files (.ANT format, Annotation Pro) and synchronized with the time-aligned transcriptions and the vowel stage boundary timestamps (Sect. 2.2). This way, the data originally coming from the acoustic field energy distribution analysis and formant values were saved within the same workspace. An example view of the imported data is shown in Fig. 3 (formant values are represented by the dotted lines, the software enables also numerical display).

The Praat formant listing files provided information about the frequencies of the formants F1, F2, F3, F4 for the whole utterances (container words). We then used a C# plugin scripts for Annotation Pro [9] to link the obtained formant frequency values to the respective vowel stages, and also to calculate mean formant values per stage. Subsequent statistical analysis using both the measurement results and the mean values was carried out with Statistica software package [17]. In this contribution, we focus on the analysis of F1 and F2 frequency values within the vowel realization stages.

3   Results

3.1   The Number of Vowel Stages

According to the results generated based on the spatial acoustic field distribution in the function of time, the majority of the realizations of the vowel [ẽw̃] were determined as 2-stage ones. Altogether, 57 out of 76 instances of [ẽw̃] were produced with two subsequent stages: an oral stage and a stage where both oral and nasal resonances were detected. 14 instances were produced with the use of just one, oral resonance. The remaining 5 vowels were identified as: 3-stage (four occurrences), 4-stage (one occurrence) or 5-stage (one occurrence) realizations.

In the case of the vowel [õw̃], out of the total 85 instances, only 22 were produced as 2-stage realizations (again, in the 2-stage variant, one of the stages was always produced with an oral resonance, and the second one with both oral and nasal resonances active). 25 realizations were identified as 3-stage (the third stage was based on a nasal resonance without any oral component). Further 27 vowels were realized using a 4-stage structure (based on alternately activating the three types of resonance, usually either the oral or oronasal stage occurred for the second time). In 9 cases five stages were detected, and only two realizations were produced with a single type of resonance, i.e. with a 1-stage structure.

3.2   Formant Analysis

The Nasalized Vowel [ẽw̃]. The mean values of the first two formant frequencies for the nasalized front vowel [ẽw̃] are presented in Table 1.

Only the data for the 1-stage and 2-stage realizations have been included in the Table because these two types of structure represent the vast majority of cases in the present material. The formant frequency values observed for these two stages confirm the distinction in case of 2-stage realizations. The differences between means are statistically significant according to ANOVA ($p \leq 0.001$).

The frequency values in 1-stage realizations are close to Polish non-nasalized [e] reported by Jassem [6].

Figure 4 displays scatter-plots of the mean formant frequency values in the F2-F1 space for the 2-stage realizations of [ẽw̃] (inverted axes were used as in typical vowel charts). Although the values obtained for the subsequent stages overlap to a certain extent, two different areas can still be distinguished.

Table 1. Mean F1 and F2 values in 1-stage and 2-stage realizations of the vowel [ẽw̃] (F- female, M-male).

Structure	Resonance type	Speaker sex	Mean F1 [Hz]	Mean F2 [Hz]
1-stage	Oral	F	635	1821
		M	475	1486
2-stage	Oral	F	672	1768
		M	494	1490
	Both	F	435	1673
		M	365	1459

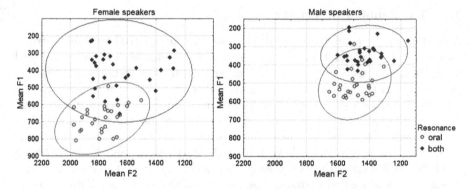

Fig. 4. Scatter-plots of the mean formant values in the F2-F1 space for 2-stage [ẽw̃] (left: female speakers, right: male speakers).

In 2-stage realizations of [ẽw̃], the stage based on both oral and nasal resonances was characterized by lower F1 and F2 median (and mean) values when compared to the stage based on exclusively oral resonance, which is in line with the results of earlier studies (see Fig. 5), assuming a decrease in the openness and frontness of the vowel in the course of its realization.

As expected, formant values for men are lower on average than those obtained for women. When looking at the differences between formants for male and female voices, similar tendencies can be observed with regard to the differences between stages, as well as slightly smaller dispersion around the middle value for male voices.

The Nasalized Vowel [õw̃]. Table 2 provides mean formant frequency values obtained for 2-, 3-, and 4-stage realizations of [õw̃] (i.e. the most frequent types of structures detected for this vowel) (Fig. 6).

The majority of multiple-stage realizations of [õw̃] began with the oral resonance, however, in certain cases at the beginning, both oral and nasal cavities were active. Table 2 includes data for all the observed stages with regard to resonance types for [õw̃] but it does not fully account for the order of appearance of these stages inside the vowel as it varied across speakers. The labels 'both2'

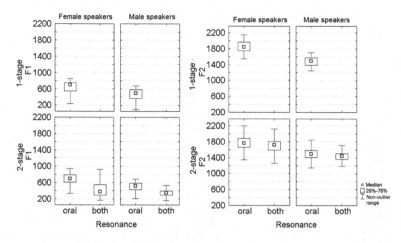

Fig. 5. Median values of F1 (left) and F2 (right) in 1-stage and 2-stage realizations of [ew̃] by male and female speakers.

Table 2. Mean F1 and F2 values in 2-stage, 3-stage and 4-stage realizations of the vowel [ow̃] (F- female, M-male).

Structure	Resonance type	Speaker sex	Mean F1 [Hz]	Mean F2 [Hz]
2-stage	Oral	F	644	1206
		M	558	1008
	Both	F	542	1404
		M	497	1161
3-stage	Oral	F	717	1129
		M	519	977
	Both	F	708	1364
		M	365	1043
	Nasal	F	399	1410
		M	398	1193
	Both2	F	450	1225
		M	391	1251
	Oral2	F	285	1268
		M	315	1304
4-stage	Oral	F	635	1821
		M	475	1486
	Both	F	616	1183
		M	602	1176
	Nasal	F	551	1470
		M	453	1343
	Both2	F	242	1305
		M	507	1579

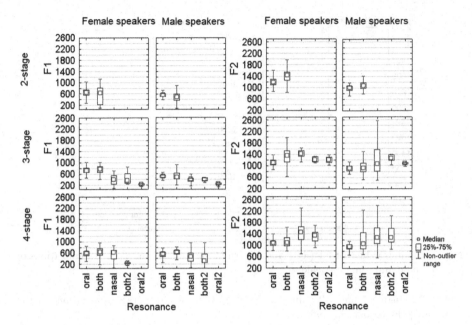

Fig. 6. Median values of F1 (left) and F2 (right) in 2-stage, 3-stage and 4-stage realizations of [ow̃] by male and female speakers.

and 'oral2' used in Table 2 denote the second occurrence of a stage based on the respective resonance (both or oral) in the course of the vowel realization. The median values of F1 and F2 are shown in the box-whiskers plots in Fig. 6.

As it might be seen, the case of the nasalized back vowel [ow] appears to be more sophisticated than [ew̃] in terms of the number of realization stages as well as by the results of formant frequency values measurements.

The average and median values of F1 were the highest for the oral stage in 2-stage realizations, however, for female voices, the dispersion around the middle value was significant. Furthermore, in 3-stage and 4-stage structures, F1 was similar or even higher for the oronasal stage (realized with both resonators) than for the oral stage, i.e. unlike for [ew̃] where F1 was systematically higher in the oral stage.

The values of F2 in 2-stage realizations were higher for stages produced with both resonances than for the oral ones. In 3-stage and 4-stage realizations, the frequency values appeared to be even higher but also more dispersed around the median.

4   Conclusions and Future Work

In this contribution, we presented preliminary results of the analysis of Polish nasalized vowels structure using a combined methodology. Formant frequencies were measured and analyzed for the subsequent stages of vowel realization. The boundaries of the stages were established based on time-aligned spatial energy distribution data obtained from acoustic camera [11,14].

We provide new empirical input with regard to the structure of the vowels, which may be useful both for basic research and application purposes, cf. the difficulties reported for handling the two sounds in speech technology (such as speech segmentation tasks or acoustic modeling), and the on-going discussion in the subject literature.

Based on the findings, it may be concluded that the two nasalized vowels should not be treated in the same way when considering their internal structure, even if they are realized in the same preceding and following context within the utterance. The front vowel denoted in orthography by the grapheme ę might be seen as a prevalently two-stage, diphthong-like vowel (in agreement with many earlier studies). Notably, many of the remaining realizations of ę were produced as single-segment vowels. However, a different situation occurs with regard to the back vowel orthographically spelt with ą, where 2-, 3-, or even 4-stage realizations are equally common. The mean formant values do differ between these stages, but much overlapping and dispersion of the frequency values occur. That is consistent with the results obtained in our earlier production studies [13]. Consequently, the case of ą should be seen as much more sophisticated and prone to individual differences than ę which might have implications for both fundamental studies and practical applications in speech therapy or speech technology, e.g., acoustic modelling tasks.

Future work will include more detailed investigation of formant frequency variability, such as identification of F1 and F2 values at potential steady states in the course of particular vowel production stages or the differences in minimum vs. maximum values of the formant frequencies. As observed by Goldstein, speaker-identifying features based on formant tracks relate to the F1 and F2 variability [5]. Another step will be the inspection of other related parameters, e.g., formant frequency bandwidths that have been reported to influence speech intelligibility, and to enhance vowel identification processes [2,12]. The same parameters will be studied with regard to higher formant frequency values (F3, F4) that in turn might be associated with individual or more fine-grained differentiation of speech sounds. The differentiation might follow varying patterns depending on the speech sound category,cf. the differences in the importance of the higher formants in two vowel groups for Japanese (/o/ and /a/ as compared to /u/ and /e/) found by Fujisaki and Kawashima [4].

As a follow-up to the present contribution, we will report in more detail on the levels and potential differences between the formant frequencies within the stages produced with the same active resonances but at different positions in the course of the vowel, with a view to closer investigate the role of the order of appearance of particular resonances as well as the actual variability of formant trajectories.

References

1. Boersma, P., Weenink, D.: Praat: Doin (2014). http://www.praat.org. Accessed 20 Dec 2017
2. De Cheveigné, A.: Formant bandwidth affects the identification of competing vowels. ICPhS: International Congress of Phonetic Sciences, 2093 2096 (1999)
3. Dukiewicz, L.: Polskie głoski nosowe: analiza akustyczna. PWN, Warszawa (1967)
4. Fujisaki, H., Kawashima, T.: The roles of pitch and higher formants in the perception of vowels. IEEE Trans. Audio Electroacoust. **16**(1), 73–77 (1968)
5. Goldstein, U.G.: Speaker-identifying features based on formant tracks. J. Acoust. Soc. Am. **59**(1), 176–182 (1976)
6. Jassem, W.: Podstawy fonetyki akustycznej. Państwowe Wydawnictwo Naukowe (1973)
7. Jassem, W.: Illustrations of the IPA: Polish. J. Int. Phonetic Assoc. **33**(1), 103–107 (2003)
8. Karaś, M., Madejowa, M. (eds.): Słownik wymowy polskiej PWN. PWN, Warszawa (1977)
9. Klessa, K., Karpiński, M., Wagner, A.: Annotation Pro-a new software tool for annotation of linguistic and paralinguistic features. In: Proceedings of the Tools and Resources for the Analysis of Speech Prosody (TRASP) Workshop, Aix en Provence, pp. 51–54 (2013)
10. Klessa, K., Szymanski, M., Breuer, S., Demenko, G.: Optimization of Polish segmental duration prediction with cart. In: SSW, pp. 77–80 (2007)
11. Król, D., Lorenc, A., Święciński, R.: Detecting laterality and nasality in speech with the use of a multi-channel recorder. In: 40th IEEE International Conference on Acoustics, Speech and Signal Processing (ICASSP), pp. 5147–5151. IEEE, Brisbane (2015)
12. Kuwabara, H., Ohgushis, K.: Role of formant frequencies and bandwidths in speaker perception. Electron. Commun. Jpn. **70**(9), 11–21 (1987)
13. Lorenc, A.: Wymowa normatywna polskich samogłosek nosowych i spółgłoski bocznej (En. Normative pronunciation of Polish nasalized vowels and the lateral consonant). Dom Wydawniczy Elipsa, Warszawa (2016)
14. Lorenc, A., Król, D., Klessa, K.: An acoustic camera approach to studying nasality in speech: the case of Polish nasalized vowels. J. Acoust. Soc. Ame **144**(6), 3603–3617 (2018)
15. Puppel, S., Nawrocka-Fisiak, J., Krassowska, H.: A Hand-Book of Polish Pronunciation for English Learners. PWN, Warszawa (1977)
16. Rocławski, B.: Podstawy wiedzy o języku polskim dla glottodydaktyków, pedagogów, psychologów i logopedów. Glottispol, Gdańsk (2010)
17. StatSoft, I.: Statistica (data analysis software system), version 6. Tulsa, USA 150 (2001)
18. Steffen-Batogowa, M.: Automatyzacja transkrypcji fonematycznej tekstów polskich. PWN, Warszawa (1975)
19. Wells, J.C.: SAMPA computer readable phonetic alphabet. In: Gibbon, D., Moore, R., Winski, R. (eds.) Handbook of Standards and Resources for Spoken Language Systems, Part IV, Section B. Mouton de Gruyter, Berlin and New York (1997)
20. Wierzchowska, B.: Wymowa polska. PZWS, Warszawa (1971)
21. Ziółko, B.: Speech Recognition of Highly Inflective Languages. Ph.D. thesis, Department for Computer Science, University of York (2009)

Morphology

RNN Language Model Estimation
for Out-of-Vocabulary Words

Irina Illina$^{(\boxtimes)}$ and Dominique Fohr

MultiSpeech Team, Université de Lorraine, CNRS, Inria, LORIA, 54000 Nancy, France
`illina@loria.fr`

Abstract. One important issue of speech recognition systems is Out-of Vocabulary words (OOV). These words, often proper nouns or new words, are essential
for documents to be transcribed correctly. Thus, they must be integrated in the
language model (LM) and the lexicon of the speech recognition system. This article proposes new approaches to OOV proper noun probability estimation using
Recurrent Neural Network Language Model (RNNLM). The proposed approaches
are based on the notion of closest in-vocabulary (IV) words (*list of brothers*) to
a given OOV proper noun. The probabilities of these words are used to estimate
the probabilities of OOV proper nouns thanks to RNNLM. Three methods for
retrieving the relevant list of brothers are studied. The main advantages of the
proposed approaches are that the RNNLM is not retrained and the architecture of
the RNNLM is kept intact. Experiments on real text data from the website of the
Euronews channel show relative perplexity reductions of about 14% compared to
baseline RNNLM.

Keywords: Speech recognition · Neural networks · Vocabulary extension ·
Out-of-vocabulary words · Proper names

1 Introduction

Voice is seen as the next big field for computer interaction. From *Statista Research
Department*, as of 2019, there are an estimated 3.25 billion digital voice assistants being
used in devices around the world. Global smart speaker sales hit a record high in 2019
with shipments of 146.9 million units, up 70% over 2018, according to a recent report
on the state of the smart speaker market from *Strategy Analytics*. Google reports that
27% of the online global population is using voice search on mobile.

Dictating e-mails and text messages works reliably enough to be useful. In this
context, an automatic speech recognition system (ASR) should accommodate all voices,
all topics and all lexicons.

The proper nouns (PNs) play a particular role: they are often important to understand a
message and can vary enormously. For example, a voice assistant should know the names
of all your friends; a search engine should know the names of all famous people and
places, names of museums, etc. For the moment, it is impossible to add all existing proper
nouns into a speech recognition system. A competitive approach is to dynamically add

© Springer Nature Switzerland AG 2020
Z. Vetulani et al. (Eds.): LTC 2017, LNAI 12598, pp. 199–211, 2020.
https://doi.org/10.1007/978-3-030-66527-2_15

new PNs into the ASR system. It implies knowing where to look for them, and knowing how to introduce them into the lexicon and into the language model. *Updating the language model of the ASR system with a list of retrieved OOV PNs* is the central point of this article.

Although the LM adaptation to contextual factors (style, genre, topic) [2, 17] has been well studied, there is little work done on integration of new words in language model. Traditionally, integration of new words is performed implicitly by using the *'unk'* word and *back-off* probability. Open vocabulary ASR represents an OOV word by a sub-lexical model [1] or as sub-word units [10, 12]. [11] proposed to estimate n-gram LM scores for OOV words from syntactically and semantically similar in-vocabulary (IV) words. In class-based approaches [9], an OOV is assigned to a word class and the OOV LM probability is taken from this class.

In our previous works, we proposed several approaches to estimate the bigram probability of OOV proper nouns using word similarity [3]. In our current work, we propose new methods for estimating OOV proper noun probability using Recurrent Neural Networks-based language model (RNNLM). The main advantage of RNNLM is a possibility of using arbitrarily long histories [5, 8]. Using classes at the output layer allows to speed-up the training [6]. *A novel aspect of the proposed methodology is the notion of brother words:* for each OOV PN we look for a list of "similar" in-vocabulary words, called a *list of brothers*, and we use their RNNLM probabilities to estimate the OOV PN probabilities. The main advantage of our methodology is the fact that the RNNLM is not modified: no retraining of the RNNLM is needed and the RNN architecture is not modified, there are the same number of layers and the same number of nodes. The proposed method can be applied for other neural network LMs, such as Long Short-Term Memory model or Gated Recurrent Units model. Indeed, we do not modify the internal architecture of the model.

2 Proposed Methodology

The naive solution for taking into account OOV PNs would consist in integrating all PNs contained in the available corpus in the lexicon and LM of the ASR. This solution is not feasible for several reasons: using corpus, like newswire or Wikipedia, will result in adding millions of OOV PNs [11]. The ASR would become very slow. Moreover, it would increase acoustic confusability: many PNs could have pronunciations close to common names. For instance, adding the names of all English footballers is useless to recognize a document that talks about war in Syria. In our work, we want to add to the ASR only OOV PNs relevant to the document to be transcribed. In this article, we focus on dynamic updating of the language model.

In our methodology we assume that we have a list of retrieved OOV proper nouns and we want to estimate their language model probability using a previously trained RNN LM. The list of OOV PNs can be retrieved according to the semantic context modeling of OOVs [13]. This list will be added to the original lexicon of ASR. In this paper, we want to integrate the list of OOVs in RNNLM using a contemporary corpus. It is important to notice that the RNNLM is *not retrained*, it is used to estimate the probabilities of OOV words. Therefore, as inputs we have a previously trained RNNLM, the original lexicon, the list of OOV proper nouns and some text data, called *contemporary corpus*. As output, we want to estimate LM probability for OOV proper nouns using RNNLM.

We assume that the topology of RNN used for LM consists of three layers. The input layer consists of a vector *w(t)* that represents the current word w_t encoded as *1* (size of *w(t)* is equal to the size of the vocabulary *V*), and a context vector *h(t − 1)* that represents values of the hidden layer from the previous time step (see Fig. 1). The output layer represents $P(w_{t+1}|w_t, h(t − 1))$. The aim of RNNLM is to estimate the probability $P(w_{t+1} |w_t, h(t − 1))$.

To take into account OOV words, we have two problems:

- w_t (previous word) can be an OOV;
- or w_{t+1} (predicted word) can be OOV.

For the first case, the difficulty is how to find a relevant representation of OOV at the RNNLM input. One classical solution is to add a specific neuron for all OOVs [16], but all OOVs will be treated in the same way, which is not optimal. We propose to introduce a specific representation for each OOV using the similar in-vocabulary words (*brother list*).

For the second case, we propose to estimate the probability $P(OOV|w_t, h(t − 1))$ using the probabilities (given by the RNNLM) of the in-vocabulary words of the brother list.

The main idea of our method is to build a list of similar in-vocabulary words for each OOV PN. The similarity can be modeled at the syntactic/semantic level. It means that the in-vocabulary brother words will play the same syntactic or/and semantic role as the corresponding OOV PNs. For instance, for the OOV proper noun *Fukushima*, the brother word can be another Japan city, like *Tokyo*. The list of similar in-vocabulary words will be used to generate the input of RNNLM or to use the RNNLM output probabilities to compute probabilities for each OOV PN. The structure of the RNNLM and the weights are neither modified nor retrained.

The approaches proposed in this article include the following steps:

- Finding a list of in-vocabulary words similar to OOVs, called *list of brothers*, using a contemporary corpus (see Sect. 2.1).
- Using the brother lists of in-vocabulary words, estimating the probabilities $P(w_{t+1}|OOV, h(t − 1))$ and $P(OOV|w_t, h(t − 1))$ for each OOV using RNNLM (see Sect. 2.2).

In the following sections we will present these two steps.

2.1 Brother List Generation

For each OOV from the list of OOVs, we want to generate a list of size *M* containing a ranked in-vocabulary words called *brother list:*

$$BrotherList(OOV) = \{(IV_1, v_1), (IV_2, v_2), \dots (IV_M, v_M)\} \qquad (1)$$

$$g(OOV, IV_i) = v_i \tag{2}$$

where v_i corresponds to the similarity value of ith IV. Each word of this list is *similar* in some sense to the OOV PN. As similarity values, some *distance information* from in-vocabulary word to OOV can be used. All similarity values for a given OOV proper noun sum to *1* (linear combination). The brother list will be used to estimate the OOV PN probability thanks to the RNNLM.

We propose three approaches for the generation of the list of brothers:

- *Similarity-based approach*: to generate an IV brother list for a given OOV PN, we use a similarity measure based on word embedding *word2vec* [8]. We trained a skip-gram model with a context window size of two on a large text corpus (we assume that the OOV PN is present in this corpus). According to *word2vec*, we compute the *cos*-distance between the OOV embedding vector and the in-vocabulary embedding vectors. We choose the top M in-vocabulary words and put them in the brother list for this OOV PN. We propose to use the corresponding *cos*-distance as v_i (after normalization).
- *k-gram counting approach*: in this approach we assume that if one in-vocabulary word w occurs in the same context as that an OOV PN, then w can be used as a similar word for this OOV proper noun. To find the brother list for one OOV PN, we propose to count all k-grams $<w_1, ... w, ..., w_k>$ corresponding to k-grams $<w_1, ..., OOV, ..., w_k>$ where the central OOV proper noun is replaced by w. The preceding words and the following words being the same. The N central words with the highest counts will be put in the brother list for this OOV proper noun. For a small value of k *(2, 3)*, it is possible to find a large number of central words w. For large value of k, the number of k-grams can be very small and so, we can have few brothers.
- *Wikipedia-based approach:* we take into account only OOVs that are the last names of a person name. We assume also that the persons are famous and that a Wikipedia page exists for them. In this aim, we have collected all Wikipedia webpage titles. For an OOV word, we search for all titles of Wikipedia containing this OOV. From these titles, we choose all fist names of this OOV word. After this, we search all last names of these first names from Wikipedia titles and put them in the brother list for this OOV. For instance, for OOV word *Kaymer* we find the title webpage *Martin Kaymer* (professional golfer). Then we search for webpage titles with *Martin* as first name and we find *Martin Scorsese, Martin Luther, Martin Malvy*, etc. Therefore, the brother list of the OOV word *Kaymer* will contain *Scorsese, Luther, Malvy*, etc.

2.2 OOV PN Probability Estimation Using RNNLM

For computing the probability of a sentence containing OOV PNs, we propose to use the brother list of each OOV PN.

Computing $P(w_{t+1} \mid OOV, h(t-1))$

As OOV proper noun is not in the lexicon, RNNLM has no corresponding input neuron for it. We propose to represent each OOV PN by a linear combination of in-vocabulary

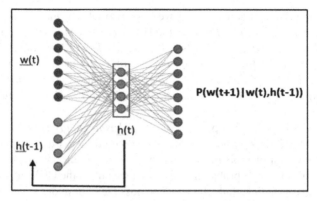

Fig. 1. Schematic representation of RNNLM.

words from the brother list of this OOV. For instance, if the brother list of an OOV proper noun contains *2* IVs:

$$BrotherList(OOV) = \{(IV_1, 0.6), (IV_2, 0.4)\} \qquad (3)$$

the RNN input vector for this OOV proper noun will be:

$$w(t) = (0 \dots 0 \; 0.6 \; 0 \dots 0 \; 0.4 \; 0 \dots 0) \qquad (4)$$

where *0.4* and *0.6* correspond to the similarity values of two IV words and their positions (instead of a single 1 in a classical one-hot representation). In this case, the OOV can be seen as a linear combination of IV words of the brother list. If brother list contains *M* words, all *M* in-vocabulary words can be used. After this, the input is propagated through the RNNLM. At the output, we will obtain probabilities $P(w_{t+1}|OOV, h(t-1))$.

$$class(IV) = \{IV \text{ and all } OOV \text{ PNs such as this } IV \text{ is in the brother list of the } OOV \text{ PNs}\} \qquad (5)$$

$$BrotherList(Fukushima) = \{(Tokyo, 0.6), (Nagasaki, 0.4)\} \qquad (6)$$

$$BrotherList(Sendai) = \{(Tokyo, 0.5), (Nagasaki, 0.3), (Nagoya, 0.2)\} \qquad (7)$$

$$class(Tokyo) = \{Tokyo, Fukushima, Sendai\} \qquad (8)$$

$$class(Nagasaki) = \{Nagasaki, Fukushima, Sendai\} \qquad (9)$$

$$P\big(Fukushima \mid w_t, h(t-1)\big) = P\big(class(Tokyo) \mid w_t, h(t-1)\big) * P\big(Fukushima \mid class(Tokyo), w_t, h(t-1)\big)$$

$$+ \; P\big(class(Nagasaki) \mid w_t, h(t-1)\big) * P\big(Fukushima \mid class(Nagasaki), w_t, h(t-1)\big) \qquad (10)$$

$$P(Fukushima \mid class(Tokyo, w_t, h(t-1)) = (1-\alpha) * g(Fukushima, Tokyo)/(g(Fukushima, Tokyo) + g(Sendai, Tokyo)) \qquad (11)$$

$$P(Fukushima \mid class(Nagasaki, w_t, h(t-1)) = (1-\alpha) * g(Fukushima, Nagasaki)/(g(Fukushima, Nagasaki) + g(Sendai, Nagasaki)) \qquad (12)$$

Computing $P(OOV \mid w_t, h(t-1))$

As OOV PN is not in the lexicon, RNNLM has no corresponding output neuron for it. The probability of OOV will be estimated using the probabilities of in-vocabulary proper nouns from the brother list. For each IV, we define a class containing the in-vocabulary word itself and all OOV proper nouns for which this IV is a brother (cf. Eq. (5)).

As an example, let us consider that we have two OOVs: *Fukushima* and *Sendai*. The obtained brothers for *Fukushima* are the IVs *Tokyo* and *Nagasaki* (cf. Eq. (6)). The obtained brothers of *Sendai* are the IVs *Tokyo, Nagasaki* and *Nagoya* (cf. Eq. (7)). We can define the classes of *Tokyo* and *Nagasaki* according to Eq. (8) and (9). We compute the probability of OOV PN *Fukushima* $P(Fukushima|w_t, h(t-1))$ as defined by Eq. (10). $P(class(Tokyo)|w_t, h(t-1))$ and $P(class(Nagasaki)|w_t, h(t-1))$ are computed by the RNNLM.

We   can   compute $P(Fukushima|class(Tokyo, w_t, h(t-1)))$ and $P(Fukushima|class(Nagasaki, w_t, h(t-1)))$ according to Eq. (11) and (12). α represents the proportion of probability mass that we put on the IV of *class(IV)*. *(1- α)* represents the proportion of probability mass that we put on the OOV of *class(IV)*. This weight is adjusted experimentally. It should be possible to have one α per *class(IV)*, but it would be difficult to accurately estimate these parameters. We chose to estimate only one α for all words.

$$(Tokyo|w_t, h(t-1)) = P(class(Tokyo)) * \alpha \tag{13}$$

This ensures that the sum of probability of all words is one:

$$\sum_{m \in IV} P(m|w_t, h(t-1)) + \sum_{m \in OOV} P(m|w_t, h(t-1)) = 1 \tag{14}$$

3   Experimental Setup

3.1   Data Description

Training Textual Corpora
We used the following corpora for training our language model, OOV PN retrieval system and brother list's generation:

- *Le Monde:* textual data from the French newspaper *Le Monde* (*200*M words; corresponding to 1988–2006, only eleven years);
- *Le Figaro*: textual data from the French newspaper *Le Figaro* (*8*M words, 2014);
- *L'Express*: textual data from the French newspaper *L'Express* (*51*M words, 2014).

The original LM was trained using the *Le Monde* corpus. The lists of OOV PNs to add were created using the *l'Express* corpus. The *Le Figaro + l'Express* corpus was used as the *contemporary corpus* for estimating word embeddings and for generating brother lists. These corpora correspond to the same time period as the development and test data.

Development and Test Textual Corpus
The development and test corpus come from the website of the *Euronews* channel: textual news articles from January 2014 to June 2014 [14]. We selected only the sentences containing at least one OOV word. For the development and test we used the same number of sentences *1148* sentences (about *29*K words per corpus, different sets of sentences for development and test corpus). The development corpus is used to evaluate the

methodology proposed in this paper and to adjust the involved parameters. the evaluation is performed on the test corpus using the adjusted parameters. The results will be presented in term of *word perplexity*.

Test Audio Corpus

The test audio corpus consists of video files reports from the *Euronews* website and their accompanying transcripts (2014). It could be noted, that the reference transcriptions for the recognition experiments are the transcripts provided with the news videos, which may not always be an exact match to the audio. The test audio corpus consists of *300* articles (60K words) and the OOV rate is about *2%*. The number of retrieved OOV PNs is *9300* OOVs. Confidence interval is *±0.3%*.

3.2 RNNLM

The lexicon contains about *87*K words. The RNNLM is trained with the toolkit developed by Mikolov [7] with *310* classes and *500* hidden nodes. The standard backpropagation algorithm with stochastic gradient descent is used to train the network.

3.3 OOV Proper Noun List

The *original lexicon* of *87*K words is augmented by adding the retrieved OOV proper noun word list as follows:

- For each development/test file, we create a ranked list of OOV proper nouns according to the methodology presented in [13];
- From each list we keep only top *128* words;
- All lists from the development set are merged into one list; all lists from the test set are merged into one list.

Finally, we obtain the extended lexicon of *95*K words.

3.4 Language Model

In our experiments, different language models are evaluated. It is important to notice that all the language models contain the same vocabulary: the extended lexicon (*95*K words).

- The baseline RNNLM language model is built as follows: it is trained using the original lexicon (*87*K words) on the train corpus (*Le Monde* corpus). The probability of an OOV from the retrieved OOV proper noun list is computed using the probability of *unk* (unknown word) estimated by the RNNLM. We consider *unk* as a class corresponding to all OOV proper noun words.

$$P(OOV|w_t, h(t-1)) = P(class(unk)) * P(OOV|class(unk)) \qquad (15)$$

where $P(class(unk))$ is computed by the RNN (output neuron corresponding to unk). To estimate $P(OOV|class(unk))$, we assume that all OOVs are equiprobable:

$$P(OOV|class(unk)) = 1/NbrOOV_{train} \tag{16}$$

where $NbrOOV_{train}$ is the number of OOV PNs in the training corpus. A similar approach was used in [16].

- **The modified** RNNLM is the same as the baseline LM and corresponds to the extended lexicon, but the probabilities are estimated according to the proposed methodology.

Note that these LMs have *the same number of words*, corresponding to the extended lexicon, and so the computed perplexities will be comparable.

During brother generation, we removed stop words (articles, adverbs, adjectives) from the brother list, because it is unlikely that these words appear in the same context as the proper nouns. So they cannot be used as brother words.

4 Results

As usual, the development corpus is used to tune the parameters and to find the best configuration for each method. After this, the best configuration is evaluated on the test corpus.

4.1 Results on the Development Corpus

Table 1 gives examples of brother list generation for some OOVs using similarity-based and Wikipedia-based approaches. We can observe that the brother choice seems to be reasonable. We would like to note, that the brother lists generated by these methods

Table 1. Examples of brother list generation for some OOV words using similarity-based and Wikipedia-based approaches

OOVs	Brother words
Similarity-based approach	
CEZ	*Microsoft, KPN, Vivendi*
Bouar	*Donetsk, Kidal, Kharkiv, Kayes, Tripoli, Lucerne, Brno, Paris*
Randstad	*Areva, CNPC, Dassault, Boursorama, MSF, Dongfeng, Ikea*
Kaymer	*Andre, Martin, Citroen, Nestle*
Heslov	*Bollore, Nestle, Lagardere, Kevin*
Wikipedia-based approach	
Kaymer	*Scorsese, Luther, Malvy, Bouygues, Bangemann, Marietta, Walser, Heidegger*
Heslov	*Dalton, Fox, Hackett, Hill, Wood*

are different because the brother choice criterions are different. For example, for OOV *Kaymer,* similarity-based method proposes 4 words (*Andre, Martin, Citroen, Nestle)* chosen according to Mikolov similarity. While Wikipedia-based method proposes (*Scorsese, Luther, Malvy, Bouygues, Bangemann, Marietta, Walser, Heidegger, Hirsch, Winckler*) because these family names have the same first name *Martin,* as OOV name *Kaymer.*

Parameter Choice

Figure 2 shows the evolution of the word perplexity in function of the brother number for the similarity-based approach. This number represents the maximal size of every brother list and corresponds to M (it is possible to have less brothers that this number). This number of brothers is used to compute $P(w_{t+1} \mid OOV, h(t-1))$ and $P(OOV \mid w_t, h(t-1))$. We can observe that using only 1 or 2 brothers gives a high word perplexity. Using more brothers is better. The best value of the brother number is around 26 brothers for similarity-based approach. In the following experiments, we will use 26 brothers for this approach. For n-gram counting approach, the best value is 28 and for Wikipedia-based approach 5 is optimal.

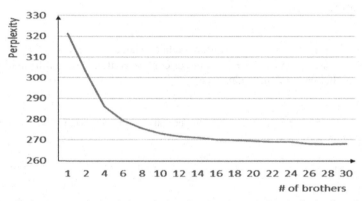

Fig. 2. Perplexity versus maximal size of every brother list (M) for similarity-based approach. Development text corpus, $\alpha = 0.6$.

Figure 3 presents the word perplexity evolution in function of the coefficient α (cf. Eq. (11)–(13)) for similarity-based approach. *(1 − α)* can be seen as the probability mass that is removed from the IV words to be given to the OOV words. The perplexity decreases when coefficient α increases until *0.6*. After this value, the perplexity begins to increase. We decided to use this value of *0.6* for this method in the following experiments This means that for this method the probability mass that we put on the IV of *class(IV)* is *0.6*. For other brother generation methods this coefficient is adjusted experimentally, method per method.

Word Perplexity Results

Table 2 presents the perplexity results of experiments on the development data. In this table, as previously, *#brothers* represents the maximal size of every brother list and corresponds to M. It is important to note that in these experiments the extended lexicon

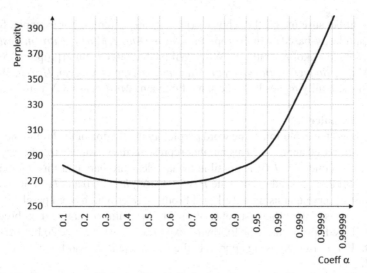

Fig. 3. Perplexity versus coefficient α for similarity-based approach. Development text corpus, brother number (M) is 26.

is used. For the k-gram brother generation method, a larger context ($k = 5$) gives a better result than a smaller context ($k = 3$): a larger context contains more information about the similarity between IV and OOV words.

Table 2. Word perplexity results for OOV proper noun's probability estimation in the RNNLM on the development text corpus.

Language models	#Brothers (M)	α	PPL
Baseline RNNLM			311.4
Modified RNNLM, similarity-based	26	0.6	**267.9**
Modified RNNLM, n-gram counting, $k = 5$	28	0.9	299.0
Modified RNNLM, Wikipedia-based	5	0.9	295.5

The best result is obtained by the similarity-based method: we obtained the perplexity of *267.9* compared to the perplexity of *311.4* for the baseline method. We note an important difference between two brother generation method results: PPL of *267.9* for similarity-based and *299.0* for n-gram-based methods. This can be explained by the fact that Mikolov's word embedding allows to better model the word contexts. We tried to mix the two best approaches, but no word perplexity improvement was observed.

In conclusion, from this table we observe that the proposed method for OOV integration in the RNNLM using similarity-based brother generation gives a good perplexity reduction over the baseline: the reduction is *14%* for the best configuration, compared to the baseline RNNLM (*267.9* versus *311.4*).

4.2   Results on the Test Text Corpus

The best-performing configuration of brother selection methods from the experiments on the development data is applied to the test data. For similarity-based brother selection method, we use the list of 26 brothers, $\alpha = 0.9, k = 5$.

Table 3 displays the word perplexity results on the test data. The results are consistent with the results obtained on the development data. The proposed methods improve the perplexity compared to the baseline system. As previously, n-gram count and Wikipedia-based methods perform worse than the similarity-based method. The best perplexity reduction is *14%* relative compared to the baseline RNNLM (*258.6* versus *299.5*). This improvement is consistent to the one obtained on the development set.

Table 3. Word perplexity results for OOV proper noun's probability estimation in the RNNLM on the test text corpus.

Language models	PPL
Baseline RNNLM	299.5
Modified RNNLM, similarity-based, 26 brothers, $\alpha = 0.6$	**258.6**
Modified RNNLM, n-gram count, $k = 5$, 28 brothers, $\alpha = 0.9$	291.4
Modified RNNLM, Wikipedia-based, $k = 5$, 28 brothers, $\alpha = 0.9$	283.2

4.3   Recognition Results on the Test Audio Corpus

After finding the best parameters and algorithms on the text corpus, we use the test audio corpus to further examine speech recognition system performance.

The Kaldi-based Automatic Transcription System (KATS) uses context dependent DNN-HMM phone models. These models are trained on 250-h broadcast news audio files. Using the SRILM toolkit [15], a pruned trigram language model is estimated on the *le Monde + Gigaword* corpus and used to produce the word lattice. From lattice, we extracted 200-best hypotheses and we rescored them with the RNNLMs (baseline RNNLM and modified RNNLM using similarity-based approach).

We computed the *Word Error Rate* (WER) for three language models: RNNLM with original lexicon (*87*K words); baseline RNN language model with extended lexicon (*95*K words); modified RNNLM using similarity-based approach with the best parameter set and using extended lexicon (*95*K words). The last two RNNLM correspond to the models used in the previous sections. All these models are used to rescore *200*-best hypotheses.

The results for the recognition experiments on the audio corpus are shown in Table 4. The baseline RNNLM with original lexicon gives *20.2%* WER. Using the extended lexicon with the baseline RNNLM or with the modified RNNLM gives similar results: *18.7%*. Thus, extended lexicon yielded a statistically significant improvement over the original

Table 4. WER results using different lexicons and RNN language models on the audio corpus.

Lexicons and language models	WER (%)
Original lexicon and rescoring with baseline RNNLM	20.2
Extended lexicon and rescoring with baseline RNNLM	**18.7**
Extended lexicon and rescoring with modified RNNLM, similarity-based, 26 brothers, $\alpha = 0.6$	**18.7**

lexicon. In contrast, no improvement is observed for the proposed method (*18.7%* WER) compared to the baseline RNNLM with the extended lexicon. However, the proposed similarity-based method obtained a good perplexity improvement compared to the baseline RNNLM on the development and test corpus (cf. Sect. 4.1 and 4.2). This can be due to the fact that reducing perplexity does not always imply a reduction of WER.

5 Conclusion

In this paper we explore different ways of adding OOVs to the language model of ASR. We propose new approaches to OOV proper noun probability estimation using RNN language model. The key ideas are to use similar in-vocabulary words, word-similarity measures, *n*-gram counting and Wikipedia. The main advantage of our methodology is that the RNNLM is not modified and no retraining or adaptation of the RNNLM is needed. The proposed methods can be applied for other NN LMs (more hidden layers or LSTM/GRU layers), because we do not modify the internal architecture of the model.

Experimental results show that the proposed approaches achieve a good improvement in word perplexity over the baseline RNNLM system, and that the similarity-based approach gives the lowest perplexity.

Acknowledgements. This work is funded by the ContNomina project supported by the French National Research Agency (ANR) under contract ANR-12-BS02-0009. Experiments presented in this paper were carried out using the Grid'5000 testbed, supported by a scientific interest group hosted by Inria, CNRS, RENATER and other Universities and organizations (https://www.grid5000.fr).

References

1. Bisani, M., Ney, H.: Open vocabulary speech recognition with flat hybrid models. In: Proceedings of Interspeech (2005)
2. Chen, X., et al.: Recurrent neural network language model adaptation for multi-genre broadcast speech recognition. In: Proceedings of Interspeech (2015)
3. Currey, A., Illina, I., Fohr, D.: Dynamic adjustment of language models for automatic speech recognition using word similarity. In: Proceedings of IEEE Workshop on Spoken Language Technology (SLT) (2016)
4. Mikolov, T.: Statistical language models based on neural networks. Ph. D. thesis, Brno University of Technology (2013)

5. Mikolov, T., Karafiát, M., Burget, L., Cernocký, J., Khudanpur, S.: Recurrent neural network based language lodel. In: Proceedings of Interspeech (2010)
6. Mikolov, T., Kombrink, S., Burget, L., Cernocky, J., Khudanpur, S.: Extensions of recurrent neural network language model. In: Proceedings of IEEE ICASSP, pp. 5528–5531 (2011)
7. Mikolov, T., Kombrink, S., Deoras, A., Burget, L., Cernocky, J.: RNNLM - recurrent neural network language modeling toolkit. In: Proceedings of IEEE ASRU (2011)
8. Mikolov, T., Sutskever, I., Chen, K., Corrado, S., Dean, J.: Distributed representations of words and phrases and their compositionality. In: Advances in Neural Information Processing Systems, pp. 3111–3119 (2013)
9. Naptali, W., Tsuchiya, M., Nakagawa, S.: Class-based N-gram language model for new words using out-of-vocabulary to in-vocabulary similarity. IEICE Trans. Inf. Syst. **E95-D**(9), 2308–2317 (2012)
10. Parada, C., Dredze, M., Sethy, A., Rastrow, A.: Learning sub-word units for open vocabulary speech recognition. In: Proceedings of ACL (2011)
11. Qin, L.: Learning Out-of-Vocabulary Words in Automatic Speech Recognition. Ph. D. thesis, CMU University (2013)
12. Shaik, A., Mousa, E.-D., Schlüter, R, Ney, H.: Hybrid language models using mixed types of sub-lexical units for open vocabulary German LVCSR. In: Proceedings of Interspeech, pp. 1441–1444 (2011)
13. Sheikh, I., Fohr, D., Illina, I., Linares, G.: Modelling semantic context of OOV words in large vocabulary continuous speech recognition. IEEE/ACM Trans. Audio Speech Lang. Process. Inst. Electr. Electron. Eng. N **25**(3), 598–610 (2017)
14. Sheikh, I., Illina, I., Fohr, D.: How diachronic text corpora affect context based retrieval of OOV proper names for audio news. In: Proceedings of LREC (2016)
15. Stolcke, A., Zheng, J., Wang, W., Abrash V.: SRILM at sixteen: update and outlook. In: Proceedings of IEEE Automatic Speech Recognition and Understanding Workshop (2011)
16. Sundermeyer, M., Oparin, I., Gauvain, J.-L., Freiberg, B., Schlüter, R., Ney, H. Comparison of Feedforward and recurrent neural network language models. In Proceedings of IEEE ICASSP (2013)
17. Wen, T.-H., Heidel, A., Lee, H.-Y., Tsao, Y., Lee, L.-S.: Recurrent neural network based personalized language modeling by social network crowdsourcing. In: Proceedings of Interspeech (2013)

Automatic Pairing of Perfective and Imperfective Verbs in Polish

Zbigniew Kaleta(⊠) (iD)

AGH University of Science and Technology,
al. Mickiewicza 30, 30-059 Kraków, Poland
zkaleta@agh.edu.pl

Abstract. The information about verb's aspect and its aspectual pair should be available to algorithms for languages where this feature is present. This paper presents an algorithm that automatically detects morphological dependencies between verbs in Polish and uses them to match corresponding perfective and imperfective verbs. The results of the algorithm are tested against *Słownik Gramatyczny Języka Polskiego*.

Keywords: Aspectual pair · Polish verb morphology

1 Introduction

Verbs in some inflectional languages e.g. Polish, Russian and Greek, have a lexico-semantic feature of aspect, which is not present in positional languages. There are two aspects: perfective and imperfective.[1] [2] The imperfective verbs are used to emphasize the duration or repetitiveness of an action. The perfective verbs emphasize the completion of an action or its one-time character. E.g. the sentence *Czytałem książkę* ('I was reading a book') describes an action that was stretched over time, while *Przeczytałem książkę* ('I have read a book') puts emphasis on the fact that the action has been completed. Two verbs of different aspects but the same meaning are called an aspectual pair. E.g. imperfective *czytać* and perfective *przeczytać* both mean 'to read'. Verbs forming the aspectual pair enter into a morphological dependency, e.g. *przeczytać* was created from *czytać* with prefix *prze-*.

There is also a subset of the imperfective verbs which put emphasis on repetitiveness of an action. They are called the iterative verbs. Pairing of the iterative and noniterative verbs, such as *czytać* (to read) – *czytywać* (to read from time to time), is also an interesting task, but it is not in the scope of this paper.

There are two major morphological differences between the perfective and imperfective verbs that reflect the feature. The first difference is that the perfective verbs lack present tense forms. The imperfective ones have the present tense forms but no future tense forms.[2] The second difference is reflected in participles,

---

[1] From here on this article refers to Polish, unless stated otherwise.

[2] They can be used in future tense using a grammar construct consisting of verb *to be* and infinitive or past tense form.

© Springer Nature Switzerland AG 2020
Z. Vetulani et al. (Eds.): LTC 2017, LNAI 12598, pp. 212–224, 2020.
https://doi.org/10.1007/978-3-030-66527-2_16

namely present participles formed with suffix *-ąc*, and past participles formed with *-łszy* or *-wszy*. The imperfective verbs have present participles and do not have past participles, exactly opposite to the perfective verbs. E.g. imperfective *iść* (to go) has present participle *idąc* but no past participle and its aspectual pair *pójść* has past participle *poszedłszy* but no present participle.

To complete the Polish verb description two more groups have to be mentioned: dual-aspect verbs and impersonal verbs.

As for the dual-aspect verbs, they are both perfective and imperfective at the same time. Such verbs are *abdykować* (to abdicate), *aresztować* (to arrest), *emitować* (to broadcast/to issue/to emit) and so on. "Słownik fleksyjny języka polskiego"[3] contains 39 such verbs, Perlin [5] reports 73 of them based on "Słownik Języka Polskiego PWN" by M. Szymczak. These verbs have simple future tense forms equal to corresponding present tense forms. They also have both present and past participles, e.g. *abdykować* have *abdykując* and *abdykowawszy*.

The impersonal verbs do not have a subject but may have an object. "Słownik fleksyjny..." has 99 of such verbs, including *brak* (stating that something is missing), *czuć* (something can be felt or smelled), *padać* (to rain), *ściemniać się* (to get dark) and so on. They also do not have present nor past participle and have only a handful of forms.

Verbs from both of these groups do not have aspectual pairs, so they are of no further interest to this algorithm or paper.

1.1 Goal and Motivation

As can be easily noticed simply changing the aspect of the verb significantly changes the meaning of the sentence. Therefore recognizing the aspect of the verb and being able to change it may be crucial. Examples of such situations are translation and natural language interface.

Additionally, as both members of an aspectual pair share meaning and are interchangeable in contexts, statistical algorithms should treat them equally.[4]

There is no clear rule on how to generate the aspectual pair for a given verb. Therefore if an algorithm needs such information, it must be available in a dictionary.

The aim of this article is to create an algorithm that would retrieve all aspectual equivalents from a given set of correct verbs using morphological rules. For every verb in a given set the algorithm should return its aspectual pair or information that this verb has no pair.

2 Polish Verb Morphology

Aspectual pairs are closely related in terms of word formation. Therefore discovering possible aspectual equivalents for a given verb requires recognition of morphological structure of the verb in question.

---

[3] see Sect. 3.

[4] As they actually do in English and other languages where the aspect is associated with a grammatical construction rather than a verb itself.

In Polish verb morphology there are two major ways of forming a new verb on the base of existing one: a prefix addition and a suffix substitution. [3] Please note that multiple derivatives might be created from a single verb in one or both of these ways and only one of these derivatives enters into imperfective - perfective opposition, if the aspectual pair exists.

2.1 Prefix Addition

The prefix addition is a way of creating a perfective verb from an imperfective one. A new verb is formed by simply adding a prefix while the verb's root or suffix are not changed. E.g. *myć* → *umyć* or *zmyć* (to wash). There are 22 valid prefixes. 7 of them may optionally contain an *e*, depending on the phonological structure of the verb.[5] In all cases but one this optional *e* appears at the end of the prefix, e.g. *nadrywać* and **naderwać**. The remaining case is prefix wz-, which becomes wez- in some cases, e.g. **wezbrać** and **wzbierać**. This is a phonetic phenomenon and the choice of prefix form with or without *e* does not affect its semantics.

All verb prefixes are listed in Table 1 (please see also [3]). The optional *e* is denoted by parenthesis.

Table 1. Prefixes

bez	do	na	nad(e)	o	ob(e)
od(e)	po	pod(e)	prze	przy	roz(e)
s	ś	u	w(e)	ws	wy
w(e)z	z(e)	za			

Example derivation using the prefix addition is shown in Fig. 1. The ϕ symbol denotes no prefix.

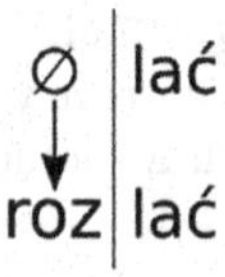

$$\varnothing \,|\, \text{lać}$$
$$\downarrow$$
$$\text{roz} \,|\, \text{lać}$$

Fig. 1. Example of prefix derivation

---

[5] Namely: if attaching the prefix would create difficult to pronounce conglomerate of consonants.

2.2   Suffix Substitution

The suffix substitution is a bit more complex than the prefix addition and may be used to create both perfective and imperfective verbs. A perfective verb may only be created from an imperfective one in this way, whereas imperfective may also be created from another imperfective. This is the case for iterative verbs.

Most verbs end with a verb forming suffix. Please note, that the verb forming suffix is not the same as the inflection ending. Therefore a part of a verb after removing the suffix and the prefix is called a root here, not a stem. There are about 30 verb forming suffixes. A new verb may be formed by changing the suffix. However, not every suffix change is legal. All changes where the new suffix ends with the previous one, such as *-ywać*→*-owywać*, are legal. This is the way how *zmy|wać* was created from *zmy|ć*[6] – the pipe symbol denotes the end of the root and start of the suffix. The suffixes are listed in Table 2 in order in which they replace each other [3].

Table 2. Suffixes

ć	ać	jać		
ć	ać	wać		
ć	ać	ować	yzować	
ć	ać	ować	izować	
ć	ać	ywać	owywać	izowywać
ć	ać	ywać	owywać	yzowywać
ć	ać	iwać	owiwać	izowiwać
ć	ać	iwać	owiwać	yzowiwać
ć	ać	otać		
ć	ić	olić		
ć	ić	ocić		
ć	eć			
ć	yć			
ć	nąć			

Other legal changes are listed in Table 3. Verb's suffix listed in the first column may be replaced with any one suffix from the right column of the same row, as in *wybebesz|yć* → *wybebesz|ać* (to gut). The substitutions listed here are in the same form as used by the algorithm – some of the suffixes in this table are not actual verb forming suffixes but pseudo-suffixes. They also contain information about the context in which given substitution may occur or the alteration that always co-occurs. E.g. *-gnąć* → *-gać* is actually substitution *-nąć* → *-ać* which may occur only if the root ends with *g*. This is where the algorithm does not strictly represent the linguistic analysis for the sake of simplicity.

---

[6] Please note, that *zmyć* has already been created from *myć* with a prefix.

Table 3. Suffix substitutions

Substituted	Substitution
ć	ć
eć	ać
ąć	ijać, ynać
ić	ać, ewać, eć, iać, ijać, iwać, jać, ować, ywać
ać	ać
yć	ać, ewać, eć, iwać, ować, ywać
nąć	eć, jać, kać, wać, ynać
gnąć	gać
snąć	stać
znąć	zgać
owć	owiwać, owywać

Example of derivation using the suffix substitution is shown in Fig. 2. The vertical line denotes the start of the suffix.

Fig. 2. Example of suffix derivation

When the suffix substitution occurs there may be also an alteration in the verb's root. The root alteration is when one letter or group of letters changes into another or when a letter is added or removed from the root. Some alterations are typical and consist of one phone[7], like o:a and ź[8]:ż in *zamroz|ić → zamraż|ać* (to freeze). Another example is shown in Fig. 3. The ϕ symbol denotes empty suffix. Others are unique, such as gią:gin in *zgią|ć → zgin|ać* (to bend). They are deeply rooted in language history and can take place only in certain roots. They may consist of more than one phone. If the unique alteration occurs, no other alteration, typical nor unique, can occur. A change between aforementioned prefix forms is not considered an alteration and may occur alongside unique alteration as in *zebra|ć → zbier|ać*. Another example of unique alteration is shown in Fig. 4.

---

[7] The algorithm works only on orthographic forms and so all examples in this paper are also in the orthographic forms.

[8] *Ż* is a palatal *z*. In ortographic form of *zamrozić* there is *z* but it becomes palatal due to the following *i*.

3   Data Sources

There are two main sources of data for this algorithm: "Słownik fleksyjny języka polskiego" (Inflectional Dictionary of Polish) as an input data source and Grammatical Dictionary of Polish as a reference data source for the purpose of testing.

Fig. 3. Example of suffix derivation with typical alteration

Fig. 4. Example of suffix derivation with unique alteration

Inflectional Dictionary of Polish is a dictionary and a programming library (named CLP) by Lubaszewski et al. [1,4,6] It is an automatically generated inflectional dictionary of Polish with all the data checked by linguists. It also contains morphosemantic relations between words. CLP was used as a source of verbs and corresponding participles.

Grammatical Dictionary of Polish or SGJP (short for its Polish name: *Słownik Gramatyczny Języka Polskiego*) is an inflectional dictionary created by Saloni et al. [7] It contains morphosyntactical tag or tags for every word and morphosemantic connections between words, including the connection between a verb and its aspectual counterpart. The aspectual counterpart for the purpose of testing of this algorithm is considered an aspectual pair, regardless of slight differences.

4   The Algorithm

The whole algorithm can be divided into 3 steps: determining the aspect of every verb, discovering all derivations which are candidates for the aspectual pair and finally selecting correct aspectual pairs from these derivations.

4.1 Determining the Aspect

This part is fairly easy, when correct participles for every verb are given. Determining the aspect of a verb boils down to checking which participles it has, based on rules described in Sect. 1.

4.2 Discovering Possible Aspectual Pairs

This part of the algorithm bases on prefix addition and suffix substitution described in Sect. 2. It searches for all connections $verb_1 \rightarrow verb_2$, where $verb_2$ can be legally derived from $verb_1$ in terms of the word formation. The left-hand side verb in this relation will be called the source verb and right-hand side verb – the destination verb or the derived verb.

Since only potential aspectual pairs are of interest, the algorithm is looking for the derivations of a perfective verb from an imperfective one and then vice versa. This allows the algorithm to avoid the problem with the iterative verbs, because they are imperfective derived from another imperfective. The order of search is dictated by the fact, that finding derivations by prefix addition adds information useful for splitting the suffix – see subsection Splitting the Suffix below.

Since reflexive verbs can only have reflexive aspectual pairs, they are separated from non-reflexive verbs. Both groups are treated the same way, so there will be no distinction between them further. This separation is done purely for the optimization of computation time.

The algorithm works in three runs in each direction. Each run it iterates over all pairs (source verb, destination verb) $\in$ source set $\times$ destination set and checks given condition for this pair. If the condition is true, the pair is accepted as legally derived. At the start of the first run source set consists of all verbs of one aspect and destination set of all verbs of the other aspect. After each run all source verbs of relations detected in this run are removed from source set and all derived verbs are removed from the destination set. This prioritizes the conditions in order to avoid false derivations.

The runs are as follows:

1. Prefix addition or suffix substitution without root alteration
   Connects pairs where destination verb consists of only one of the prefixes and the source verb without any change, like *gotować→ugotować* (to cook). Also connects pairs derived with the suffix substitution where both verbs have exactly the same root, such as *da|ć→da|wać* (to give).
2. Suffix substitution with unique root alteration
   Connects pairs where the suffixes can be legally changed using the suffix substitution where one root is obtained from the other with exactly one unique root alteration. It allows the change of prefix's form. E.g. *zebra|ć→zbier|ać* (to gather, to harvest) with alteration *bra:bier*.
3. Suffix substitution with typical root alteration

Connects pairs derived with the suffix substitution where one verb's root can be created from the other using up to two typical alterations, like *krzyk|nąć→krzycz|eć* with *k:cz* (to shout, to scream).

Completing this part of the algorithm creates derivative chains – sequences of verbs, where each one is derived from the previous one, starting with the base verb which is not derived from any other. There is no rule as to whether the base verb is perfective or imperfective.

Since those chains intersect with each other – a verb can have multiple derivations – it is useful to group them together in a derivation tree. Figure 5 shows an example derivation tree starting with an imperfective verb *cumować* (to moore) and Fig. 6 a derivation tree starting with perfective *plusnąć* (to splash).

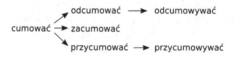

Fig. 5. Derivation tree sample

Fig. 6. Derivation tree sample

In the case of the algorithm described here, the generated graph is not always a tree, due to incorrectly discovered derivations. E.g. the algorithm discovers correct derivation *akumulować → zakumulować*, as well as incorrect *kumulować → zakumulować*, which causes *zakumulować* to have two parents. Nonetheless the term derivation tree will be used throughout this paper.

Splitting the Suffix. In order to correctly recognize verbs derived using the suffix substitution, the suffix must be identified first. The most basic approach to this problem is listing all suffixes in order of descending length.[9] The first suffix that exactly matches at the verb's end and meets the condition that verb's root consists of at least one syllable is the correct one. Therefore it is *my|ć* (to wash) and *pi|ć* (to drink), not *m|yć* and *p|ić*.

This approach does not take into consideration prefixes. Adding a prefix to a verb does not change its root and suffix. E.g. correct suffix of *myć* and *pić*

---

[9] Order of suffixes with the same length is irrelevant, since two different suffixes cannot match at the same time.

is -*ć* and so it is for their perfective counterparts: *umyć* and *wypić*. This naive algorithm would return -*yć* and -*ić* respectively, since *um* and *wyp* are correct syllables.

This problem could be solved by recognizing the prefix of a verb, but it is a difficult task. The reason are pseudo-prefixes. Some verbs start with a sequence of letters equal to one of the prefixes, but it is not an actual prefix. E.g. *zakumulować* is *z+akumulować* not *za+kumulować* (although *kumulować*, *akumulować* and *zakumulować* all mean 'to accumulate'). Also *dostać* (to receive) is not *stać* (to stand) with a prefix *do-*.

The algorithm described in this paper first discovers all derivations using the prefix addition and stores information about the prefix in the derived verb. When analyzing two verbs for possible suffix substitution, if one of the verbs has a prefix denoted then the other must have the same or equivalent[10] prefix. If neither has a prefix denoted, then they are only checked for the equivalent prefixes. In case of exactly the same prefix, when neither was discovered during earlier analysis, it is assumed to be the part of the root.

4.3   Selecting the Correct Aspectual Pair

This part of the algorithm works under the assumption that each verb has zero or one aspectual pair. There are some cases when it is not true. First, there are cases of concurrent verbs in a single derivative chain, e.g. *giąć* → *zgiąć* → *zginać* (to bend), where *zgiąć* is perfective, two others are imperfective. There is no reason to prefer *giąć* over *zginać* or vice versa as a pair for *zgiąć*. Second, there are homographic verbs, where one meaning has one aspectual pair, and the other has another. For example there are pairs: *rozstrzeliwać* – *rozstrzelić* (to scatter) and *rozstrzeliwać* – *rozstrzelać* (to shoot, to execute). Unfortunately homonymy cannot be detected on the morphological level. SGJP has 940 verbs with more than one aspect counterpart, three of them have four counterparts.

Input data for this part of the algorithm is a tree, similar to the one depicted in Figs. 5 and 6, but each node consists of not only a verb but also its aspect. This avoids the problem with homographic verbs, where one of them is perfective and the other is imperfective, such as *cisnąć* (perfective of 'to throw' or imperfective of 'to push') or *trącić* (perfective of 'to smack' or imperfective of 'to smell'). Edges represent the derivation relation determined by the previous part of the algorithm.

The term *unambiguous sequence* will be used to denote a path of at least two nodes where each node has no more than one predecessor and one successor.

When the algorithm matches an aspectual pair it removes both nodes from the tree along with all adjacent edges. Verbs can only be aspectual pair if they are connected with an edge.

The algorithm repeats all following steps in a loop until no new aspectual pair is matched in a pass.

---

[10] with *e* added or removed.

1. All verbs that are neither left nor right side of any relation are designated as having no aspectual pair, e.g. *administrować* (to administer). Note that this can happen because the verb has no derivatives or that they were removed in previous runs.
2. If two verbs are in relation only with each other then they are matched, like *dezaktywować – zdezaktywować* (to deactivate).
3. If there is an unambiguous sequence where the first verb has no predecessor and the last verb has no successor it is split into pairs.

   If the length of such sequence is even, then it is split in twos so that every verb has an aspectual pair. E.g. *klepnąć się→klepać się→poklepać się→poklepywać się* (to clap, to pat) creates two pairs: *klepnąć się – klepać się* and *poklepać się – poklepywać się*.

   If the length is odd then the first three verbs are analyzed. Derivation due to the suffix substitution is preferred over prefix addition. If both derivations are based on prefix addition then the first one is chosen. If both are based on the suffix substitution then the pair with longer common substring (starting at the beginning of a string) is chosen. If said length is equal, the first pair is chosen. E.g from derivative chain *dostawić→dostawiać→podostawiać*, the aspectual pair is *dostawić – dostawiać* because it is based on suffix substitution and *podostawiać* is left without a pair.

   If the second and third verbs were determined to be a pair, all three verbs are removed from further analysis and the first is assumed to have no pair. If the first two, then only they are removed. If the third verb has any successors – i.e. the path had length of at least five nodes – it will be analyzed again in the next run.
4. If there is an unambiguous sequence where the last node has no successors and predecessor of the first node does not belong to an unambiguous sequence (i.e. has more than one successor) it is analyzed as in the previous step.

Taking for example the derivation tree from Subsect. 4.2 the algorithm would operate in the following way:

The first three steps yield no results. In the fourth step unambiguous sequences *odcumować → odcumowywać* and *przycumować → przycumowywać* are analyzed. Since they have only two verbs each, these verbs are paired and removed from the tree. This leaves only *cumować → zacumować* in the tree, which are paired in the second step of the second run and the tree is then empty. Therefore third run yields no new results and the algorithm ends. Figures 7 and 8 show the process described above.

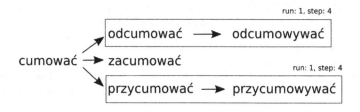

Fig. 7. Example of pair selection - 1st run

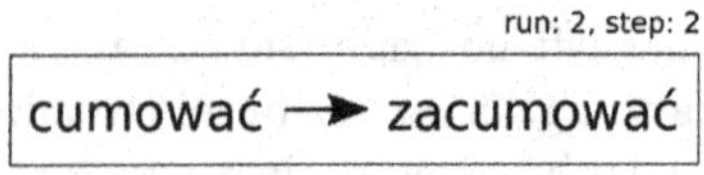

run: 2, step: 2

cumować ⟶ zacumować

Fig. 8. Example of pair selection - 2nd run

The algorithm has been implemented in Python.

5 Tests

The algorithm has been tested using two metrics.

The first one was the accuracy of the final result – assigned aspectual pairs, which tests the algorithm as a whole. Since SGJP may have more than one aspect counterpart for a single verb the answer is considered correct if it matches any of those. E.g. *zatykać* have three aspectual pairs depending on the meaning: *tykać* (to tick), *zatkać* (to plug, to cork) or *zatknąć* (to stick, to shove). The algorithm's answer for *zatykać* is *zatkać*, so it is correct. Also, since not all verbs available for the algorithm were present in the reference data and vice versa, only the results on the common part are shown below. SGJP has a total of 29950 verbs and CLP has 20200. 14302 of these verbs are present in both dictionaries.[11]

The second metric tests only discovering of candidates, checking how often the correct aspectual pair of a verb was classified to the set of verbs derived from given one or vice versa.

SGJP does not have separate entries for reflexive verbs as CLP does. Even for verbs like *awanturować się* (to brawl) which can only be reflexive, SGJP consists entries without *się* and includes information about reflexiveness in the description. For the purpose of testing it was assumed that a reflexive verb behaves exactly as its non-reflexive counterpart. So particle *się* was removed from both the verb and its pair for testing.

6 Results

As can be seen in Table 4 the algorithm yielded an answer for over 80% of all verbs. This seems to be as far as an algorithm based purely on morphology can go. In order to improve this result contexts should be taken into consideration. For example for a verb *pisać* (to write) after pairing there are two options left: *napisać* and *popisać*. They are both generated using the prefix addition and from a morphological point of view they are equally good candidates. Knowledge of semantics would lead to a decision that *napisać* is a pair for *pisać*, while *popisać*

---

[11] CLP does not include so called potential verbs, i.e. verbs which can be created by regular morphological processes, e.g. from a noun, by replacing the inflection ending with suffix -ować. For example the verb *bistrować* may be created in this way from the noun *bistro*. *Bistrować* is not registered in dictionary of Polish but many potential verbs enter to dictionaries each year.

has no pair. It has to be noted that in some cases choosing one aspectual pair is disputable. That is why SGJP for some verbs lists more than one aspect counterpart, e.g. *zbiegać – zbiec, zbiegnąć* (to run down).

Table 4. Pairing ratio

Solved	16081	81.7%
Unsolved	3605	18.3%

Table 5 shows the accuracy of the algorithm. Errors are split into three classes: verbs that have a pair according to SGJP but not according to the algorithm (falsely without pair), have a pair according to the algorithm but not SGJP (falsely with pair) and that have different pair according to both sources (wrong pair). Correct answers have also been split into two classes: verbs without and with a pair.

Table 5. Algorithm's accuracy

Correct	without pair	1435	12601 (84.4%)
	with pair	11166	
Incorrect	falsely without pair	1019	2330 (15.6%)
	falsely with pair	937	
	with wrong pair	374	

As can be seen the biggest part of incorrect answers are verbs that the algorithm wrongly marked as having no pair. This indicates that too little connections between verbs are discovered.

Verbs that were wrongly assigned a pair when they should not have any are also a major group of errors. However, closer analysis shows that some of these cases seem to be correct assignments. In order to more accurately assess the scale of this problem more extensive tests should be conducted. More on this in Sect. 7. The reason for those assignments that are truly incorrect are wrong connections between verbs, which again points to the second part of the algorithm.

This conclusion is further proven by the second test's results which are shown in Table 6. Less than 90% of correct pairs are discovered by the algorithm. This measure cannot be expected to be 100% as some of SGJP's aspect counterparts are not really derived from the verb they are assigned to. E.g. for *spoglądać* (to glance) SGJP lists 3 counterparts: *spoglądnąć, spojrzeć* and *spójrzeć* (archaic), where only *spoglądnąć* is actually derived from *spoglądać*. In this regard SGJP seems to take into consideration semantic relevancy without concern of morphology. Even so this result is not fully satisfactory.

Table 6. Candidates discovery recall

Discovered	8937	88.2%
Undiscovered	1198	11.8%

Total: 10135

7 Conclusions

The results described above are promising, but prove that there is still work to be done for this algorithm to be fully usable.

First of all more extensive tests should be conducted in order to better assess the algorithm's quality and locate its strong and weak points. Such tests should work only on data from SGJP so that input and reference data are relevant. Also tests with new verbs are needed, because those verbs are the main reason, why the automatic method of pairing is needed.

Further, in order to improve the quality of recreating the derivative chains, better recognition of a prefix, root and suffix is needed, which means that context rules should be introduced.

Also, the algorithm should use contexts to choose from among multiple candidates. Contexts or a semantic dictionary could also be used to detect synonymy and homonymy, which may cause a verb to have more than one aspectual pair.

Acknowledgment. This research was partially funded by AGH University of Science and Technology Statutory Fund.

References

1. Gajęcki, M.: Słownik fleksyjny jako biblioteka języka C. In: Lubaszewski, W. (ed.) Słowniki komputerowe i automatyczna ekstrakcja informacji z tekstu pod red. W. Lubaszewskiego, chap. 6, pp. 107–134. Uczelniane Wydawnictwa Naukowo-Dydaktyczne AGH, Kraków (2009)
2. Gołąb, Z., Heinz, A., Polański, K.: Słownik Terminologii Językoznawczej. Państwowe Wydawnictwo Naukowe, Warszawa (1968)
3. Lubaszewski, W.: Program automatycznej segmentacji morfologicznej (na materiale polskich czasowników). POLONICA II **21–32**, (1976)
4. Lubaszewski, W., Wróbel, H., Gajęcki, M., Moskal, B., Orzechowska, A., Pietras, P., Pisarek, P., Rokicka, T.: Słownik fleksyjny języka polskiego. Wydawnictwo Prawnicze LexisNexis, Kraków (2001)
5. Perlin, J.: Ile jest we współczesnej polszczyónie czasownikźw dwuaspektowych? Linguistica Copernicana **1**(3), 165–171 (2010)
6. Pisarek, P.: Słownik fleksyjny. In: Lubaszewski, W. (ed.) Słowniki komputerowe i automatyczna ekstrakcja informacji z tekstu pod red. W. Lubaszewskiego, chap. 6, pp. 37–67. Uczelniane Wydawnictwa Naukowo-Dydaktyczne AGH, Kraków (2009)
7. Saloni, Z.: Podstawy teoretyczne "Słownika gramatycznego języka polskiego", Warszawa (2012)

Computational Semantics

Transforming Syntactic Relations in Attributive Groups

Iuliia Romaniuk[1], Nina Suszczańska[2], and Przemysław Szmal[2]

[1] Institute of Ukrainian Language NAS of Ukraine,
Grushevskogo 4, Kyiv, Ukraine
ju.romaniuk@gmail.com
[2] Silesian University of Technology,
Akademicka 2A, Gliwice, Poland
n.w.suszczanski@gmail.com, przemyslaw.szmal@polsl.pl

Abstract. The article deals with the *Thetos* system that translates Polish texts into sign language. We try to significantly improve the quality of translations. Thus we have recently decided to take into account the meaning of sentences, syntactic groups (SGs) and individual words they are composed of, what we previously ignored. In the paper we describe the work on our SG-model of semantics. The foundation of the model is a predicate-argument representation, which expresses relations between the syntactic tree root, which is a verb syntactic group, and its arguments. The model is completed by semantic relations that exist between the terms occurring in separate arguments, as well as by those that are hidden in the syntactic relations. In the paper we deal with the interpretation of relations within 1st level syntactic groups, especially the attributive groups. We show transformation rules of relations properties which are taken into account. We give examples of transformations.

Keywords: NLP · Polish Sign Language · Semantic processing · Semantic relations · Interpreting relations · Thetos system

1 Introduction

The article deals with the *Thetos* system that translates Polish texts into sign language [9], and recent work aimed at improving translation quality. The ambitious goal set by the research team is to translate Polish texts into PJM, at

---

[1] In Poland there are two versions of sign language: PJM and SJM. PJM is a natural language used by Polish Deaf people. SJM is an artificial language, primarily created for educational purposes and still used in the media. For PJM carriers it is a foreign language, just like spoken Polish. Over the last few years, many discussions have taken place about the choice between PJM and SJM. In this work we only find a difference in approach to translation from Polish into PJM in relation to SJM.

This work was co-financed by SUT grant for maintaining and developing research potential.

Z. Vetulani et al. (Eds.): LTC 2017, LNAI 12598, pp. 227–241, 2020.
https://doi.org/10.1007/978-3-030-66527-2_17

least in some approximation to this natural language[1]. Research conducted by the authors since 1998 confirmed the hypothesis that for SJM translations the analysis of the predicate-argument (P-A) structure is sufficient, but for PJM a deeper semantic analysis is necessary. In our approach, semantic analysis is based on the results of syntactic analysis [5]. Syntactic groups (SGs) derived from the application of SGGP[2] formalism [8] are semantic in principle. This fact allows for transforming a syntactic representation of a Polish sentence into a semantic representation. Semantic analysis is applied both to the body of SGs and to the syntactical relations between the components.

The aim of the research being reported is "understanding" of the content of utterances. Adducing Z. Vetulani [10], for "understanding" we assume a form of translation from a natural language into some artificial, formalized one, whose units are used to store the meanings contained in syntactic structures. In addition, the formal structures of the semantic representation of the text should be convenient for further processing. With us, the mentioned above formal structures are a model of semantics. Since the semantics of a language is an extensive subject of research, it makes sense to speak of only modeling a small semantics fragment, and this is also in our case.

We model the Polish syntax on 5 levels. SGs of each level as well as relations between them are interpreted by separate grammars: for each syntax level – a separate grammar being a set of grammars to interpret each SG type and for each type of syntactic relations in them. In this paper we focus on a description of rules for interpreting the relations that occur in 1st level noun groups (NG^1).

Let us recall that the syntactic structures interpretation stage follows the stage of converting the resultant syntax representation to the P-A structure.

The work described in this paper is an extension of the works [6,7].

2   Predicate-Argument Structure

It is known that the predicate's arguments can be characterized by their type: referential, attributive or sentential [10]. The role of referential arguments is played by NGs (noun SGs), PRON_Gs (pronoun SGs) and a part of PREP_Gs (preposition SGs). Attributive arguments are ADV_Gs (adverbial SGs), MOD_Gs (modifier SGs), and the remainder of PREP_Gs. The number of arguments and argument types depend on the predicate features and are defined in the semantics grammar. One of such grammars is defined in [2]. In the algorithms, which determine the semantic structure of the sentence, we also use the grammar [3], which is an adaptation of Fillmore's case grammar [1].

Apart from ACTION (predicate), there are 16 roles taken into consideration, including AGENT (action's executor), OBJECT, INSTRUMENT, etc. We realize that this list neither includes all the P-A relations in any language, nor is required in full at any processing system, including the system *Thetos*. The list is experimental, designed to carry out the research, including the transfer of space-time relation elements in Polish to the corresponding structures of the Polish Sign Language.

---

[2] SGGP is the Syntax Groups Grammar for Polish.

In this model part we distinguish two types of semantic units: semantic elements and semantic relations. Semantic elements are notions, which represent the meaning of SG or semantic constants. They are the alphabet of the model and are contained in dictionaries. As to semantic relations, they arise during interpretation of SGs and their internal relations.

The aim of SG interpretation is to detect the semantic representative in it and to determine both its meaning and the content of the SG as a whole. Each SG has also a syntactic representative, which not always coincides with the semantic one. The latter is determined based on the SG type and internal syntactic relations in the SG. Semantic interpretation rules for the ADV_G, PREP_G, NG and VG groups are briefly described in [5].

The semantic interpretation should tell us whether there is a mapping from a particular relation in a syntactic representation into a semantic relation, and if so, how it looks. In order to find the mappings we formulated several rules. Some of them are discussed in the following sections. We use the notational convention, according to which semantic relations are written in small letters, with double '#' sign preceding the relation name.

Briefly, tasks of the *Polsem* semantic analyzer are twofold: conversion of 4th level syntactic structures to P-A form and interpretation of P-A elements. With that, the P-A elements are again considered as SGs created by the *Polsyn* parser at all levels of syntax modeling.

In the next section we discuss the syntactic structure of NG^1 and rules for creating the syntactic relations in NG^1 and its components.

3   Syntactic Structure of NG^1

3.1   General Information About NG^1s

The 1st level NG consists of a noun and its attributives. An attributive can be each of numeral (NUM_G) and attributive (ATR_G) SGs, and also selected adverbial (ADV_G), modifier (MOD_G), and negation (NEG_G) SGs. The NG shape can be represented by the following string:

$$(ADV\_G(ATR\_G(NUM\_G(\ldots(noun)\ldots)))))^3.$$

Here the ellipsis '...' indicates that theoretically, this chain can be infinite due to reuse of attributive SGs. In practice, however, the number of attributives to a noun is limited and does not exceed 7 nests [4].

If NG contains a noun or a nominal pronoun, then it is the NG representative. However, filling positions in the NG chain is not mandatory. There are structures where a noun is absent. In this case, the NG representative is instead of noun – for example – adjective or numeral. For each component of NG^1, the representative and relations between these representatives are also defined.

---

[3] MOD_G and NEG_G groups are not shown in this chain. These groups usually have an auxiliary character, while the execution of the production can be seamlessly merged with another SG, for example ADV_G.

NG grammatical features: type, case, and number are mostly defined as common features for all components. In other cases, the special circumstances of the agreement of pronouns and nouns, numerals and nouns, numerals and adjectives to the noun, etc., are taken into account. These rules in SGGP are fully compliant with the relevant rules in traditional grammar. NG also has features that indicate the class of the word being a representative of the NG, which may be important for semantic interpretation. NG may also have semantic features, such as *vitality – non-vitality*. Semantic features are inherited from the main element or calculated on the basis of special rules.

3.2   Modifier SG (MOD_G) and Negation SG (NEG_G)

MOD_G consists of a string of words, among the attributes of which is the *modifier* feature. Most often it is a single word, but there may be more, then each of them affects the meaning and characteristics of the next. Then there is the #*mod* relation between the elements of the group. MOD_G type groups are usually of an auxiliary character, during production execution they can be merged without a trace with another SG, for example ADV_G or VG. Once merged, only the relations that have been generated so far remain. At times, however, the MOD_G may remain an "unreclaimed", independent group. Then it occurs as a coherent group and aspires to a semantic role in P-A.

The NEG_G group is a MOD_G variant: the first NEG_G component is always the particle *nie (not)*. If the group contains additional words or modifiers, between them there are relations #*not* or #*not* and #*mod*. Groups of type NEG_G are auxiliary. Just like MOD_G they can be merged with another SG. Relations generated during grouping NEG_G, after merging, are part of syntactic interpretation of merged SG. Occurrence of NEG_G as an independent SG is a special case that happens in elliptic contexts.

3.3   Adverbial SG (ADV_G)

The ADV_G group consists of a string of words: an adverb and, possibly, words that modify its meaning. For example, *szybko (fast), bardzo szybko (very fast), nie bardzo (not really), nie bardzo szybko (not very fast)*. ADV_G includes the following constructions:

adverb + adverb,	(i)
adverbial operator + adverb,	(ii)
ADV_G + adverb,	(iii)

where the adverbial operator is an expression suitable for the role of a semantic modifier for the adverb. The list of these operators is given in a special dictionary. The need for specialized semantic modifier dictionaries is that not all modifiers can act as operators for all word classes. For example, the word *wszystko (everything)* can be a modifier for a verb, the word *więcej (more)* – for a noun, but none of them can be a modifier for an adverb.

In the structures given above, the adverb standing to the right of the '+' sign is the main component and represents ADV_G, the left component is subordinate to the main. The syntactic relation between the ADV_G elements is modification and is denoted #mod. There is also a denial relation (#not) between the operator-particle *nie (not)* and the adverb.

In the rules (i) and (iii), the elements that stand to the left of the '+' sign also act as semantic modifiers. This means that rules (i)–(iii) can be rewritten as follows:

$$[operator] + adverb, \tag{1}$$

where parentheses indicate operator's optionality, and the operator is:

1. operator from the list of adverbial modifiers,
2. an adverb whose grammatical features indicate the possibility of being a semantic modifier,
3. ADV_G, whose representative is an adverb having the grammatical characteristics of a semantic modifier.

The condition for grouping words into ADV_G is the fact that these words are placed side by side in the sentence and fulfill one of the conditions 1–3.

ADV_G components can also be connected using relation of linking in series:

$$ADV\_G \oplus ADV\_G, \tag{2}$$

where $\oplus$ is a comma or one of serial conjunctions. In this case, the main component is not defined. Relation between the components is series zależą od kontekstu, od tego, czym jest wyrażony szereg i jest is denoted as #conj_and, #conj_or, and #acc_comma, for example *szybko i bezpiecznie (quickly and safely)*. Condition of realization of the rule (2) is the presence of common grammatical features in ADV_Gs from both sides of the sign $\oplus$.

ADV_G grammatical features, including semantic, are inherited from the representative, or determined as the result of the unification and generalization of features of the serial components (see [8]).

3.4 Numeral SG (NUM_G)

The basic rule for grouping NUM_G is as follows. The NUM_G consists of simple or complex numerals: *dwa (two)*, *dwieście dwadzieścia dwa (two hundred twenty-two)*, *dwudziestego drugiego (of twenty second)*. The SGGP rule can be written as follows:

$$numeral \ [+ \ numeral[+ \ldots [+ \ numeral]]], \tag{3}$$

where brackets indicate the facultative nature of the component.

The NUM_G representative is the last word in the sequence. The properties of coordination of NUM_G of this type with a noun and its adnominal are inherited from NUM_G's representative. NUM_G's grammatical features are calculated as follows: gender and case are taken from representative as well as the NUM_G type. The category *number* is inherited from the representative in the

case of a simple numeral, in other cases it is set to *plural*. This feature is passed
to NG, what is important for the semantic characterization of NG. Components of NUM_G formed from complex numerals do not form a tree. Therefore,
between NUM_G elements only the relation of preserving their position in SG is
generated, namely #*nextto*.

In more complex cases, to NUM_G creation apply the following rules:

$$[operator] + NUM\_G, \tag{4}$$

$$NUM\_G \oplus NUM\_G. \tag{5}$$

The term 'operator' is defined here as in the case of ADV_G. The list of numeral
modifiers should be provided in the form of a dictionary. The fact of a change
in the numeral status caused by a semantic operator is noted by the syntactic
relation #*mod*; by this – similarly to the case of ADV_G – the negation relation
#*not* is distinguished. The serial relations corresponding to the operator '⊕' are
the same as in case of production (2) for ADV_G.

NUM_G's grammatical features are inherited from the main component or
designated as common to the constituents when the rule (5) is executed. The
requirements for NUM_G components are similar to the requirements for ADV_G
group components.

3.5   Attribute SG (ATR_G)

The ATR_G consists of an adjective, adjective pronoun or an adjective participle,
which may be preceded by an adnominal operator:

$$[operator] + adjective, \tag{6}$$

$$[operator] + adjectival\ pronoun, \tag{7}$$

$$[operator] + adjectival\ participle. \tag{8}$$

The adnominal modifiers are in the dictionary, the operator can also be ADV_G,
MOD_G or NEG_G. The syntactic relation between ATR_G elements when
applying rules (6)–(8) is *modification* or *denial*. For example, #*mod(bardzo,
dobry) (very, good)*.

ATR_G components can also be linked through a series relation:

$$ATR\_G \oplus ATR\_G, \tag{9}$$

where the sign ⊕ means the same as for ADV_G and NUM_G. ATR_G's representative is the main component if the rules (6)–(8) are executed, or the last
component in a series. ATR_G's grammatical features are inherited from the
main component or are calculated as common to the series constituents. Syntactic relations generated between ATR_G members using rule (9) are the same as
when applying rules (2) and (5).

For example, ATR_G *białego, żółtego i niebieskiego (white, yellow and blue)*
consists of five tokens, each of which plays a role in the grouping process. In

Table 1. Step 1 of grouping the ATR_G *białego, żółtego i niebieskiego (white, yellow and blue)*

Token no	1	2	3	4	5
Token	białego	,	żółtego	i	niebieskiego
	(white)	*(comma)*	*(yellow)*	*(and)*	*(blue)*
Result of	ATR_G1	$\oplus_{comma}$	ATR_G2	$\oplus_{and}$	ATR_G3
grouping	(rule 06)		(rule 06)		(rule 06)

step 1, three ATR_G groups are recognized by rule (6). The comma ',' and the conjunction *i (and)* are recognized as two different grouping operators: $\oplus_{comma}$ and $\oplus_{and}$ (see Table 1).

In Step 2, based on rule (9) the group ATR_G4 is formed:

$$ATR\_G4 = ATR\_G1 \oplus_{comma} ATR\_G2.$$

The representative of ATR_G4 is *żółty (yellow)*. The relation, which is generated when grouping is #*acc_comma(biały, żółty)* (01).

In Step 3, based again on rule (9) the entire ATR_G group is formed:

$$ATR\_G = ATR\_G4 \oplus_{and} ATR\_G3.$$

The representative of ATR_G is *niebieski (blue)* (02). The relation, which is generated during this operation is #*conj_and(żółty, niebieski)*.

As you can see, the role of the grouping operator $\oplus$ in ATR_G4 is played by the comma, and in ATR_G – by the conjunction *and*. Respectively, the generated relations are different.

According the SGGP grammar, all the SG's features are transferred to the representative of the SG. If necessary, however, these features can be multiplied and transferred to each SG component. This fact will be helpful in transforming the syntactic relations between component representatives in a complex ATR_G and between an ATR_G's representative and a noun.

After the transformations that will be discussed in Sect. 4, the following sequence of relations can be get: #*acc_comma(biały, żółty) (white, yellow)*, #*conj_and(biały, niebieski) (white, blue)*, #*conj_and (żółty, niebieski)*.

For the NG group: *białego, żółtego i niebieskiego koloru (of white, yellow, and blue color)* the #atr relation between the ATR_G representative and the noun, #*atr(niebieski, kolor)* (03), is created. The interpretation of relations in NGs is discussed in Sect. 4.

4   Grammar for Interpreting Internal Relations of 1st Level NG Groups

4.1   Relations in *Polsyn*

The *Polsyn* parser can potentially create 38 relations (PR, potential relations). A partial list of PRs and their arguments is shown in Table 2.

Table 2. Names and arguments of potential relations

No	PR name	PR symbol	PR arguments				
			$SG^{0\ \&\ 1}$	SG^2	SG^3	SG^4	Relations
1	Attribute	#atr	+	+			
2	Quantitative attribute	#q_atr	+				
3	Numeric attribute	#n_atr	+				
...		...					
21	"i" class series *(and)*	#conj_and	+	+	+	+	
22	"lub" class series *(or)*	#conj_or	+	+	+	+	
23	',' series (including comma)	#acc_comma	+	+	+	+	
...		...					
38	Forbidden link	#N_link				+	+

4.2   Interpreting Relations in Attributive Groups

Between the components of an attributive group (ATR_G) and its parent noun some syntactic relations may occur. A set of such potential relations is shown in Table 3.

Table 3. Names and arguments of potential relations for ATR_Gs

No	PR name	PR symbol	PR arguments				
			$SG^{0\ \&\ 1}$	SG^2	SG^3	SG^4	Relations
1	Attribute	#atr	+	+			
2	Quantitative attribute	#q_atr	+				
3	Numeric attribute	#n_atr	+				
4	Modifier	#mod	+	+			
5	Negation/denial modifier	#not	+	+	+		
6	Side by side	#side_by_side	+	+			
7	Denotation	#denot	+	+			
8	Comparison	#comp	+	+			
9	"i" class series *(and)*	#conj_and	+	+	+	+	
10	"lub" class series *(or)*	#conj_or	+	+	+	+	
11	',' series (including comma)	#acc_comma	+	+	+	+	
12	Negation	#neg	+	+	+	+	
13	Accord	#accord	+	+	+		
14	Anaphoric relation	#anaphora	+	+	+	+	

The ATR_G syntactic relations are divided into two types: atomic and non-atomic. Only atomic relations are subject to semantic interpretations. For example, #atr is an atomic relation. Non-atomic relations should be decomposed or transformed, i.e. presented as a set of atomic relations. In the paper, the transformation operations will be recorded as an arrow with a lower index, for example $\rightarrow_1$.

Let's recall that ATR_G is a component of a NG group and is used to describe, characterize the NG representative, which usually is a noun or – if not present – a substitute word. In this work we assume that a NG group contains a noun. As a rule, in absence of a noun the rules of interpretation remain the same.

Relations in Table 3 are binary and may occur both between the ATR_G and the NG representative and between the ATR_G components. In any case, the relation has the general form $rel(a,b)$.

Another note concerns the list of relations discussed in this article. We do not consider relations no. 7–8 and 12–14 of Table 3. They arise when the ATR_G or its components are complex constructions, such as a sentence or a participle group. These complex aspects are not dealt with here.

Before starting interpretation of the ATR_G syntactic relations set, do the following:

- Find the relations in which b is the NG representative. For example, #atr(czerwony, jabłko) (red, apple) for NG czerwonych jabłek (red apples).
- Analyze the list of such relations to distinguish between atomic and non-atomic ones.
- Perform operations $\rightarrow_n$ for non-atomic relations.
- To do this, find the relations in which a occurs as the second argument. For example #accord(dobry, czerwony) (good, red) in NG dobrych czerwonych jabłek (good red apples).
- Perform a proper transformation of each relation; several grammar rules are given in the next section. The result is a list of atomic relations that undergo interpretation.
- Interpret each of the relations in this list.

For the purposes of research a classification system for ATR_Gs was adopted, containing the semantic classes color, shape, quantifier, number, belongingness, question, size, and property.

The specificity of ATR_G is that if an ATR_G component belongs to a specific semantic class then it is the basis for generating in the parent NG the relation ##class_name(A, B), where class_name is an identifier – item in the above-mentioned list of classes for ATR_G, A is the ATR_G component, and B – the parent NG representative.

Our semantic dictionary of attributes contains dozens of entries extracted from test examples. An example semantic dictionary for ATR_G is provided in Table 4.

Note that these relations may also exist between the elements of ATR_G; consider for instance the NG group kilka bardzo dobrych czerwonych jabłek (a few very *good red apples)* and its component ATR_G *kilka bardzo dobrych czerwonych (few very good red).*

In our example, ATR_G contains a numeral, an adverb, and two adjectives. NG in our example is a component of the test sentence *Na stole leży kilka bardzo dobrych czerwonych jabłek. (On the table lie a few very good red apples).* In the P-A structure, this NG was assigned the role of ACTOR. This fact will be important when generating utterances in the target language.

Table 4. Example semantic dictionary for ATR_G

Lexeme	Feature
niebieski *(blue)*	color
dobry *(good)*	property
kilka *(some)*	quantifier
mało *(few)*	quantifier
mały *(little)*	size
nasz *(our)*	belongingness

As a result of automated syntactic analysis for this NG, the following set of relations was generated:

$$\{ \#accord(kilka,\ dobry), \tag{1}$$
$$\#mod(bardzo,\ dobry), \tag{2}$$
$$\#n\_atr(kilka,\ dobry), \tag{3}$$
$$\#accord(dobry,\ czerwony), \tag{4}$$
$$\#n\_atr(kilka,\ czerwony), \tag{5}$$
$$\#atr(czerwony,\ jabłko)\}. \tag{6}$$

In the following subsections, we explain the principles of semantic interpretation on the example of these relations.

4.3   Transformation Rules – Relation Properties

Attributiveness relation $\#atr(a,b)$ is atomic one. Interpretation of $\#atr$ is based on a semantic dictionary.
Conformity relation $\#accord(a,b)$. Its properties are:

1. Non-atomic.
2. Symmetric, that is, the formula applies:
   $$\#accord(a,b) \equiv \#accord(b,a),$$
   and in the place where stands a can appear b.
3. Can enter into operation with atomic relation $\#atr$, then creates a new relation $\#atr$:
   $$\#accord(a,b)\ \&\ \#atr(b,c) \rightarrow_2 \#atr(a,c)$$

Modification relation #*mod(a,b)* and its properties:

1. Non-atomic. Let's recall that in this work we only deal with relations arising in ATR_Gs.
2. Quasi-transitive in reference to the relation #*atr*. That is, the formula applies

    #*mod(a,b)* & #*atr(b,c)* →₃ #*mod(a,c)*

    and the transformation operation generates a new, additional #*mod(a,b)* relation.
3. Interpretation of #*mod* is dictionary based. The interpretation algorithm takes into account the fact that under the #*mod* relation may be hidden the relation #*q_atr*. For example, in the ATR_G *mało dobrych książek (few good books)* the adverb *mało (few)* enters into the #*mod(małodobrych) (few, good)* relation, while the lexeme *mało* has the semantic feature *quantifier*. Then further transformations are already applied to #*q_atr*. This fact will be discussed more extensively when describing the #*q_atr* relation.

Numeric relation #*n_atr*. Its properties are:

1. Relation #*n_atr* is interpreted as ##*number*.
2. It occurs in NG class groups and is obligatory in them.
3. In the absence of #*n_atr* in the NG relation list, it is generated based on the grammatical representation of NG. This happens when the ATR_G does not contain an explicitly indicated numeral, compare NGs *jabłka* and *dwa jabłka (apples, two apples)*.
4. The relation #*n_atr* can conceal the quantifier relation #*q_atr*. Then more than one relation ##*number* can be created. This fact is discussed below.
5. It is a quasi-transitive relation in reference to the relation #*atr*: if the set of relations of a given ATR_G or NG contains relations #*n_atr(a,b)* and #*atr(b,c)*, then a new, implicit relation #*n_atr(a,c)* arises. We'll write it like this:

    #*n_atr(a,b)* & #*atr(b,c)* →₃ #*n_atr(a,c)*.

Sometimes the #*n_atr* relation may occur in a "strange", as if a "not matching" place, as in the example with *kilka dobrych czerwonych jabłek (a few good red apples)* – relation (5). This is because "on the occasion of grouping NG" was formed a numeral group NUM_G *kilka bardzo dobrych (few very good)* with the representative *dobrych (good)*, an attributive group ATR_G *czerwonych (red)* with the representative *czerwonych*, which before merging with the noun *jabłek (apples)* was merged with NUM_G, leaving its relations unchanged, and in doing so, organizing new relations with each other. So relations (4) and (5) appeared.

As mentioned above, there are IDs of group representatives in the relation record, that is, there occurs so-called delegation of properties of a group to its representative. There is always the possibility to "unscrew" the delegation and go back to the source.

Quantifier relation #*q_atr*. Its properties are:

1. Non-atomic; as a rule, it transforms into #*n_atr*, but theoretically it can be ##*quantifier*, which may be in NG, but not ATR_G.

2. Hidden in NUM_G, MOD_G, and ADV_G groups due to morphology. For example, the word *mało (little)* after the morphological analysis occurs as an adverb, *wiele (many)* – like a numeral, not a quantifier. That is to say, the relation #q_atr must first be "discovered", recognized in other relations.
3. The #q_atr interpretation is dictionary based. The quantifier dictionary contains transformation patterns #$q\_atr(a,b)$. The quantitative treatment of #q_atr attributes is compatible with Polish grammar. An example of the quantifier dictionary is given in Table 5.

Table 5. Quantifier dictionary

#q_atr(**a**,**b**)		
Class **a**	Transformer **b**	Example
k1	>2	wiele, niemało *(many, quite a lot)*
k2	<3	mało, niewiele *(little, not much)*
k3	>2 & <11	kilka *(several)*

Relation #*conj_and* – series of class *'i' (and)*. Its properties are:

1. Non-atomic.
2. Symmetric, that is, the formula is used: #$conj\_and(a,b) \equiv$ #$conj\_and(b,a)$, and in this place where a stands, there may be b.
3. Can enter into operation '&' with atomic relation #atr, then a new relation #atr is created:
    $$\#conj\_and(a,b) \ \& \ \#atr(b,c) \rightarrow_2 \#atr(a,c)$$
4. Quasi-transitive in reference to #atr. That is, the formula applies:
    $$\#conj\_and(a,b) \ \& \ \#atr(b,c) \rightarrow_3 \#conj\_and(a,c)$$
    and the transformation operation generates a new, additional relation #$conj\_and(a,b)$.

Relation #*conj_or* – the series of class *'lub' (or)* has the same properties as #*conj_and*. In addition, it is interpreted as an ##*alternative*.
Relation #*acc_comma* – series with *','* *(comma)* has the same properties as #*conj_and*. An additional feature – quasi-transitivity refers not only to #atr, but also to #$conj\_and$. It means, it is performed:
$$\#acc\_comma(a,b) \ \& \ \#atr(b,c) \rightarrow_3 \#acc\_comma(a,c),$$
$$\#acc\_comma(a,b) \ \& \ \#conj\_and(b,c) \rightarrow_3 \#acc\_comma(a,c)$$
As mentioned, the relations *Denotation, Comparison, Negation, Accord*, and *Anaphoric* are not discussed.

4.4   Examples of Transformations

We will explain the rules of transformations and algorithms on the example of the NG group from Subsect. 3.5: *białego, żółtego i niebieskiego koloru (of white, yellow, and blue color).* The result of automated syntactic analysis of this NG are three relations: (01), (02) and (03) also mentioned in Subsect. 3.5. They are collected in Table 6.

Table 6. Relation set for the NG group *białego, żółtego i niebieskiego koloru (of white, yellow, and blue color)*

Relation id	Relation
(01)	#acc_comma(biały, żółty) *(white, yellow)*
(02)	#conj_and(żółty, niebieski) *(yellow, blue)*
(03)	#atr(niebieski, kolor) *(blue, color)*

Ordering relations

Command 1. Search for NG representative.

Result: *kolor.* We denote this result as **R**.

Command 2. Search for relations *r(x, b)* where **R** is the second argument, *b* = **R**.

Find *r(x, **kolor**)*

Result: One relation: (03) – #*atr(niebieski, **kolor**)*. Put it on the stack!

Command 3. Search relations where the first argument of (03) occurs as the second argument.

Find *r(x, niebieski)*

Result: One relation: (02) – #*conj_and(żółty, niebieski)*. Put it on the stack!

Find *r(x, żółty)*

Result: One relation: (01) – #*acc comma(biały, żółty)*. Put it on the stack!

Find *r(x, biały)*

Result: No relations. Stop iteration.

As a result of ordering, all relations have been accepted. The relations appear on the stack as shown in Fig. 1.

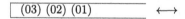

Fig. 1. Relation stack

Stack analysis

Analysis of relations from the stack consists in verifying in turn whether there is a processing chain for it to the **R** representative (in other words, if there is a derivation). We will give the analysis briefly:

Transformation of relation (01) – #acc_comma(biały, żółty).

Find *r(biały, **kolor**)* – not found

Find *r(biały, x)* – found: #*acc_comma(biały, żółty)* – error: looping.

Is it possible to transform the relation? Yes: symmetry transformation

   #*acc_comma(biały, żółty)* ≡ #*acc_comma(żółty, biały)*

We are analyzing the new relation #*acc_comma(żółty, biały)*:

Find *r(żółty,* **kolor**) – not found.

Find *r(żółty, x)* – found: #*conj_and(żółty, niebieski)*

Is it possible to transform the relation? Yes: use of quasi-transitivity

   #*acc_comma(biały, żółty)* &

        #*conj_and(żółty, niebieski)* $\rightarrow_3$ #*acc_comma(biały, niebieski)*

We are analyzing the new relation #*acc_comma(biały, niebieski)*:

Find *r(biały,* **kolor**) – not found.

Is it possible to transform the relation? Yes: symmetry transformation

   #*acc_comma(biały, niebieski)* $\equiv$ #*acc_comma(niebieski, żółty)*

We are analyzing the new relation #*acc_comma(niebieski, żółty)*:

Find *r(niebieski,* **kolor**) – found: #*atr(niebieski,* **kolor**); there is a path derivation to the representative. As a result of feedback transformations the following symmetry is determined:

   #*acc_comma(niebieski, żółty)* $\equiv$ #*acc_comma(biały, niebieski)*

Then, after the transitivity operation, we have:

   #*acc_comma(biały, niebieski)* &

        #*atr(niebieski,* **kolor**) $\rightarrow_2$ #*atr(biały,* **kolor**)

– success. A new relation #*atr(biały,* **kolor**) arose between the first argument of the analyzed relation and the NG representative. This relation is atomic, it can be interpreted semantically.

Similarly we proceed with further relations from the stack.

The resulting set of atomic relations has the form:

#*atr(biały, kolor)*,
#*atr(żółty, kolor)*,
#*atr(niebieski, kolor)*.

Using the dictionary from Table 4 we get a semantic interpretation of relations in the group:

##*color(biały, kolor)*,
##*color(żółty, kolor)*,
##*color(niebieski, kolor)*.

5   Final Remarks

Let's get a few facts about our experiments in the field of semantics:

1. Computer experiments were conducted only with the relations #*atr*, #*mod*, and #*accord*.
2. Interpreted are those groups that occur somewhere in the P-A structure.
3. For translation purposes in *Thetos*, a full semantic interpretation of the relations in NG is not needed. It's enough to get a "light" interpretation in the form of atomic relations, which are the basis for generating complex gestures. For sample NGs, such interpretation allowed to generate text very close to Polish Sign Language PJM, and not to SJM sign language, as has been until

now. Original NG text: *kilka bardzo dobrych czerwonych jabłek (a few very good red apples)* was translated into two texts for two sequences of gestures: *bardzo_dobry_jabłko (very_good_apple)* and *czerwony_jabłko (red_apple)*. The underscores determine the difference in these gestures: *jabłko (apple)* and *dobre_jabłko (good_apple)* are in PJM signed differently. The fact that this NG (consisting of two texts) got the semantic role AGENT, allowed to place it in the first position in the newly generated sentence: *bardzo_dobry_jabłko czerwony_jabłko leżeć na stół (very_good_apple red_apple lie on table)*.

Polish Sign Language does not contain complex structures, which in the spoken Polish and in many other natural languages appear in abundance. We are dealing with a fairly common situation, when the translation should convey the content of input utterance, without being suggested by its syntactic structure in the input language. Interpretation of semantic structures allows to recount an utterance that reflects the original content in the target language in the form of single sentences with a fixed word order.

References

1. Fillmore, C.J.: The case for case. In: Bach, E., Harms, R. (eds.) Universals in Linguistic Theory, pp. 1–88. Holt Rinehart and Winston, Chicago (1968)
2. Grund, D.: Computer implementation of syntactic-generative dictionary of Polish verbs. Studia Informatica **21**(3), 243–256 (2000). (in Polish)
3. Kozielski, S., Świderski, M., Bach, M.: The use of natural language as an intuitive semantic integration system interface. In: Tkacz, E., Kapczynski, A. (eds.) Internet - Technical Development and Applications. AINSC, vol. 64, pp. 51–58. Springer, Heidelberg (2009). https://doi.org/10.1007/978-3-642-05019-0_7
4. Miller, G.: The magical number seven, plus or minus two: some limits on our capacity for processing information. Psychol. Rev. **63**, 81–97 (1956)
5. Romaniuk, J., Suszczańska, N., Szmal, P.: Semantic analyzer in the Thetos-3 system. In: Vetulani, Z. (ed.) LTC 2009. LNCS (LNAI), vol. 6562, pp. 234–244. Springer, Heidelberg (2011). https://doi.org/10.1007/978-3-642-20095-3_22
6. Romaniuk, J., Suszczańska, N., Szmal, P.: Interpreting the predicate-argument representation for the needs of the Thetos translator. In: The 7th Language & Technology Conference, LTC 2015, Poznań, Poland, pp. 175–180 (2015)
7. Romaniuk, J., Suszczańska, N., Szmal, P.: Interpreting syntactic relations in attributive groups. In: The 8th Language & Technology Conference, LTC 2017, Poznań, Poland, pp. 194–200 (2017)
8. Suszczańska, N., Szmal, P., Simiński, K.: The deep parser for Polish. In: Vetulani, Z., Uszkoreit, H. (eds.) LTC 2007. LNCS (LNAI), vol. 5603, pp. 205–217. Springer, Heidelberg (2009). https://doi.org/10.1007/978-3-642-04235-5_18
9. Szmal, P., Suszczańska, N.: Selected problems of translation from the Polish written language to the Sign Language. Archiwum Informatyki Teoretycznej i Stosowanej **13**(1), 37–51 (2001)
10. Vetulani, Z.: Man-machine communication. In: Computer Modeling of Linguistic Competence. Exit, Warszawa (2004). (in Polish)

Syntactic-Semantic Classes
of Context-Sensitive Synonyms Based
on a Bilingual Corpus

Zdeňka Urešová⬤, Eva Fučíková⬤, Eva Hajičová⬤, and Jan Hajič⁽✉⁾⬤

Faculty of Mathematics and Physics, Institute of Formal and Applied Linguistics
(ÚFAL), Charles University, Malostranské nám. 25, 11800 Prague 1, Czech Republic
{uresova,fucikova,hajicova,hajic}@ufal.mff.cuni.cz

Abstract. This paper summarizes findings of a three-year study on
verb synonymy in translation based on both syntactic and semantic criteria and reports on recent results extending this work. Primary language resources used are existing Czech and English lexical and corpus
resources, namely the Prague Dependency Treebank-style valency lexicons, FrameNet, VerbNet, PropBank, WordNet and the parallel Prague
Czech-English Dependency Treebank, which contains deep syntactic
and partially semantic annotation of running texts. The resulting lexicon (called formerly CzEngClass, now SynSemClass) and all associated resources linked to the existing lexicons and corpora following from
this project are publicly and freely available. While the project proper
assumes manual annotation work, we expect to use the resulting resource
(together with the existing ones) as a necessary resource for developing
automatic methods for extending such a lexicon, or creating similar lexicons for other languages.

Keywords: Verbal synonymy · Multilingual synonym Lexicon ·
Semantic lexicon · Valency · Language resources · Czech · English

1 Introduction

The goal of the project is to group verbs used as synonyms in Czech and English
into (cross-lingual) synonym classes representing a cross-lingual meaning of the
state or event expressed by the set of verbs assigned to that class. Each class
thus represents a "concept" of an event or state type, expressed by a set of
synonyms in that particular class, their syntactic behavior and their links to
existing lexical-semantic resources.

For the purpose of this work, we use the term "synonym" in the "loose" interpretation [15], i.e., the necessary semantic equivalence takes also wider context
into account.

The novel feature is the use of a richly annotated bilingual corpus to get
more insight into the usage of verbs (together with their arguments) in translation. In the present paper, we have extended the discussion as presented in an

© Springer Nature Switzerland AG 2020
Z. Vetulani et al. (Eds.): LTC 2017, LNAI 12598, pp. 242–255, 2020.
https://doi.org/10.1007/978-3-030-66527-2_18

earlier version as published at the LTC 2017 conference [27], where the initial results have been discussed based on a sample of 60 classes manually processed and linked to the existing resources. The current size covers about 200 classes processed in several steps. The relevant features of the classes and their verbal members are also described.

While not being the goal of this very project, the ultimate use of such resource is both for followup linguistic studies and for use in natural language applications. The resulting lexicon, together with the existing resources to which it will be linked, will be used as a "gold standard" for evaluating automatic methods that should mimic the laborious manual work performed in this project (and possibly also as training data for systems based on deep learning, depending on its final extent). That way, it will serve as a seed resource for future, automatically extracted, cross-lingual lexicons with the same properties.

Since the publication at LTC 2017, partial findings have been published at various workshops, some results of which are summarized here too [28–33].

2   Resources Used

While the corpora (Sect. 2.2) are the basis for providing the lexicon annotators with examples of real use of the verbs in question, as well as material for automatic pre-extraction of the classes, the existing lexicons (Sect. 2.1) are used to link the new resource to, in various ways described later in the article.

2.1   Lexical Resources

The lexical resources used are of two types: the valency lexicons associated with previously annotated corpora, the Prague Dependency Treebank [10] and the Prague Czech-English Dependency Treebank [8], which serve as the sense identification source, and the external lexicons (VALLEX, FrameNet, VerbNet, OntoNotes, PropBank and WordNet), which are being linked to from the individual SynSemClass Entries.

The actual versions of these lexical resources are as follows:

– PDT-Vallex (Czech) is a Czech valency lexicon used for the annotation of the Prague Dependency Treebank [11] family of treebanks [9,25][1] This lexicon is also published as part of the Prague Dependency Treebank 2.0 by the Linguistic Data Consortium.[2] PDT-Vallex is based on the Functional Generative Description theory [24]. PDT-Vallex contains 7,121 verbs structured into 11,933 valency frames (roughly corresponding to verb senses), and the latest version is available as part of the PDT 3.5 distribution [10]. For a detailed information about the actual structure of the PDT-Vallex lexicons and its entries, see [25].

---

[1] https://lindat.mff.cuni.cz/services/PDT-Vallex/.
[2] http://www.ldc.upenn.edu/LDC2006T01.

- EngVallex (English)[3] is an English valency lexicon with 7,148 valency frames for 4,337 verbs, using the same valency framework as PDT-Vallex. It was built by an (largely manual) adaptation of the PropBank Lexicon [19] to the PDT labeling standards and principles [2].
- CzEngVallex (Czech-English) [6,26,34] is a Czech-English bilingual valency lexicon. It contains 20,835 explicitly linked verb senses (frame-to-frame pairs) and their aligned arguments (argument-to-argument pairs). It is linked, entry by entry and frame by frame, to the Prague Czech-English Dependency Treebank [7][4] and to the two monolingual valency lexicons mentioned above: PDT-Vallex and EngVallex.
- Berkeley FrameNet (English) [1,22] is a lexical database of English[5], containing about 13,000 word senses from more than 200,000 manually annotated sentences linked to more than 1,200 Semantic Frames. FrameNet is based on the Frame Semantics theory [5]; each lexical unit evokes a Semantic Frame (SF) which lists relevant Frame Elements (FEs), or Semantic Roles (SRs).
- OntoNotes [20,21,23] is a large-scale, multi-genre, multilingual corpus manually annotated with syntactic, semantic and discourse information. In the SynSemClass lexicon, we have used the OntoNotes Sense Groupings as published e.g., in the SemLink database (see below). They provide verb sense resource at a middle level of granularity. In these Groupings, each verb sense is identified and marked with a numeric index (1, 2, etc.). These indexes then serve as the distinguishing factor in the links from the individual class members in SynSemClass to the OntoNotes Grouping sense.[6]
- VerbNet (English) [3,13,23] is a class-based verb lexicon[7] with mappings to other lexical resources such as WordNet or FrameNet. VerbNet contains syntactic and semantic information on English verbs. It extended Levin [14] verb classes by refinement and addition of subclasses [13]. Each verb class is described by thematic roles, selectional restrictions on the arguments, and frames. Currently, VerbNet contains about 5,257 verb senses structured in 274 classes.
- PropBank (English) [19] is not only a lexicon but also a corpus[8] of one million words of English text, annotated with argument role labels for verbs (113,000 tokens, 3,324 frames files/types). Arguments are linked to their semantic roles [19].[9]

---

[3] http://hdl.handle.net/11858/00-097C-0000-0023-4337-2.

[4] PCEDT 2.0 is available from the LINDAT/CLARIAH-CZ repository at http://hdl.handle.net/11858/00-097C-0000-0015-8DAF-4).

[5] https://framenet.icsi.berkeley.edu.

[6] OntoNotes Sense Groupings are also viewable at the Unified Verb Index Reference Page at http://clear.colorado.edu/compsem/index.php?page=lexicalresources&sub=ontonotes.

[7] http://verbs.colorado.edu/~mpalmer/projects/verbnet.html.

[8] http://propbank.github.io/.

[9] The PBs semantic roles are not the same as SynSemClass lexicon roles.

- SemLink (English) [18][10] links together different lexical resources (PropBank, VerbNet, FrameNet) through sets of mappings. The Semlink lexicon can be browsed online using the Unified Verb Index.[11]
- WordNet(s) [4,16] is a semantic network[12] of English. Words (nouns, verbs, adjectives, adverbs) are hierarchically grouped into sets of synonyms (117,000 "synsets"). Each synset contains word forms (referring to a given concept), a definition gloss and an example sentence. Czech WordNet 1.9[13] [17] will be used in future work when extending the classes on the Czech side.

2.2 Corpus Resources

The Prague Czech-English Dependency Treebank 2.0 (PCEDT 2.0) [8] is a parallel treebank with over 1.2 million tokens in almost 50,000 sentences for each side. The PCEDT is based on the texts of the Wall Street Journal part of the Penn Treebank and their manual translations. Each language part is annotated in the Prague Dependency Treebank style, i.e., the annotation is dependency-style with argument structure of verbs (syntactic and semantic labeling), which corresponds to the associated valency lexicons for both languages: the PDT-Vallex (for Czech) and the EngVallex (for English); see Sect. 2.1.

In addition, we also use various monolingual corpora, such as the COCA corpus[14], corpora available in the SketchEngine[15] and corpora accessible and searchable through the KonText tool in the LINDAT/CLARIAH-CZ repository.[16]

3 Structure of the SynSemClass Lexicon

As the first thing in building the SynSemClass lexicon (called CzEngClass at that time), a structure of the lexicon (Fig. 1, from [27]) has been designed.

The SynSemClass lexicon builds upon the existing resources, as described above: CzEngVallex, PDT-Vallex and EngVallex lexicons and the PCEDT parallel corpus. In addition, the other lexicons listed (FrameNet, VerbNet, PropBank, OntoNotes and WordNet) are used as additional sources and links will be kept between their entries and the SynSemClass entries.

At the core of the SynSemClass lexicon, there are Synonym Classes, which are, for the purpose of this project, defined as (multilingual, or rather cross-lingual)[17] groups of verb senses (of different lexemes/words) that have the same meaning *and* the arguments of which can be mapped to a common set of semantic

---

[10] https://verbs.colorado.edu/semlink/.
[11] http://verbs.colorado.edu/verb-index/.
[12] https://wordnet.princeton.edu/.
[13] http://hdl.handle.net/11858/00-097C-0000-0001-4880-3.
[14] http://corpus.byu.edu/coca/.
[15] https://www.sketchengine.co.uk/.
[16] http://lindat.mff.cuni.cz/services/kontext/.
[17] For the time being, bilingual: in Czech and English.

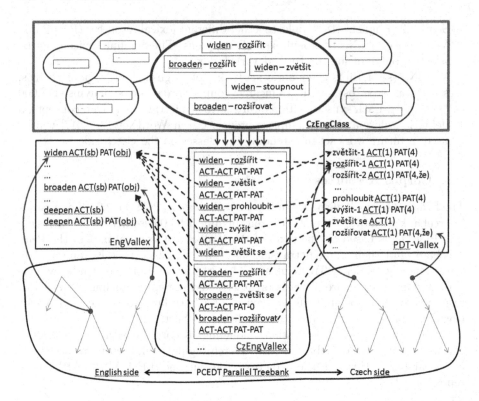

Fig. 1. SynSemClass lexicon & relation to core resources (Color figure online)

roles.[18] The term "same meaning" is understood with regard to a context, the relevant information about which is expected (at least in some cases) to be part of the argument mapping in the form of certain restrictions (lexical, syntactic, semantic) put on the arguments or even on a wider context. This is the case of most light verb constructions (*hold a meeting - meet*), idiomatic verbal MWEs (*cut loose - sever*), in some cases clear cases of hyperonymy (*pay [back] - repay*), and more (e.g., *return [call] - call back*), with clear patterns emerging.

In Fig. 1, the red box shows the SynSemClass proper, with its links to external lexicons. Below the red box, the resources directly used for its creation, extraction of examples, etc. are shown: in the middle, the CzEngClass lexicons links the original valency frame pairs, taken from the two monolingual valency lexicons, PDT-Vallex and EngVallex. These two lexicons are in turn linked to the parallel Czech-English corpus (see the bottom of the figure), the PCEDT 2.0.

Another view of the resulting lexicon in tabular form, depicting (a small sample of) one synonym class ("complain") is shown in Fig. 2 (taken from [33]).

---

[18] The term "sense" is used here for the differentiation of a single verb lexeme ("word") into one or more senses, represented technically by its valency frame ID, as it is done in the original valency lexicons (PDT-Vallex and EngVallex).

Fig. 2. SynSemClass class example ("complain"), taken from [33]

In this example, the class "complain" (ID `vec00132`) contains five verbs, identified in turn by their PDT-Vallex or EngVallex frame IDs. For each of these verbs (verb senses), the argument mapping between their syntactic properties (the valency frame slots, labeled by the PDT/FGD-style arguments such as `ACT, PAT, EFF` and others) and the unique set of Semantic Roles defined for the whole class (here: `Complainer, Addressee, Complaint`). The argument mapping part is followed by links to external lexicons, for each class member separately, since the nuances in the meaning of the particular senses can lead to quite different targets in the external lexicons. While in this case, the FrameNet frame is the same for the three English members of the class, it often happens that several FrameNet classes are linked to from a single SynSemClass class. Finally, several example sentences from the PCEDT parallel corpus are attached too (here identified by their WSJ sentence IDs).

4   Data Preparation

The work so far has been done in several steps. First, we have randomly selected a set of 200 Czech verbs (verb senses) from three categories based on their frequency in the PCEDT corpus (high, medium, low). We have used the bilingual valency lexicon CzEngVallex [34] to determine a set of candidate English verbs for one synonym class, based simply on their pairings with the original Czech verb. Since CzEngVallex is linked to the PCEDT corpus, this gave us also a set of usage examples of these verb pairs, i.e., the context in which they have been used in the original English sentence and in its Czech translation.

For each of the English verbs in the candidate synonym group we have extracted links from the Unified Verb Index.[19] leading to PropBank, VerbNet, FrameNet and WordNet. These links (readily available to the annotators in the annotation software - cf. e.g., [31]) have been used for guiding the subsequent manual annotation pass.

There have been three manual annotation passes, gradually expanding the classes by new verbs, again taken from the PCEDT corpus using the parallel links in the CzEngVallex lexicon and the PCEDT corpus. After each expansion, a manual pass followed (Fig. 3), as described below.

---

[19] https://verbs.colorado.edu/verb-index.

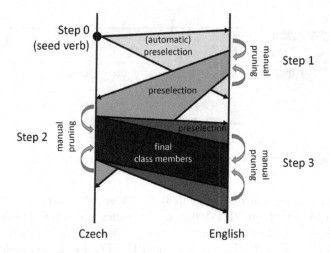

Fig. 3. Scheme of the data preprocessing and the manual annotation passes, taken from [33]

5 Annotation

5.1 Class Membership

At the beginning of each pass, including the initial one, the automatically pres-elected verbs (class members) had to be examined in order to determine if they actually are synonymous with each other. The reasons for exclusions have been numerous: bad automatic alignment in the corpus, very free or liberal translation leading to good sentence-level translation, but not word- (verb-) level transla-tion, translation using hyperonyms or hyponyms, translation using negation of antonyms, etc.

Of course, the inclusion of any verb has been subject to further annotation, as described below; even if a verb has been retained after the first pruning based on intuitions about synonymy and verb senses, it could be excluded later, e.g., because argument mapping between valency slots and the Semantic Roles was deemed impossible.

5.2 Sense Determination

We have linked each English verb in the initial sample (in all three passes) to the OntoNotes sense as available from OntoNotes Groupings, in order to get more precise (even if not unique) links to the corresponding entry(ies), PropBank ID(s), VerbNet ID(s), FrameNet frame(s) and WordNet synset(s).[20] For example, for the English verb *set up* in a group extracted from the translation pairs linked to the Czech verb *budovat* (lit. *build*), OntoNotes sense No. 4 of set-v has

---

[20] Using the Unified Verb Index, http://verbs.colorado.edu/verb-index.

been assigned ("prepare (something) for a particular purpose"), attaching it to FrameNet frames ARRANGING and INTENTIONALLY_CREATE, VerbNet classes braid-41.1.2 and preparing-26.3-2, PropBank IDs set.03 and set.08 and WordNet senses 6, 7, 21, 22, 25 (of *set*).

5.3   Common Semantic Roles (SRs)

In the next phase of the initial step,[21] we have devised a common set of SRs for each group (candidate synonym class) and mapped them to the valency frame slots from the PDT-Vallex and EngVallex lexicons for the Czech source verb and all the English verbs. In devising these roles, we have used FrameNet's Frame Element (FE) labels and descriptions as the initial pool of roles.[22] Many of the verbs (verb senses) in our candidate synonym sets have been found in FrameNet, so that we have started with the core FEs in the frame(s) associated with these verb senses (to be pruned later during a reconciliation phase also with VerbNet thematic roles). For example, for the *set up* example, we have first listed Agent, Configuration and Theme, Creator and Created_Entity from the ARRANGING and INTENTIONALLY_CREATE frames, finaly the SRs Agent, Components and Created_Entity are used; see Sect. 5.4. In Fig. 2, class "complain", the SRs are, as has already been mentioned, Complainer, Addressee and Complaint.

5.4   Argument - Role Mapping

For each verb in the candidate class, we have then paired each of the SRs to a valency slot as found in PDT-Vallex (for the Czech verb) and to EngVallex (for the English verbs). This has not, as expected, been straightforward, as will be described in more details in the next section. We have also used the other English resources to help clarify the relations if necessary. For example, some of the SRs initially listed have to be merged or deleted; in our "set up" example, it is clear that Agent and Creator are to be merged to one role.

6   Analysis of the Current Version

This section is based on a (sub)sample of 60 candidate synonym groups that have been created and mapped initially so far during the annotation process. More thorough examination will follow when all the 200 classes are fully checked.

---

[21] With possible modifications in the subsequent two steps.

[22] Using FrameNet v1.7, there are 1,168 different FE labels available across all frames. Later, VerbNet's thematic roles will be compared with the selected FEs and a common set used, provided a suitable common theoretical framework can be found.

6.1 Synonym Classes Composition

While the translation pairs extracted from the parallel corpus should have been clear synonyms, in some cases, even if the particular context has been taken into account, verbs had to be deleted from the group. For example, sometimes the parallel corpus correctly identified hyperonyms as translation equivalents, but there was no specific context that would restrict the hyperonym to the particular sense on the other side of the translation (this has happened in both directions):

- *That would hold.*PRED *spending.*PAT *on the program at about the previous year's level.*
- *To by znamenalo.*PRED *investice.*PAT *do programu přibližné ve stejné výši jako loni.*

In the above example, the Czech verb *znamenat* (form *znamenalo*) (in the sense of lit. *mean, imply, indicate*) is aligned with English *hold* (... *hold spending ... at about the same level*, Czech lit. ... *mean spending ... etc.*). This is considered a functional equivalent translation in this context,[23] but since the context cannot be described just in terms of verb arguments, it has been decided to delete *hold* from this synonym class.

6.2 Roles and Argument Mapping

The initial set of roles for each class has been a union of FrameNet's Frame Elements (FEs) of all frames in which the appropriate English verbs have been found (see also Sect. 5.4). The goal was to establish a common set of roles for a given class, carefully considering both the FrameNet's FEs and the corresponding valency slots in the valency lexicons associated with the parallel corpus, including their use in the bilingual texts themselves.

Merging FrameNet-provided (or -inspired) SRs has been the most frequent operation, even within the same frame. For example, verbs inheriting from the STATEMENT frame might in some cases have merged the Topic SR (the subject matter to which the Message pertains) with the Message FE (what the Speaker is communicating to the Addressee)[24], since in our view, these typically occupy just one "slot" [12], with Topic being often part of the Message.[25] For example: *She said about her past* (Topic) *that it was wild* (Message) - *She said that her past was wild* (Topic+Message). Similarly, SRs differing only in animateness

---

[23] We can only speculate why the translator has used *znamenat* here; possibly because literal translations of *hold* are awkward in Czech (in this context), and the translator also determined that in fact the semantics of *hold* is already contained in the phrase *previous year's level*, and thus a translation of a hyperonym of *hold* can be used instead.

[24] https://framenet2.icsi.berkeley.edu/fnReports/data/frameIndex.xml? frame=Statement.

[25] In fact, the "Topic" part should be annotated within the information structure "layer" (topic/focus), not using semantic roles.

have often been merged, such as in the `Agent/Cause` case in the class represented by "widen" in Fig. 1.

The mapping of SRs to valency slots is mostly 1:1, as in the example of the synonym class corresponding most closely to the FrameNet's COMMERCE_PAY frame (Table 1): the arguments of *cover, pay, reimburse* are mapped 1:1 (if we tentatively add some of the missing ones into EngVallex, e.g., EFF/Effect to *reimburse* and `EFF/Effect` and `BEN/Benefactor` to *settle*). `Buyer` typically maps to the valency frame argument `Actor`, `Goods` to `Effect`, `Seller` to `Addressee` and `Money` to `Patient`.[26]

However, the correspondence of SRs and valency arguments is not necessarily always 1:1 - SRs have been occasionally merged (cf. `Topic` and `Message`) or split. For example, in BECOMING_AWARE, `Phenomenon` is mapped either to valency frame argument `Effect` or to valency frame argument `Patient` as shown in the following example *...to know details.*Effect *of one side only*, where `Phenomenon` is mapped to `Effect` and for another class member of the same class, the verb *hear*, `Phenomenon` is mapped to `Patient`: *she heard about the artery-clogging hazards.*Patient. Similarly, the mapping of `Goods` in the *pay* class is either to `Patient` (for *cover*) or `Effect` for PAY and other verbs; see also Table 1).

There are also examples with some specific context restrictions when a mapping can be applied. E.g., for the idiomatic verb *foot [the bill]* of the PAY synonym class (Table 1), the restriction to this idiomatic meaning (using *bill*) must be recorded. As another example, the `Patient` mapping of the verb *drill* in the BUILDING-related synonym class must be restricted to *drill a well (or other [large] hole-like thing)*. For light verb constructions, the nominal argument (labeled `CPHR` in the valency lexicons) often maps to the same role for the light verb argument as the `Patient` argument does for "non-light" verbs.

7   SynSemClass Size and Other Statistics

The current version of SynSemClass (version 2.0) contains 200 classes.[27] As of end of April, 2020, it contains 200 classes, which has been processed at least by 2 steps of the manual annotation (or better to say, they also contain additional annotation so that they are close to be through all three steps). Finished are 157 classes with 1567 Czech and 2836 English verbs (both relatively small as well as quite rich classes, e.g., "build", "wait", "allow", "learn", "invest", "think",

---

[26] Later investigation, as well as testing the addition of new verbs into this class, however revealed that most of the verbs should have had only three valency slots in EngVallex: ACT, PAT and ADDR, where PAT corresponds to either "payment" or "obligation", but not both. In addition, the EFF does not seem to be core for the "reimbursement" concept. Therefore the Roleset has been reduced (or, generalized) to three SRs only, namely `Payee` (mapped to ADDR), `Obligation_Payment` (mapped to PAT or EXT) and `Payer` (mapped to ACT). In any case, this is still an example of a prevailing 1:1 valency slot:SR mapping.

[27] http://hdl.handle.net/11234/1-3215; previous version was available as SynSemClass 1.0, http://hdl.handle.net/11234/1-3125.

Table 1. Argument mapping for PAY class

Roles				
	Buyer	Goods	Seller	Money
Hradit	ACT	EFF	ADDR	PAT
Cover	ACT	PAT	BEN?	MANN
Foot	ACT	DPHR(*bill*)	BEN?	MANN
Pay	ACT	EFF	ADDR	PAT
Reimburse	ACT	EFF?	ADDR	PAT
Settle	ACT	EFF?	BEN?	PAT

etc.). In the last phase of step 3 annotation (i.e., almost finished) are another 37 classes, with 404 Czech and 208 English verbs, such as "hold", "speak", "watch", "announce", etc.). The remaining six classes from the 200 have been in fact merged with others, since after the expansion and annotation, they grew to be in fact identical with another class (created initially from a different seed verb).

These numbers are summarized in Table 2.

Table 2. SynSemClass 2.0 size and statistics

	In Step 3	Finished	Merged	Total
Classes	37	157	(6)	194
Czech verbs	404	1567	(110)	1971
English verbs	208	2836	(92)	3044

8   Conclusions and Next Steps

We have described some findings on synonymy of verb senses of generally different verbal lexemes in a bilingual setting, and specifically, we focused on their valency behavior and common Semantic Roles. Our future research will be aimed at extending it to more verbs, at further refinement of our semantic roles and their explicit mappings from valency arguments to the semantic roles, and at formalizing the additional restrictions. We will analyze in more detail the relation of valency and semantic roles also from their morphosyntactic realization point of view. We will also confront the findings as supported by the corpus material to the underlying theoretical framework(s), in order to possibly refine them in their approach to verb sense distinctions, valency and argument description. We will also compare our results with automatic approaches to cross-lingual semantic similarity detection, such as in [35], which is very much related to our work.

Finally, the resulting lexicon is now openly available in the LINDAT/ CLARIAH-CZ repository at http://lindat.cz[28].

Acknowledgments. This work has been supported by the grants No. GA17-07313S and GX20-16819X of the Grant Agency of the Czech Republic, and it uses resources hosted by the LINDAT/CLARIAH-CZ Research Infrastructure, project No. LM2018101, supported by the Ministry of Education, Youth and Sports of the Czech Republic.

References

1. Baker, C.F., Fillmore, C.J., Lowe, J.B.: The berkeley FrameNet project. In: Proceedings of the 36th Annual Meeting of the Association for Computational Linguistics and 17th International Conference on Computational Linguistics, ACL 1998, vol. 1, pp. 86–90. Association for Computational Linguistics, Stroudsburg (1998). https://doi.org/10.3115/980845.980860. http://dx.doi.org/10.3115/980845.980860
2. Cinková, S.: From propbank to engvallex: adapting the propbank-lexicon to the valency theory of the functional generative description. In: Proceedings of the 5th International Conference on Language Resources and Evaluation (LREC 2006), pp. 2170–2175. ELRA, Genova (2006)
3. Duffield, C.J., et al.: Criteria for the manual grouping of verb senses. In: Proceedings of the Linguistic Annotation Workshop, LAW 2007, pp. 49–52. Association for Computational Linguistics, Stroudsburg (2007). http://dl.acm.org/citation.cfm?id=1642059.1642067
4. Fellbaum, C. (ed.): WordNet: An Electronic Lexical Database. Language, Speech, and Communication. MIT Press, Cambridge (1998)
5. Fillmore, C.J.: Frame semantics and the nature of language. Ann. New York Acad. Sci.: Conf. Origin Dev. Lang. Speech **280**(1), 20–32 (1976)
6. Fučíková, E., Hajič, J., Šindlerová, J., Urešová, Z.: Czech-English bilingual valency lexicon online. In: 14th International Workshop on Treebanks and Linguistic Theories (TLT 2015), pp. 61–71. IPIPAN, Warszawa (2015)
7. Hajič, J., et al.: Announcing Prague Czech-English dependency treebank 2.0. In: Proceedings of the 8th LREC 2012), pp. 3153–3160. ELRA, Istanbul (2012)
8. Hajič, J., et al.: Prague Czech-English dependency treebank 2.0 (2012). https:// catalog.ldc.upenn.edu/LDC2004T25. http://hdl.handle.net/11858/00-097C-0000-0015-8DAF-4
9. Hajič, J., Panevová, J., Urešová, Z., Bémová, A., Kolářová, V., Pajas, P.: PDT-VALLEX: creating a large-coverage valency lexicon for treebank annotation. In: Nivre, J., Hinrichs, E. (eds.) Proceedings of The Second Workshop on Treebanks and Linguistic Theories. Mathematical Modeling in Physics, Engineering and Cognitive Sciences, vol. 9, pp. 57–68. Vaxjo University Press, Vaxjo (2003)
10. Hajič, J., et al.: Prague dependency treebank 3.5 (2018). http://hdl.handle.net/11234/1-2621
11. Hajič, J., et al.: Prague dependency treebank 2.0 (2006)
12. Kettnerová, V.: Konstrukce s rozpadem tématu a dikta v češtině (constructions with topic and message separation in Czech). Slovo Slovesnost **70**(3), 163–174 (2009)

---

13. Kipper, K., Korhonen, A., Ryant, N., Palmer, M.: Extending VerbNet with novel verb classes. In: Proceedings of LREC, p. 1 (2006)
14. Levin, B.: English verb classes and alternations. The University of Chicago Press, Chicago and London (1993)
15. Lyons, J.: Introduction to Theoretical Linguistics. Cambridge University Press, Cambridge (1968)
16. Miller, G.A.: WordNet: a lexical database for English. Commun. ACM **38**(11), 39–41 (1995). https://doi.org/10.1145/219717.219748. http://doi.acm.org/10.1145/219717.219748
17. Pala, K., Smrz, P.: Building Czech Wordnet. Roman. J. Inf. Sci. Technol. **7**(1–2), 79–88 (2004). http://nlp.fi.muni.cz/publications/romjist2004_pala_smrz/
18. Palmer, M.: Semlink: linking PropBank, VerbNet and FrameNet. In: Proceedings of the Generative Lexicon Conference, pp. 9–15 (2009)
19. Palmer, M., Gildea, D., Kingsbury, P.: The proposition bank: an annotated corpus of semantic roles. Comput. Linguist. **31**(1), 71–106 (2005). https://doi.org/10.1162/0891201053630264. http://dx.doi.org/10.1162/0891201053630264
20. Pradhan, S.S., Hovy, E., Marcus, M., Palmer, M., Ramshaw, L., Weischedel, R.: OntoNotes: a unified relational semantic representation. Int. J. Semant. Comput. **01**(04), 405–419 (2007). https://doi.org/10.1142/S1793351X07000251. https://www.worldscientific.com/doi/abs/10.1142/S1793351X07000251
21. Pradhan, S.S., Xue, N.: OntoNotes: the 90% solution. In: Proceedings of Human Language Technologies: The 2009 Annual Conference of the North American Chapter of the Association for Computational Linguistics, Companion Volume: Tutorial Abstracts, pp. 11–12. Association for Computational Linguistics, Boulder, May 2009. https://www.aclweb.org/anthology/N09-4006
22. Ruppenhofer, J., Ellsworth, M., Petruck, M.R.L., Johnson, C.R., Scheffczyk, J.: FrameNet II: extended theory and practice. Unpublished Manuscript (2006). http://framenet.icsi.berkeley.edu/
23. Schuler, K.K.: VerbNet: a broad-coverage, comprehensive verb lexicon. Ph.D. thesis, University of Pennsylvania (2006). http://verbs.colorado.edu/~kipper/Papers/dissertation.pdf
24. Sgall, P., Hajičová, E., Panevová, J.: The Meaning of the Sentence in Its Semantic and Pragmatic Aspects. D. Reidel, Dordrecht (1986)
25. Urešová, Z.: Valence sloves v Pražském závislostním korpusu. Studies in Computational and Theoretical Linguistics. Ústav formální a aplikované lingvistiky, Praha, Czechia (2011)
26. Urešová, Z., Fučíková, E., Hajič, J., Šindlerová, J.: Czengvallex - Czech English valency lexicon (2015)
27. Urešová, Z., Fučíková, E., Hajičová, E., Hajič, J.: Syntactic-semantic classes of context-sensitive synonyms based on a bilingual corpus. In: Vetulani, Z., Mariani, J. (eds.) Proceedings of 8th Language and Technology Conference, pp. 201–205. Fundacja Uniwersytetu im. Adama Mickiewicza, Fundacja Uniwersytetu im. Adama Mickiewicza w Poznaniu, Poznań (2017)
28. Urešová, Z., Fučíková, E., Hajičová, E., Hajič, J.: Creating a verb synonym lexicon based on a parallel corpus. In: Proceedings of the 11th International Conference on Language Resources and Evaluation (LREC 2018). European Language Resources Association (ELRA), Miyazaki, May 2018
29. Urešová, Z., Fučíková, E., Hajičová, E., Hajič, J.: A Cross-lingual synonym classes lexicon. Prace Filologiczne **LXXII**, 405–418 (2018)

30. Urešová, Z., Fučíková, E., Hajičová, E., Hajič, J.: Defining verbal synonyms: between syntax and semantics. In: Haug, D., Oepen, S., Øvrelid, L., Candito, M., Hajič, J. (eds.) Proceedings of the 17th International Workshop on Treebanks and Linguistic Theories (TLT 2018), pp. 75–90. Universitetet i Oslo, Linköping University Electronic Press Pub No. 155, Linköping (2018)

31. Urešová, Z., Fučíková, E., Hajičová, E., Hajič, J.: Tools for Building an Interlinked Synonym Lexicon Network. In: Proceedings of the 11th International Conference on Language Resources and Evaluation (LREC 2018). European Language Resources Association (ELRA), Miyazaki, May 2018

32. Urešová, Z., Fučíková, E., Hajičová, E., Hajič, J.: Meaning and semantic roles in CzEngClass lexicon. Jazykovedný časopis/J. Linguist. **70**(2), 403–411 (2019)

33. Urešová, Z., Fučíková, E., Hajičová, E., Hajič, J.: SynSemClass linked lexicon: mapping synonymy between languages. In: Proceedings of the Globalex 2020 Workshop at the 12th International Conference on Language Resources and Evaluation (LREC 2020). European Language Resources Association (ELRA), Marseille, May 2020

34. Urešová, Z., Fučíková, E., Šindlerová, J.: CzEngVallex: a bilingual Czech-English valency lexicon. Prague Bull. Math. Linguist. **105**, 17–50 (2016)

35. Wu, S., Choi, J.D., Palmer, M.: Detecting cross-lingual semantic similarity using parallel PropBanks. In: Proceedings of the 9th Conference of the Association for Machine Translation in the Americas, AMTA 2010, Denver, CO (2010). https://amta2010.amtaweb.org

Towards the Evaluation of Feature Embedding Models of the Fusional Languages

Alina Wróblewska$^{(\boxtimes)}$ ⓘ, Katarzyna Krasnowska-Kieraś, and Piotr Rybak

Institute of Computer Science, Polish Academy of Sciences,
Jana Kazimierza 5, 01-248 Warsaw, Poland
alina@ipipan.waw.pl, kasia.krasnowska@gmail.com,
piotr.cezary.rybak@gmail.com

Abstract. An important component of a NLP system with the neural network architecture is an encoder that represents word features as dense vector representations, i.e. feature embeddings. According to the concept of feature embeddings, features sharing common linguistic information should have similar vectors and thus feature similarities can be captured. In this paper we investigate which features should be used in estimating NLP models of the fusional languages – tokens or lemmata. Furthermore, we research the methodological question whether the results of the intrinsic evaluation of feature embeddings are informative for downstream applications, or feature embedding models should be evaluated extrinsically. The presented evaluation experiments are conducted on Polish – a fusional Slavic language with a relatively free word order. However, the evaluation results can be approximately generalised to other Slavic languages, because the studied problems are common to them.

Keywords: Feature embedding · Evaluation · Morphosyntactic disambiguation · Dependency parsing · Fusional language

1 Introduction

Neural networks are very successful in various language processing tasks, e.g. language modelling [7,21], neural machine translation [2,6,29], dependency parsing [4,8,15,28], sentiment analysis [14]. An important component of a system with the neural network architecture is an encoder that represents word features[1] (e.g. word forms, part-of-speech tags) as dense vectors, i.e. feature embeddings. The particular feature types are represented as vectors of presumably different number of dimensions (more dimensions for word features and less for part-of-speech features). Features represented as d-dimensional vectors are embedded into a d-dimensional Euclidian space ($\mathbb{R}^d$). Different feature types require different vector spaces.

---

[1] Not only words but also texts, e.g. sentences or documents, or even images, can be represented as feature embeddings. However, the research presented in this article is limited to word features.

Z. Vetulani et al. (Eds.): LTC 2017, LNAI 12598, pp. 256–270, 2020.
https://doi.org/10.1007/978-3-030-66527-2_19

According to the concept of feature embeddings, some features sharing common information should have similar vectors and thus feature similarities can be captured. For example, some Slavic languages make a clear distinction between perfective and imperfective aspects, and contain pairs of the imperfective and perfective verbs, e.g. in Russian СЪЕДАТЬ vs. СЪЕСТЬ, in Polish ZJADAĆ vs. ZJEŚĆ ('to eat'). As the verbs marked for different aspects tend to share most of their valence and selectional preferences, they should be represented with similar vectors.

From the perspective of foregoing research and provided experimental results, it is indisputable that embedding models of the isolating languages, such as English, can be trained on tokens, e.g. [21,25]. Other languages largely adopt procedures of estimating and evaluating feature embeddings proposed for English. However, it is not obvious what should be used as features in estimating embeddings of the fusional languages[2] – tokens, subwords, lemmata, or maybe stems.

In experiments on Russian, the embedding models are estimated on stems [19] or tokens simplified with some language-specific rules [30]. The embedding models are then intrinsically evaluated on a dataset[3] which is a translation of English SimLex999 [13] and consists of pairs of lemmata re-scored by Russian speakers. In experiments on Polish [23], the embedding models are trained on tokens[4] or lemmata, and tested on pairs of lemmata from the publicly available thesauri or on a lemma-based analogy dataset.

Therefore, the first research question addressed in our paper is what should be used as features in training the embedding models of the fusional languages – lemmata or tokens[5] (see Sect. 2)? Lemma embeddings seem to be advantageous from the point of view of the intrinsic evaluation. Token embeddings, in turn, seem to be helpful for downstream applications. The second research problem concerns the validation methodology. Are the results of the intrinsic evaluation of embeddings sufficiently informative for downstream applications (see Sect. 4)? Or should the embedding models be evaluated *in vivo* in the realistic scenarios (see Sects. 5 and 6)? The presented evaluation experiments are conducted on Polish – a fusional Slavic language with a relatively free word order. However, the evaluation results can be approximately generalised to other Slavic languages, because the studied problems are common to them.

---

[2] According to Aikhenvald (2007:4), in *"fusional* – sometimes misleadingly called (in)flectional *languages* – there is no clear boundary between morphemes, and thus semantically distinct features are usually merged in a single bound form or in closely united bound forms". For example, the suffixes -АМИ in Russian ДОМАМИ ('house'.inst.pl) and -*ami* in Polish *domami* ('house'.inst.pl) fuse the case and number information.

[3] http://www.leviants.com/ira.leviant/MultilingualVSMdata.html#SimLex999.

[4] Tokens are automatically lemmatised, in order to enable the comparison with the test sets.

[5] Throughout this paper, we use the following terms: *lemma* for a word's base form (e.g. КОТ 'cat'); *token* for a string of characters in running text (e.g. *kota* 'cat'); *inflectional form/inflection* for a token assigned a morphosyntactic interpretation (e.g. *kota*.subst:acc:sg:m2, *kota*.subst:gen:sg:m2).

2   Feature Embeddings Models

When considering a fusional language such as Polish, it is important to make a distinction between possible types of embeddings, including vector representations of tokens, lemmata, inflections, and character n-grams.

2.1   Token Embedding

Token embeddings are the most straightforward to obtain: the only required resource and tool are a possibly large collection of texts and a tokeniser. Based on a tokenised text corpus, a neural NLP model (e.g. language model) can be estimated and a trainable part of the model (i.e. embedding weight matrix) can be used to map tokens onto real-valued feature vectors. In the resulting embedding space, each distinct token has its own vector. For instance, there is no single vector corresponding to the 'concept' of POCIĄG ('train'), but separate vectors for each distinct token of POCIĄG that appears in the training data: *pociąg* ('train'.sg:nom|acc), *pociągu* ('train'.sg:gen|loc), *pociągami* ('train'.pl:inst) etc. Note that under this approach, all syncretic inflectional forms of the same lemma (as well as those homonymous between distinct lemmata) receive a common embedding.

2.2   Lemma Embedding

A lemma embedding model that embeds lemmata into a vector space is more tool/resource-dependent, i.e. either an existing large lemmatised corpus or a lemmatiser is necessary. The lemma model is trained on a sequence of lemmata assigned to consecutive tokens of a text corpus and an embedding weight matrix of the model is used to map lemmata onto lemma embeddings. The resulting embedding model shares vectors between text occurrences belonging to the same lemma rather than to syncretic/homonymous word forms, but its quality will depend on the quality of the corpus annotation or the lemmatiser.

2.3   Inflectional Form Embedding

A third possible approach to training an embedding model for a fusional language is to employ a morphosyntactically annotated corpus or a tagger in order to induce vectors for inflections, i.e. tokens paired with their respective morphosyntactic interpretations. This approach is the most resource/tool costly and, since automatic morphosyntactic tagging of Polish is a difficult task that still requires improved solutions [16], it seems the most prone to propagation of errors.

2.4   Subword Embedding

Apart from models trained on word forms, there are some experiments on estimating real-valued vectors of parts of words, e.g. subword embeddings by [3]. Subword embedding models are dedicated for morphologically rich languages with large vocabularies and a high rate of rare words. Subword embedding models are also useful for estimating vectors of unknown words (i.e. words outside the dictionary). To estimate a subword embedding model, a large collection of tokenised texts is required. The subword embedding model represents tokens as bags of character n-grams, i.e. it computes vectors both for tokens from the dictionary and for tokens' parts. An embedding of a token is estimated as the sum of the appropriate n-gram embeddings, e.g. for $n = 5$, the token *pociągami* ('train'.pl:inst) is represented as the sum of the vectors of the following n-grams: *<poci, pocią, ociąg, ciąga, iągam, ągami, gami>*, and the vector of the entire token *<pociągami>* if available.

3   Experimental Setup

We examine token embeddings $\mathcal{T}$, lemma embeddings $\mathcal{L}$, and subword embeddings $\mathcal{S}$ in our studies. All tested vectors are estimated on the same collection of textual data from Polish Wikipedia and National Corpus of Polish [26]. The details of preprocessing (tokenisation and lemmatisation) the dataset are described in [23]. As training data are either tokenised or lemmatised, the vocabulary sizes of feature embedding models estimated on these datasets differ considerably. The vocabulary sizes of token, lemma, and subword embeddings are 2.12M, 1.55M and 5M, respectively.

A subword embedding model for Polish [12] was recently trained on Common Crawl[6] and Polish Wikipedia[7], and publicly distributed. This subword embedding model (hereafter $\mathcal{F}$) is tested in our additional evaluation experiment.

3.1   Token and Lemma Embeddings

We use the pre-trained token $\mathcal{T}_d$ and lemma $\mathcal{L}_d$ embeddings[8] for d being the vector size. The vectors were derived with Gensim[9] [27] – a Python implementation of CBOW and Skip-gram models [21]. The vectors selected for our experiments were estimated with the following parameters:

- dictionary parameters: trimming the items occurring less than 5 times in training data,
- training parameters: the vector size of 100 or 300; the context window of 5; 10 training epochs; training procedures: hierarchical softmax or negative sampling (with 5 negative samples).

---

[6] https://commoncrawl.org.
[7] https://pl.wikipedia.org/wiki/Wikipedia.
[8] dsmodels.nlp.ipipan.waw.pl.
[9] https://radimrehurek.com/gensim.

3.2 Subword Embeddings

The subword vectors S_d are estimated with fastText library[10] [3]. FastText is an extension of the continuous Skip-gram model [21]. The subword embeddings are trained with the following parameters:

- dictionary parameters: trimming the items less frequent than 5; character n-grams for $3 \leq n \leq 6$,
- training parameters: the vector size of 100 or 300; the context window of 5; 10 training epochs; negative sampling training method (with 5 negative samples).

The subword embeddings $\mathcal{F}_{300}$ were trained with fastText library using CBOW procedure. The word vectors are available in the binary format[11] and in the text format[12]. The subword embeddings were trained with the following parameters:

- dictionary parameters: character n-grams of length 5,
- training parameters: the vector size of 300; negative sampling training method (with 5 and 10 negative samples).

4   Evaluation Methodology

There are two standard ways of evaluating the embedding models: intrinsic evaluation and extrinsic evaluation.

4.1 Intrinsic Evaluation

Intrinsic evaluation consists in testing a model against a dedicated test set. There are two standard benchmarks for word embeddings: determining word similarity/relatedness, e.g. [11,13], and word analogy solving, e.g. [22]. The word similarity evaluation estimates the semantic proximity of two words (e.g. with Cosine similarity[13]), and correlates this score with the human judgements of the word similarity. The methodology of similarity testing has recently been criticised, e.g. [5,10], because of the uncertain indication of the impact of embeddings on downstream applications.

The word analogy method, in turn, assumes that linear relations between pairs of words (e.g. *king–man* and *woman–queen*) are indicative of the embedding quality. According to this method, a vector of one word (e.g. *queen*) can be estimated of the vectors of the remaining words (i.e. **queen = king − man + woman**). This methodology is questioned as well, e.g. [20], because "information not detected by linear offset may still be recoverable by a more sophisticated search method, and thus is actually encoded in the embedding", cf. [9].

---

[10] https://github.com/facebookresearch/fastText.
[11] https://dl.fbaipublicfiles.com/fasttext/vectors-crawl/cc.pl.300.bin.gz.
[12] https://dl.fbaipublicfiles.com/fasttext/vectors-crawl/cc.pl.300.vec.gz.
[13] Cosine similarity is a measure of similarity between two words that measures the cosine of the angle between embeddings of these words.

Despite criticism the intrinsic evaluation method is still widely used. Regarding intrinsic evaluation of Polish embeddings, an extensive study of synonymy and analogy recognition is presented in [23]. According to their results, intrinsic evaluation is appropriate for investigated tasks. However, there are no clues that it is informative for other NLP tasks which to a lesser extent rely on synonymy and analogy recognition.

4.2 Extrinsic Evaluation

Extrinsic evaluation consists in integrating a model into a sophisticated NLP task and verifying the impact on the results of this task. According to our knowledge, there are no experiments on evaluating Polish embeddings extrinsically. Two extrinsic evaluation settings are thus considered here: morphosyntactic disambiguation (Sect. 5) and dependency parsing (Sect. 6).

5 Morphosyntactic Disambiguation

The first extrinsic evaluation context considered in this paper is that of morphosyntactic disambiguation (MD). The task of morphosyntactic disambiguation is formulated as follows: given a sequence of tokens together with their respective sets of possible morphosyntactic tags (most commonly obtained from a morphosyntactic analyser), select a single correct tag for each token. For tokens unknown to the analyser, the set contains a special ign tag and the task amounts to generating a tag instead of selecting one from a provided set. Table 1 shows an example token sequence corresponding to the sentence *Mieszka w Kotkowicach.* '(S)he lives in Kotkowice.', together with possible tags and correct tags that a disambiguator should choose for each token. The performance of a morphosyntactic disambiguation tool is measured in terms of accuracy, i.e. the percentage of tokens assigned the same tag as in gold standard data.

Table 1. A token sequence with possible and correct tags.

Token	Possible tags	Correct tag
Mieszka	subst:sg:gen:m3	fin:sg:ter:imperf
	fin:sg:ter:imperf	
	subst:sg:acc:m1	
	subst:sg:gen:m1	
w	prep:acc:nwok	prep:loc:nwok
	prep:loc:nwok	
Kotkowicach	ign	subst:pl:loc:n
.	interp	interp

In order to examine the impact of embedding vectors on MD accuracy, the vectors from $\mathcal{T}$ models[14] are used as input features for a morphosyntactic disambiguator – Toygger [18]. The basic disambiguator has at its core a two-layer, bi-directional LSTM network taking as input a sequence binary-valued vectors representing the sets of possible morphosyntactic tags of consecutive tokens. For a more detailed description of Toygger, see [18]. For each considered token embedding model, the basic input for each token is extended by concatenation with its embedding vector.

The PolEval task 1(A) training dataset is used [17] and following measures are calculated for each feature configuration:

– accuracy in 10-fold cross-validation (91.59% for the basic model),
– Δ: difference wrt. accuracy on folds' training data (a measure of overfitting; 2.6 for the basic model).

The results of the experiment are given in Table 2.

Table 2. Toygger's accuracy and Δ in 10-fold cross-validation for 100- and 300-dimensional embeddings estimated with Skip-gram (skipg) or CBOW (cbow) and optimised with hierarchical softmax (hs) or negative sampling (ns).

Toygger	100		300	
	Acc	Δ	Acc	Δ
$+\mathcal{T}$ (skipg-hs)	94.54%	2.6	94.97%	2.7
$+\mathcal{T}$ (skipg-ns)	94.77%	2.5	95.15%	2.7
$+\mathcal{T}$ (cbow-hs)	95.01%	2.2	**95.22%**	2.4
$+\mathcal{T}$ (cbow-ns)	94.90%	**1.9**	95.14%	2.2
$+\mathcal{S}$ (skipg-ns)	94.36%	2.5	94.81%	2.7

Augmenting the morphological information with vector embeddings brings about a substantial gain in accuracy, resulting in error reduction wrt. the basic model ranging from 35% to 43%. For each tested embedding model and configuration (rows in Table 2), the performance of the disambiguator with $\mathcal{T}_{300}$ vectors is consistently better than with $\mathcal{T}_{100}$ vectors, whereas the shorter vectors yield smaller values of Δ. In the morphosyntactic disambiguation context, the Skip-gram models perform better when trained with negative sampling while the CBOW ones work better when hierarchical softmax is employed, the latter yielding the best accuracy of 95.22% with $\mathcal{T}_{300}$ vectors.

6   Dependency Parsing

Another task in which feature embeddings are extrinsically evaluated is dependency parsing. The main goal of dependency parsing is to derive a syntactic

---

[14] The $\mathcal{L}$ models are not considered here since correct lemmata for tokens are typically not yet determined at the MD stage of text processing for Polish.

analysis (represented as a directed tree) of a sequence of words, based on their features, i.e. tokens, lemmata, part-of-speech tags, morphological characteristics, and eventually pre-trained token embeddings. We verify two scenarios: 1) the syntactic analysis of Polish sentences provided by a dependency parser with a linear classifier (baseline) vs. dependency parsers with neural network architectures; 2) the syntactic analysis of Polish sentences provided by dependency parsers integrated with pre-trained feature embeddings.

Recent study on dependency parsing of Polish [32] indicates that graph-based parsers, even those without any neural component, are better suited for Polish than transition-based parsing systems. Therefore, we conduct an additional experiment on the impact of various pre-trained token and subword embeddings on the quality of a state-of-the-art graph-based dependency parser.

The quality of the dependency parsers tested in these three scenarios is measured with two standard metrics:

- unlabeled attachment score (UAS) – the average accuracy[15] of assigning the correct head for each token in each sentence,
- labeled attachment score (LAS) – the average accuracy of assigning both the correct head and label for each token in each sentence.

6.1 Dependency Parsing Systems

Three dependency parsing systems are tested in our experiments: MaltParser [24], SyntaxNet [1], and BIST parser [15]. All of them are transition-based parsers, but they differ in internal classifiers (linear vs. non-linear) and in feature encoders (hand-crafted vs. feed-forward neural network vs. bidirectional long short term memory network). Apart from transition-based parsers, we also test the state-of-the-art graph-based dependency parser COMBO [28].

MaltParser's classifier is based on the logistic regression model (i.e. a linear model). Its hand-crafted feature model is built of the single features (e.g. tokens, lemmata, part-of-speech tags) and double or triple combinations thereof.

SyntaxNet uses a simple feed-forward neural network with two hidden layers of 1,024 dimensions each to make the transition decisions (i.e. a non-linear classifier). A transition is based on a rich set of discrete features (i.e. tokens, part-of-speech tags, dependency arcs and labels in the surrounding context of a token, k-best tags) encoded in the current parse configuration. SyntaxNet is an extension of the approach by [4].

BIST parser uses a biLSTM encoder that is jointly trained with the parser's decoder (end-to-end training). The encoder takes concatenations of the vector representations of tokens and their part-of-speech tag as input and outputs biLSTM embeddings (i.e. contextualised word embeddings) of these tokens. The feature vectors are passed to the non-linear scoring function (multi-layer perceptron) of BIST parser.

---

[15] We provide both the micro- and marco-average scores.

COMBO[16] is a natural language pre-processing system that implements not only a graph-based dependency parser, but also a tagger and a lemmatizer which can be jointly trained with the parser. In the current research, we only train the dependency parsing model. The arc scoring function is simple dot product of the dependent embedding and its governor embedding. The feature embeddings are estimated with COMBO's biLSTM encoder based on input features, i.e. tokens, lemmata, part-of-speech tags and morphological features in this case.

6.2   First Evaluation Experiment

In the first experiment, the parsing results provided by the baseline parser – MaltParser – are compared with the results of the parsers with neural components – SyntaxNet and BIST parser. All parsers are trained on the version of Polish Dependency Bank [31], which consists of over 16K dependency trees. The results of 5-fold cross-validation can be found in Table 3. BIST parser achieves the best results and outperforms two other parsers. Since SyntaxNet performs only slightly better than MaltParser, it is not used in our further experiments.

Table 3. Parsing results provided by MaltParser with a linear classifier, and SyntaxNet and BIST parser with neural network classifiers.

PARSER	LAS		UAS	
	Macro	Micro	Macro	Micro
MALT	81.56%	77.78%	86.07%	82.56%
SyntaxNet	81.64%	77.91%	87.55%	84.23%
BIST	**83.01%**	**79.70%**	**89.07%**	**85.95%**

6.3   Second Evaluation Experiment

In this experiment, we test whether training a parser with pre-trained feature embeddings increases the parsing quality. In order to augment MaltParser with token embeddings, the feature model of MaltParser is extended in a naive way: for the top token on the stack and the first token in the buffer we add a number of additional features corresponding to the dimensionality of the token embedding (each value of the token embedding as a separate feature). The results can be found in Table 4. There is a negligible increase in the parsing quality, but only if 100-dimensional token embeddings are used. We suspect that due to its linearity the logistic regression model cannot fully take advantage of token embeddings.

---

[16] https://github.com/360er0/COMBO.

Table 4. 5-fold cross-validation of MaltParser integrated with token embeddings trained using Skip-gram scenario and negative sampling technique.

MALT	LAS		UAS	
	Macro	Micro	Macro	Micro
BASELINE	81.56%	77.78%	86.07%	82.56%
$+\mathcal{T}_{100}$	**81.65%**	**77.97%**	**86.08%**	**82.64%**
$+\mathcal{T}_{300}$	80.75%	76.95%	85.43%	81.88%

As previously noted BIST parser uses a deep learning model to predict the correct sequence of transitions (i.e. to parse a sentence). Each word in the sentence gets its own feature embedding which corresponds to the hidden state of biLSTM. The vector which is input to biLSTM is a concatenation of the vectors of the token and its part-of-speech tag. By default both vectors are trained from scratch during the parser training. However, it is also possible to apply external feature embeddings. We test whether augmenting BIST parser with the pre-trained token, lemma, or character n-gram embeddings increases its accuracy. Results of the experiment are in Table 5.

Table 5. 5-fold cross-validation of BIST parser with internal token embeddings ($\mathcal{T}_{intern}$), 100- or 300-dimensional external token embeddings ($\mathcal{T}_{100}$, $\mathcal{T}_{300}$), internal lemma embeddings ($\mathcal{L}_{intern}$), external lemma embeddings ($\mathcal{L}_{100}$, $\mathcal{L}_{300}$), and external character n-gram embeddings ($\mathcal{S}_{100}$, $\mathcal{S}_{300}$). All external embeddings are estimated using Skip-gram and negative sampling training technique.

BIST	LAS		UAS	
	Macro	Micro	Macro	Micro
$\mathcal{T}_{intern}$	83.01%	79.70%	89.07%	85.95%
$+\mathcal{T}_{100}$	**85.70%**	**82.68%**	**90.29%**	**87.52%**
$+\mathcal{T}_{300}$	85.52%	82.46%	90.09%	87.30%
$\mathcal{L}_{intern}$	82.94%	79.85%	89.07%	86.06%
$+\mathcal{L}_{100}$	84.24%	81.27%	89.64%	86.78%
$+\mathcal{L}_{300}$	84.23%	81.16%	89.54%	86.64%
$+\mathcal{S}_{100}$	85.48%	82.48%	90.07%	87.25%
$+\mathcal{S}_{300}$	85.21%	82.24%	89.72%	86.93%

According to the results, external embeddings increase the accuracy of BIST parser. Similarly as for MaltParser, the best results are achieved with 100-dimensional token embeddings, however the improvement is now much more apparent.

Even if the results of BIST parsers with internal token and lemma embeddings are comparable, the results of the parser with external lemma embeddings are consistently lower than the results of the parser with the token embeddings. Finally, BIST parser integrated with the character n-gram embeddings performs slightly worse than the parser enhanced with the token embeddings.

Feature embeddings estimated in two scenarios (i.e. CBOW and Skip-gram) and with two training techniques (i.e. hierarchical softmax and negative samplings) are tested in this experiment. Due to lack of space, we cannot provide all results. We therefore reported only results of BIST parser integrated with the external 100-dimensional token embeddings (see Table 6). The best performing BIST parser is enhanced with token embeddings estimated using Skip-gram and negative sampling training technique.

Table 6. 5-fold cross-validation of BIST parser with external 100-dimensional token embeddings estimated in CBOW (cbow) or Skip-gram (skipg) scenarios, and with negative sampling (ns) or hierarchical softmax (hs) training techniques.

BIST $+\mathcal{T}_{100}$	LAS		UAS	
	Macro	Micro	Macro	Micro
(cbow-ns)	85.57%	82.49%	90.18%	87.36%
(cbow-hs)	85.45%	82.38%	90.02%	87.21%
(skipg-ns)	**85.70%**	**82.68%**	**90.29%**	**87.52%**
(skipg-hs)	85.25%	82.18%	89.93%	87.14%

6.4 An Additional Evaluation Experiment

In this experiment, we evaluate the impact of different pre-trained feature embeddings on the parsing quality of COMBO. Since we do not have access to data used in the previous experiments conducted in 2017, COMBO's parsing models are trained on Polish Dependency Bank 2.0.[17] We therefore cannot directly compare the results of this additional experiment with the previous results. We only compare the current results with the baseline results of MaltParser trained on PDB 2.0. Results of the experiment are in Table 7.

As expected, COMBO parser decidedly outperforms the baseline (Malt-Parser). The differences between LAS and UAS scores of COMBO's parsing models trained with or without external feature embeddings are negligible, with a small advantage of the models supported by pre-trained feature embeddings over the model estimated without external embeddings. The results suggest that COMBO model should be estimated with 300-dimensional word2vec token embeddings (COMBO+$\mathcal{T}_{300}$). However, it is important to note, that this model is also the largest one with the size of 7.3G compared to 2.5G and 70M of COMBO+$\mathcal{T}_{100}$ and COMBO ($\mathcal{T}_{intern}$), respectively. The question remains whether the dependency parsing model should be really trained with external feature embeddings.

---

[17] http://zil.ipipan.waw.pl/PDB.

Table 7. Evaluation of COMBO parser with internal token embeddings (T_{intern}), 100- or 300-dimensional external token embeddings (T_{100}, T_{300}) and external subword embeddings (F_{300}). All external vectors are estimated in Skip-gram scenario using negative sampling technique.

PARSER	LAS		UAS	
	Macro	Micro	Macro	Micro
MALT (BASELINE)	81.70%	77.20%	86.29%	82.21%
COMBO (T_{intern})	90.68%	88.19%	94.50%	92.43%
COMBO+T_{100}	**91.19%**	88.89%	94.64%	92.72%
COMBO+T_{300}	91.18%	**88.93%**	**94.68%**	**92.79%**
COMBO+F_{300}	91.10%	88.68%	94.65%	92.65%

7   Conclusions

Methods of evaluating feature embeddings intrinsically have recently been criticised by the NLP community, e.g. because of uncertain indication of the impact of embeddings on downstream applications. Intrinsic evaluation is undoubtedly appropriate for synonymy and analogy recognition. However, it does not provide clues for other NLP tasks, such as morphosyntactic disambiguation and dependency parsing, which to a lesser extent rely on synonymy and analogy recognition. For these tasks extrinsic evaluation seems to be reasonably informative.

Using external token embeddings in morphosyntactic disambiguation task brings about a substantial gain in accuracy (improvement by about 4 pp), resulting in error reduction up to 43%. However, the choice of the most appropriate token embedding is a secondary issue, because embeddings of different lengths estimated with different training methods have a comparable impact on the task.

Pre-trained feature embeddings (especially token embeddings) play an important role in transition-based dependency parsing, but they are definitely less important in graph-based dependency parsing. The transition-based parser – BIST – augmented with 100-dimensional token embeddings estimated in Skip-gram scenario using negative sampling training technique outperforms the parsing models boosted with other external embeddings, but the differences are actually negligible. The graph-based parser – COMBO – augmented with pre-trained feature embeddings slightly outperforms COMBO model trained without external feature embeddings, but the difference in parsing qualities is insignificant. Furthermore, COMBO's models boosted with pre-trained feature embeddings are much larger and probably less useful in practical applications. Finally, external lemma embeddings are useless for morphosyntactic disambiguation and less useful for dependency parsing than external token embeddings.

Acknowledgments. The presented research was supported by SONATA 8 grant no 2014/15/D/HS2/03486 from the National Science Centre Poland. The computing was performed at Poznań Supercomputing and Networking Center.

References

1. Andor, D., et al.: Globally normalized transition-based neural networks. In: Proceedings of the 54th Annual Meeting of the Association for Computational Linguistics (Volume 1: Long Papers), pp. 2442–2452. Association for Computational Linguistics, Berlin (2016). https://www.aclweb.org/anthology/P16-1231
2. Bahdanau, D., Cho, K., Bengio, Y.: Neural machine translation by jointly learning to align and translate (2014). http://arxiv.org/abs/1409.0473
3. Bojanowski, P., Grave, E., Joulin, A., Mikolov, T.: Enriching word vectors with subword information. Trans. Assoc. Comput. Linguist. **5**, 135–146 (2017). https://www.aclweb.org/anthology/Q17-1010
4. Chen, D., Manning, C.: A fast and accurate dependency parser using neural networks. In: Proceedings of the 2014 Conference on Empirical Methods in Natural Language Processing (EMNLP), pp. 740–750. Association for Computational Linguistics, Doha (2014). https://www.aclweb.org/anthology/D14-1082
5. Chiu, B., Korhonen, A., Pyysalo, S.: Intrinsic evaluation of word vectors fails to predict extrinsic performance. In: Proceedings of the 1st Workshop on Evaluating Vector-Space Representations for NLP, pp. 1–6. Association for Computational Linguistics, Berlin (2016). https://www.aclweb.org/anthology/W16-2501
6. Cho, K., et al.: Learning phrase representations using RNN encoder-decoder for statistical machine translation. In: Proceedings of the 2014 Conference on Empirical Methods in Natural Language Processing (EMNLP), pp. 1724–1734. Association for Computational Linguistics, Doha (2014). https://www.aclweb.org/anthology/D14-1179
7. Devlin, J., Chang, M.W., Lee, K., Toutanova, K.: BERT: pre-training of deep bidirectional transformers for language understanding. In: Proceedings of the 2019 Conference of the North American Chapter of the Association for Computational Linguistics: Human Language Technologies, Volume 1 (Long and Short Papers), pp. 4171–4186. Association for Computational Linguistics, Minneapolis (2019). https://www.aclweb.org/anthology/N19-1423
8. Dozat, T., Qi, P., Manning, C.D.: Stanford's graph-based neural dependency parser at the CoNLL 2017 shared task. In: Proceedings of the CoNLL 2017 Shared Task: Multilingual Parsing from Raw Text to Universal Dependencies, pp. 20–30. Association for Computational Linguistics, Vancouver (2017). https://www.aclweb.org/anthology/K17-3002
9. Drozd, A., Gladkova, A., Matsuoka, S.: Word embeddings, analogies, and machine learning: beyond king−man+woman=queen. In: Proceedings of COLING 2016, the 26th International Conference on Computational Linguistics: Technical Papers, pp. 3519–3530. The COLING 2016 Organizing Committee, Osaka (2016). https://www.aclweb.org/anthology/C16-1332
10. Faruqui, M., Tsvetkov, Y., Rastogi, P., Dyer, C.: Problems with evaluation of word embeddings using word similarity tasks. In: Proceedings of the 1st Workshop on Evaluating Vector-Space Representations for NLP, pp. 30–35. Association for Computational Linguistics, Berlin (2016). https://www.aclweb.org/anthology/W16-2506

11. Finkelstein, L., et al.: Placing search in context: the concept revisited. ACM Trans. Inf. Syst. **20**(1), 116–131 (2002). https://doi.org/10.1145/503104.503110
12. Grave, E., Bojanowski, P., Gupta, P., Joulin, A., Mikolov, T.: Learning word vectors for 157 languages. In: Proceedings of the Eleventh International Conference on Language Resources and Evaluation (LREC 2018). European Language Resources Association (ELRA), Miyazaki (2018). https://www.aclweb.org/anthology/L18-1550
13. Hill, F., Reichart, R., Korhonen, A.: SimLex-999: evaluating semantic models with (genuine) similarity estimation. Computat. Linguist. **41**(4), 665–695 (2015). https://www.aclweb.org/anthology/J15-4004
14. Iyyer, M., Manjunatha, V., Boyd-Graber, J., Daumé III, H.: Deep unordered composition rivals syntactic methods for text classification. In: Proceedings of the 53rd Annual Meeting of the Association for Computational Linguistics and the 7th International Joint Conference on Natural Language Processing (Volume 1: Long Papers), pp. 1681–1691. Association for Computational Linguistics, Beijing (2015). https://www.aclweb.org/anthology/P15-1162
15. Kiperwasser, E., Goldberg, Y.: Simple and accurate dependency parsing using bidirectional lstm feature representations. Trans. Assoc. Comput. Linguist. **4**, 313–327 (2016). https://www.aclweb.org/anthology/Q16-1023
16. Kobyliński, Ł., Kieraś, W.: Part of speech tagging for polish: state of the art and future perspectives. In: Proceedings of Computational Linguistics and Intelligent Text Processing, pp. 307–319 (2016)
17. Kobyliński, Ł., Ogrodniczuk, M.: Results of the PolEval 2017 competition: part-of-speech tagging shared task. In: Vetulani, Z., Paroubek, P. (eds.) Proceedings of the 8th Language & Technology Conference: Human Language Technologies as a Challenge for Computer Science and Linguistics, pp. 362–366. Fundacja Uniwersytetu im. Adama Mickiewicza w Poznaniu, Poznań (2017)
18. Krasnowska-Kieraś, K.: Morphosyntactic disambiguation for Polish with bi-LSTM neural networks. In: Vetulani, Z., Paroubek, P. (eds.) Proceedings of the 8th Language & Technology Conference: Human Language Technologies as a Challenge for Computer Science and Linguistics, pp. 367–371. Fundacja Uniwersytetu im. Adama Mickiewicza w Poznaniu, Poznań (2017). http://ltc.amu.edu.pl/book2017/papers/PolEval1-2.pdf
19. Leviant, I., Reichart, R.: Separated by an un-common language: towards judgment language informed vector space modeling. CoRR abs/1508.00106 (2015). http://arxiv.org/abs/1508.00106
20. Linzen, T.: Issues in evaluating semantic spaces using word analogies. In: Proceedings of the 1st Workshop on Evaluating Vector-Space Representations for NLP, pp. 13–18. Association for Computational Linguistics, Berlin (2016). https://www.aclweb.org/anthology/W16-2503
21. Mikolov, T., Sutskever, I., Chen, K., Corrado, G.S., Dean, J.: Distributed representations of words and phrases and their compositionality. In: Burges, C.J.C., Bottou, L., Welling, M., Ghahramani, Z., Weinberger, K.Q. (eds.) Advances in Neural Information Processing Systems, vol. 26, pp. 3111–3119. Curran Associates, Inc. (2013). http://papers.nips.cc/paper/5021-distributed-representations-of-words-and-phrases-and-their-compositionality.pdf
22. Mikolov, T., Yih, W.t., Zweig, G.: Linguistic regularities in continuous space word representations. In: Proceedings of the 2013 Conference of the North American Chapter of the Association for Computational Linguistics: Human Language Technologies, pp. 746–751. Association for Computational Linguistics, Atlanta (2013). https://www.aclweb.org/anthology/N13-1090

23. Mykowiecka, A., Marciniak, M., Rychlik, P.: Testing word embeddings for Polish. Cogn. Stud./Études Cogn. **17**, 1–19 (2017). https://ispan.waw.pl/journals/index.php/cs-ec/article/view/cs.1468
24. Nivre, J., Hall, J., Nilsson, J.: MaltParser: a data-driven parser-generator for dependency parsing. In: Proceedings of the Fifth International Conference on Language Resources and Evaluation (LREC 2006), pp. 2216–2219. European Language Resources Association (ELRA), Genoa (2006). http://www.lrec-conf.org/proceedings/lrec2006/pdf/162_pdf.pdf
25. Pennington, J., Socher, R., Manning, C.: GloVe: global vectors for word representation. In: Proceedings of the 2014 Conference on Empirical Methods in Natural Language Processing (EMNLP), pp. 1532–1543. Association for Computational Linguistics, Doha (2014). https://www.aclweb.org/anthology/D14-1162
26. Przepiórkowski, A., Bańko, M., Górski, R.L., Lewandowska-Tomaszczyk, B. (eds.): Narodowy Korpus Języka Polskiego. Wydawnictwo Naukowe PWN, Warsaw (2012)
27. Řehůřek, R., Sojka, P.: Software framework for topic modelling with large corpora. In: Proceedings of the Workshop on New Challenges for NLP Frameworks, pp. 45–50 (2010)
28. Rybak, P., Wróblewska, A.: Semi-supervised neural system for tagging, parsing and lematization. In: Proceedings of the CoNLL 2018 Shared Task: Multilingual Parsing from Raw Text to Universal Dependencies, pp. 45–54. Association for Computational Linguistics, Brussels (2018). https://www.aclweb.org/anthology/K18-2004
29. Sutskever, I., Vinyals, O., Le, Q.V.: Sequence to sequence learning with neural networks. In: Ghahramani, Z., Welling, M., Cortes, C., Lawrence, N.D., Weinberger, K.Q. (eds.) Advances in Neural Information Processing Systems, vol. 27, pp. 3104–3112. Curran Associates, Inc. (2014). http://papers.nips.cc/paper/5346-sequence-to-sequence-learning-with-neural-networks.pdf
30. Vulić, I., Mrkšić, N., Reichart, R., Ó Séaghdha, D., Young, S., Korhonen, A.: Morph-fitting: fine-tuning word vector spaces with simple language-specific rules. In: Proceedings of the 55th Annual Meeting of the Association for Computational Linguistics (Volume 1: Long Papers), pp. 56–68. Association for Computational Linguistics, Vancouver (2017). https://www.aclweb.org/anthology/P17-1006
31. Wróblewska, A.: Polish dependency parser trained on an automatically induced dependency bank. Ph.D. dissertation, ICS PAS, Warsaw (2014)
32. Wróblewska, A., Rybak, P.: Dependency Parsing of Polish. Poznań Stud. Contemp. Linguist. **55**(2), 305–337 (2019). https://doi.org/10.1515/psicl-2019-0012

Machine Translation

Syntactic and Semantic Impact of Prepositions in Machine Translation : An Empirical Study of French-English Translation of Prepositions 'à', 'de' and 'en'

Violaine Prince[✉][iD]

University of Montpellier and LIRMM-CNRS, Montpellier, France
`prince@lirmm.fr`

Abstract. This paper presents a study about ambiguous French prepositions, stressing out their roles as dependencies introducers, in order to derive some translation heuristics into English, based on a French-English set of parallel texts. These heuristics are formulated out of statistical observations and use some up-to-date results in Machine Translation (MT). Their originality mostly relies upon two items: (1) The importance given to syntax and dependency relations, along with lexicons, the latter being well browsed by the present literature in the domain (2) The existence of intrinsic semantics in prepositions, something rather discarded in NLP literature devoted to statistical MT, that tends to point at the most appropriate translation. An experiment has been run on corpora in both languages, using a dependency parser in the source language, and results looked to be encouraging for a "step by step approach" for MT improvement.

Keywords: Natural Language Processing · Machine aided translation · Prepositions

1 Introduction

Function words is the class of those words that are necessary for a syntactically correct text, ranging from determiners, conjunctions, prepositions, pronouns to auxiliary verbs, modals, etc. They are opposed to *content words* such as nouns, verbs, adjectives, etc. They are considered as dependent words, in the sense that they cannot bear a given meaning by themselves, but their presence contributes to decode (or build) the meaning of a complete linguistic unit, i.e. the sentence. They have been put aside in several works in Information Retrieval (IR), since they can serve neither as indexes or keywords, nor as queries. When it comes to Machine Translation (MT) or Machine Aided Translation, they cannot be neglected, because translation produces a text as an output, and this text needs

© Springer Nature Switzerland AG 2020
Z. Vetulani et al. (Eds.): LTC 2017, LNAI 12598, pp. 273–287, 2020.
https://doi.org/10.1007/978-3-030-66527-2_20

to be understandable by a reader. If content words are sufficient to bring out the general thematic context (what the text is about), translation needs to be more precise, and produce, in its target, the exact composition of content words that was present in the source text.

Researchers in Computational Linguistics (CL) and Natural Language Processing (NLP) have been working on the role of some subcategories of function words such as *pronouns* and *prepositions*. Prepositions and the issues they raise for an automated processing of language have been studied by [6]. When addressing the pair English-French, the role of pronouns has been thoroughly examined in [8] and works such as [10] and [5] have been devoted to prepositions. However, these works were not in the main stream, neither in NLP, nor in MT. It seems that prepositions have been less dealt with, when compared to the overwhelming importance given to nouns, noun phrases and adjectival phrases in lexical semantics. Electronic resources such as WordNet still assign to prepositions a secondary role, and mostly restricted to syntax, within multiword expressions. In MT, a huge project such as ARTFL (American and French Research on the Treasury of the French Language), led by the University of Chicago and the French ATILF association for Machine Translation, provides an electronic bilingual resource of 75,000 entries. Though, this resource seems to discard prepositional noun phrases acting as technical (and sometimes usual) multiword expressions. Although ontologies and web semantics have shed light on multiword expressions, several of which containing prepositions, as the lexical impersonation of some of their concepts, the true role of prepositions is still left aside, in spite of their quite crucial role in the semantic interpretation of the fragment.

In this pair of languages, the preposition gives the direction in interpretation in a prepositional noun phrase. For instance *a set of pairs* greatly differs in meaning from *a pair of sets*. The non commutativity of natural language is here typically demonstrated by the existence and/or position of prepositions [11].

Second, the preposition semantics greatly modify the verb meaning in a phrasal verb, especially in English. Whole sets of new verbs are created by combining verbs with a strong polysemous potential such as *do, put, make, take, get*, etc. with prepositions [10].

Third, prepositions are not univocal terms. They behave like nouns, adjectives and verbs by being as ambiguous as the latter [9]. For instance, if the preposition *at* seems to indicate a projection from a source towards a target in phrasal verbs such as *to look at*, a sense it shares with the preposition *to*, *at* seems a locative tag, with stability and immobility characteristics in *'he is at home'*, sharing the latter with the preposition *in* [5]. So characteristics such as polysemy and synonymy, common in nouns and verbs, also apply to prepositions.

The issue we try to address in this paper is the role and transformation of prepositions in an MT application in a restricted environment. MT applications are very largely statistically based, and make a very heavy use of learning (whether deep or shallow). However, if corpora used for such a learning are poorly written, the application output will not be better. Moreover, prepositions contexts are distributed: For instance, the preposition *out* translates quite differently if one says: *I put some weight out*, or *I put some garbage bags out*. Unless one needs to record all possible sentences (although this is what deep learning MT seems to aim at), it might be cheaper to have a few heuristics that could provide some guidance for a large scale system. In that sense, we have been looking for a few mainlines that could possibly act as a safeguard system, believing thus in the virtue of hybridation, between a large scale application needing an enhancement of its precision, and a very sketchy 'rule-based' framework which will play the watchdog. Thus, the main items that we needed to tackle were the following:

- Prepositions are often ambiguous in the *source language* (SL), but also their candidate equivalencies are ambiguous in the *target language* (TL). What are the most discriminant criteria that might help a task like MT disambiguate and choose the appropriate candidate?
- What is the syntactic role of the preposition? Does it introduce a particular complement? If yes, what is the influence of its semantics in SL on its representation in TL?
- Some prepositions belong to idiomatic multiword expressions and do not need to be examined as such. The recognition of their ineffectiveness as syntactic agents belongs to the existence (or the improvement) of existing translation memories and databases. But what is the limit between an idiomatic multiword expression and a non idiom? Is every overly popular recorded translation an idiom?

All these questions are naturally very broad in their scope. Since languages greatly vary in their use of linguistic dimensions, and in spite of an asymptotic aim at generalization, we have restricted their impact on the pair we address and all results will be considered as locally grounded in our French-English environment. They might possibly have an echo in other pairs of languages, and asserting the generality would be another very interesting issue that we cannot undertake in this study.

So in this paper we will address the goal that we aim at, i.e, a set of possible regularities for in the semantic behavior of the most used prepositions in French and their translation into English. We will first stress out the possible use of such as set in the next section. Then we will deal will corpus study and experiment in Sect. 3, the experiment having been extended with the study of verb -preposition bigrams as suggested by a reviewer of an earlier version of this work. Some rules based on semantic relations and syntactic properties have been a incorporated in the translating prototype. Then we conclude about the present step of this ongoing large project assessing its benefits for a broader frame.

2   Machine Translation and Prepositions: Is There a Most Appropriate Approach?

The ground hypothesis is that MT, a very large field with several paradigms and software contributing to its evolution, should be enhanced with tools able to detect as automatically as possible the nature of the dependency introduced by the preposition in SL, as a requirement for discriminating the proper equivalent in TL, provided that the preposition is not an element of an idiomatic multiword expression.

Idioms, as multiword colocations, can be stored in extensive lexicons, or translation memories. Their enhancement has been long studied in MT, and any survey, however extensive, will not capture all the achievements in this domain. A few works are still more dedicated to prepositions (e.g., [12,13]) or tend to stress them out (e.g, The Ultralingua dictionary at http://www.ultralingua.net provides human readable cross-lingual idioms containing prepositions. The problem is that is not machine operable as such). Others (e.g. [7]) have focused on the requirements for candidate idioms in order to enhance existing resources. Several works have thoroughly contributed to the domain, but those we cite here have been the most inspiring to us, or the closest to our specific issue.

If the fragment to be translated is not clearly an idiom, then two approaches might compete:

- One that would browse aligned corpora and automatically learn translations, storing them in translation memories thus 'lexicalizing' as much as possible the translation process. This is the actual main trend in MT. It needs important resources but few manpower. It has a quite good recall but its behavior, when encountering a new expression, would be bland.
- Another that would undergo the same browsing, but would try to semi-automatically extract patterns to be rewritten into transformation rules. A more costly approach in terms of human effort, but since prepositions seem to deal with hidden construction aspects (e.g. syntax, dependencies) and less with apparent lexical data, it seems that a more rule based approach would not be totally out of scope, in such a case.

We have chosen to explore the second (and less used) path for which we have particular opportunities. If statistical MT was all successful in properly translating prepositions in the selected pair of languages, the issue would have dropped. Let us mention an interesting study undergone by [4] analyzing the types of errors in pronouns translation in English-French (statistical) MT. Pronouns is a subcategory of function words sister to prepositions, and some French words, such as *en* could be tagged with *preposition* or *pronoun*. Of course, one could always make a corrective survey after a given translation, post-editing, and feeding the system with valid constructions (several papers account for such experiments). This would naturally enhance the manpower devoted to the task, diminishing statistical MT main benefit, but is a sound although time consuming method for translation improvement.

However, the rationale is not to 'beat' statistical MT but to contemplate a hybrid method (both statistical and rule-based) seen as a task-properties oriented approach (if the local syntactic properties of prepositions provide a true added-value to their translation then a local rule-based tool within a statistical overall frame could be envisaged). Our research team has access to a dependency parser for French [3], our source language, and to a French to English translation prototype that would implement, as a proof of concepts, some ideas about *transformation rules* as a formal rewriting of regular translation patterns. This prototype is compatible with the parser. In that sense, we have a similar methodological framework as the one used in [7] and [13], and these studies have shown the reliability of this approach. But in order to do implement transformation rules, we needed to produce a study of prepositions behavior in both SL and TL. As computational linguists, we tried the most automatic approach, by parsing an SL corpus for which we had a TL aligned equivalent, we tagged parsed chunks with dependency tags, and studied the regularity (or not) of the prepositions translation according to their role.

A more restricted and refined goal for this study would be to see if variability in translation, illustrated by the prepositions issue, could be stressed down into a few patterns, which would in turn give birth to a few transformation rules, or not. If not, then the main trend will prove to be the wisest path to run. But if patterns can be fathomed and reduced to a small number, then it would not be meaningless to include the results into a wider and multi-model translation system.

3   Looking for Translation Patterns: An Experimental Framework

For a POC (Proof of Concepts) result, we have restricted the study to three prepositions widely used, which are *à*, *de* and *en*. The three could be seen as **locative** ([5]) but are highly ambiguous in meaning, and might present a distributional similarity close to that described in [1]. Moreover, *en* is also syntactically ambiguous, it could be either a preposition or a pronoun, leading in the second case to the problem of anaphora resolution.

We have a pair of aligned French-English corpora of about 54,000 words (in French) about stock exchange and economics, extracted from companies reports. The present corpus is specialized, "clean" and its quality is sufficient to ensure a reliable study. Moreover, it is naturally rich in prepositions. It is a good candidate "to begin with". We also used bilingual economics dictionaries[1] and an access to French and English Wordnet. First subsection states the issue, the second gives the general methods we have tried to follow and the third provides the obtained results.

3.1   A Brief Overview of the Issue

This section describes the characteristics of the studied prepositions and their linguistic properties and the subsequent issues raised for computational representation. A particular light will be shed on the role of these prepositions as dependency triggers. As a consequence, we will address the particular problem of modeling translating actions for these prepositions.

---

[1] URL: http://www.e-anglais.com/ressources/glossary.html, developed by Kevin Halion. Mostly human readable, but machine operability was easy.

Table 1. Different lexical forms representing the preposition

Canonic Form	à	de
First Variant (Sing)	au	du
Second Variant (Plur.)	aux	des
Contracted	-	d'

Table 2. Different POS tags associated to lexical forms

Token	Tag 1	Tag 2
du, des	determiner	preposition
au, aux	determiner	preposition
en	pronoun	preposition

Source Language Morphological Variation. Table 1 refers to lexical variability of prepositions *de* and *à* in SL, whereas *en* is invariant. The two variants correspond to the contraction of a determiner (i.e. *le, les*) and the preposition and plays both the role of a preposition and a determiner.

Some linguists consider only the determiner tag, some others insist on the dual role invoking the **substitution theory** (if one substitutes the singular feminine form of the determiner to the contracted token, it expands into two tokens, the preposition + the determiner. Thus, theoretically, *aux* should expand into *à les, du* should expand into *de le*, etc.). With substitution, one has to read "tag 1" and "tag 2" in Table 2 not as mutually exclusive tags, but as conjugated tags. The ambiguity here is not to be solved but is intrinsic to the contracted form, whereas with *en*, it is a classical mutually exclusive ambiguity.

Dependency Roles. The different dependency roles of these prepositions are the following:

- in a noun phrase introducing a *noun complement* (**NC**)
- in a verb phrase introducing an *indirect noun phrase object complement* for verbs either transitive of intransitive (**OC**)
- in a verb phrase introducing an *infinitive proposition acting as an object complement* (**OCP**)
- in a verb phrase introducing *a noun phrase location complement* (**LC**)

Table 3 gives a few examples of translations with the three prepositions.

Multiple Equivalencies in Target Language. The ambiguity of the three studied prepositions is revealed and enhanced by the multiplicity of their equivalencies in the target language. Table 3 has already hinted at the fact that more than one equivalent is available for each of the prepositions. Table 4 shows what the bilingual dictionary we use suggests.

Table 3. Equivalencies for each preposition

Prep.	Translations
à	at, to, in, with, by, upon, about
de	of, out of, off, from, with, by, about
en	in, into, to, of, thereof, at, during

3.2 Modeling Prepositions Translation

One of the main items is the following question: Are prepositions part of a multiword expression? Do they already belong to a bilingual dictionary? If not, could they be possible candidates for a domain-oriented translation lexicon? The second question rises if prepositions are not part of multiword expressions. Some of the examples in Table 4 show that translation sometimes deletes them. For instance, the NC (*noun complement*) dependency role seems to favor deletion of the preposition in TL (cf Table 4). Could it be a transformation rule at the dependency level? Could it be flattened back at the pure component level in which the French Noun Phrase pattern *N1 Prep N2* (where *N1* and *N2* are nouns and *Prep* is a preposition), respectively *NP1 Prep NP2* (where *NP1* and *NP2* are noun phrases) would be translated in English into *TN2 ('s) TN1*, respectively *TNP2 TNP1* (where *Tx* is the translation of *x*)?

Another way to look at it is to consider the impact of dependency role on the translation. Does it reduce the translation ambiguity for a given preposition? If in the NC role, prepositions are deleted, in the LC (*locative complement*) role, is *à* always translated by *to*, *de* by *from*, and *en* by *by*? An experiment is interesting to conduct on such an issue. Object complements (both OC and OCP roles) are

Table 4. Dependency roles examples

Role	Example in SL	Translation
NC	*médecin de famille*	family doctor
	moulin à café	Coffeemill
	pot en terre cuite	terracotta pot
OC	*Je pense à lui*	I think **of** him
	Je parle de lui	I am talking **about** him
	Je pense en Franais	I think **in** French
OCP	*Je viens de manger*	I have just eaten
	Je refuse de parler	I refuse **to** talk
	Il parle en dormant	He talks **in** his sleep
LC	*Je pars en avion*	I am going **by** plane
	Je pars à Londres	I am going **to** London
	Je viens de New York	I am coming **from** New York

more difficult to stress down. Here, a more thorough empirical study might lead to a better view of the subject.

Experimental Setup. We ran an experiment by parsing the SL corpus (54,000 words) with our morphosyntactic parser, which provides a deep syntactic analysis (with a syntactic tree) and assigns dependency roles to subtrees. It also tags tokens with their semantic role, if it is present in the lexicon. The experiment was divided into three parts. It needed to answer the 'lexical or not lexical' question. Is the preposition an element of a multiword expression belonging or not to a lexical resource? For that we first checked that the parser was able to recognize lexical prepositional noun phrases. A second step aimed at tracking domain oriented multiword expressions by retrieving the most frequent subtrees containing any of our studied prepositions, in order to see if we can enhance a translation lexicon. The third investigated the remaining occurrences, focusing on dependency tags whenever they appeared.

3.3   Experiment Results

Checking the Parser. In our system, if the morphosyntactic parser recognizes the candidate as an idiom it transforms its pattern by adding '_%20' tags in the lexical string to replace the blanks. For instance, the idiom 'pomme de terre' meaning *potato* is transformed into 'pomme_%20de_%20terre'. The bilingual lexicon must contain it as a single entry. Other adjectival and verbal locutions exist in our dictionary, such as 'Beaucoup_%20de' (many), 'A_%20partir_%20de' (from... on) 'en_%20dépit_%20de' (in spite of, despite). We created the machine readable bilingual lexicon out of our bilingual resources in such a format. We ran the parser on the SL corpus and obtained the results in Table 5 according to the type of multiword expression. Two usual measures have been used: *recall* and *precision*. Recall is calculated as following:

$$\rho = \frac{nc}{nc+nf},$$

Where nc stands for the number of candidates correctly extracted and nf the number of candidates forgotten. Precision is calculated as:

$$\pi = \frac{nc}{np},$$

where np is the number of candidates extracted by the parser. Table 5 showing that idiomatic verbal locutions being less recognized by the parser, the latter was fed with this information and its abilities were thus enhanced.

Table 5. Extracted candidates for parser checking

Type	Typical Example	Recall	Precision
Idiomatic NP	Hôtel de ville (Town Hall)	0.88	0.92
Idiomatic Nominal Locutions	En raison de (because of)	0.75	0.87
Idiomatic Verbal Locutions	A partir de (from ... on)	0.56	0.52
Idiomatic Adjectival Locutions	A peu près (circa)	0.90	0.88

Extracting the Most Frequent Noun Phrases with Prepositions. The second step was quite important in the sense that we needed to isolate sentence fragments containing prepositions that had proper syntactic roles, that were properly constituted and to which dependencies roles have been assigned. This meant that the corpus needed to be totally parsed. The parser performances are not perfect: if POS tagging is over 98% in precision and recall [3], syntactic analysis and dependency assignment, although quite successful compared to the on-going state-of-the-art, are closer to a 70% value. We had to run a tedious checking on the results and then, for well parsed subtrees, we re-run an automatic counting on the corpus.

Table 6. Most Frequent Prepositional Noun Phrases in the Corpus (NC dependency role)

Rank	Contents	Translation	Frequency
1	Economie de marché	market economy	45
2	Taux d'intérêt	Interest rate	42
3	Taux de croissance	growth rate	41
4	Taux d'inflation	inflation rate	39
5	Chiffre d'affaires	Sales turnover	39
6	Marché à terme	Futures exchange	37
7	Indice de référence	Benchmark	39
8	Indice de satisfaction	Customer Satisfaction Value Index	32
9	Consommation des ménages	Private Consumption	28
10	Marchés des devises	Exchange market	25
11	Augmentation de capital	Capital growth	23
12	Option d'achat	Call	19
13	jour de liquidation	Winding-up date	12
14	jour de valeur	Overnight	12
15	Obligations à haut rendement	Junk Bonds	7
16	Frais de souscription	Subscription charges	5
17	Frais de rachat	back-end load	5
18	Ecart de suivi	tracking error	4
19	Teneur de marché	mark to market	4
20	Facturation d'entreprise	Enterprise billing	3

As a first approach, we were interested into enhancing our bilingual lexicon with domain oriented expressions, and most of these are known to be noun phrases, so we extracted the 20 most frequent prepositional noun phrases. Since we had a TL corpus that was a translation of our SL corpus, we associated the TL equivalent and obtained the results given in Table 6. The first observation was that the deletion of the preposition in the noun complement dependency role was quite regular, except in item 19 of Table 6, so it was not a negligible clue for our future transformation rules. Second, it seems that the pattern *N1 Prep N2* in French is not always transformed into *TN2 TN1*. For instance, item 8 in Table

Table 7. Prepositions translations (through alignment) and their percentages in the well parsed verbal phrases

Role	à	de	en
OC	at $(22, 8\%)$, to $(20, 2\%)$ - $(18, 1\%)$, of $(16, 2\%)$, with $(11, 3\%)$, by $(6, 6\%)$, about $(4, 8\%)$	of $(37, 2\%)$, -$(33, 5\%)$ some $(17, 9\%)$ by $(11, 4\%)$	by $(40, 3\%)$, in $(24, 5\%)$ with $(10, 4\%)$, - $(9, 8\%)$ about$(9, 5\%)$, into (5%)
OCP	to $(71, 2\%)$, - $(28, 8\%)$	- $(85, 3\%)$, to (10%) about $(3, 1\%)$, of $(1, 6\%)$	in $(45, 1\%)$ during (32%) while $(20, 3\%)$, into $(3, 4\%)$
LC	at $(50, 9\%)$ to $(49, 1\%)$	from	by

6 gives birth to four English nouns, whereas items 12 and 14 drop down to only one. Moreover, TN2 (respectively TN1) is not always the translation of French N2 (respectively French N1). 13 out of 20 expressions in Table 6 do not follow such a pattern. So, in our opinion, *Domain Idiomatic Noun Phrases* would be those prepositional noun phrases in which nouns in TL are not translations of nouns in SL and should be candidate entries to the bilingual lexicon. Thus, we incorporated items 5, 6, 7, 8, 9, 10, 12, 13, 14, 15, 17, 18, 19 to our lexicon. Other noun phrases were extracted, and we noticed that there was a distribution, for the pattern *N1 de N2* (value *de* for the preposition) between *TN2 TN1* and *TN1 of TN2*, the latter being a word to word translation. These noun phrases were not specific to the domain.

Comparing Translations of Prepositions According to Their Dependency Roles. If the NC role seems to lead to preposition deletion in quite an important number of cases and could be used as a rule if the expression is not an idiom or domain specific, and if the preposition is not *de*, we tried to investigate other roles and the variation in translation within a given role. From the parsed corpus, we extracted all subtrees (chunks) in which prepositions *à*, *de* and *en* had one of the other dependency roles (within verb phrases). It is to be noticed that not all sentences were completely parsed, but all of them had at least a partial parsing (local subtrees formed). We had, at that point, two solutions: Either we checked all occurrences of those prepositions in a window of words (with an n-gram approach) on the aligned corpora (the SL and its translation), then assigned by hand dependency roles to our prepositional phrases, or we had to rely on the parsed SL corpus with its shortcomings. We chose to examine the well-formed verbal subtrees of the parsed SL corpus, containing our prepositions. Out of 1305 verbal phrases containing any of these prepositions, 962 were correctly parsed and the subtrees having the proper information in terms of dependency role tag[2]. Some of the reasons for such a recall value (i.e 0.73) were issued due to a bad recall and precision on verbal locutions (cf Table 5) or adverbial ones. The comparison of translations was made semi-automatically (with sentence numbers as usual in aligned corpora). The results are given in Table 7.

---

[2] The parser developer has created a semi-automatic counting tool which allows him to check the parser abilities in terms of recall and precision.

What is quite interesting is the regularity of translations in the case of the locative complement (LC). *à* is translated as much by *at* as by *to*. The distribution has a semantic grounding: *at* is more static, pointing at the present place, whereas *to* is projective and points at the place to reach. In our corpus *de* is always translated by *from*, and *en* by *by*. Thus, the relationship with the semantic aspects is quite obvious: *en* designates the mean with which the movement is performed, and *de* the source. The only ambiguity is about the location, either present or future. The OCP role, or the verb or proposition acting as an object complement, is more ambiguous for all prepositions, with a deletion case (represented by '-' in Table 7) quite present. This is why we devoted our next experiment to OCP.

Let us note that preposition *en* is quite often translated in the meaning of a duration (*while, during*), especially with a present progressive verbal form. The widest distribution is for the OC (object complement) role assigned to a noun phrase. Here, it seems that the POS tag of the preposition (ambiguous as shown in Table 2) has a role to play. When the tag is that of a determiner (case of *au, aux, du, des* as shown in Table 1), then *to*, respectively - (i.e, the deletion of the preposition), are quite dominant, as well as *some*, an unlikely translation for *de*. It is also true for *en*: as a preposition, it is widely translated by *by*, whereas as a pronoun, it differs according to the nature of the object it refers to.

3.4 Studying the 'verb/preposition' Bigram

A very interesting track was suggested to us by a reviewer of the first version of this paper. To consider a 'verb/preposition' pair in French as a multiword expression *per se* and to study its semantic properties as well the dependency roles attached to it. The English dictionaries naturally provide entries as such. One can find, for instance, for the verb *to go*, entries such as *to go to, to go on, to go in, to go into*, etc. In man readable dictionaries, these entries are subcategories of the canonic classical entry *to go*. We wanted to see wether such an approach could improve the translation of OC and OCP roles for our prepositions.

Considering the Pair as an Idiom. For this, we had two possible paths: Either to modify the parser, or to enhance its lexicon with verb/preposition bi-grams using the local multiword format such as in 'Beaucoup_%20de'. We chose the latter, since it was easy to implement. The verb/preposition pair is thus seen as an idiom and recorded as such in the parser lexicon, and in the prototype translation memory. As an example for our three prepositions, we made a first experiment with the pairs provided in Table 8, relying only on the most 'popular' translation for the bigram. We tried with only five verbs, and generated the 15 pairs coupling them with our three prepositions.

One of the functionalities of Chauche's parser is to create alternative parsing trees according to the possible syntactic combinations. Thus, when augmenting its dictionary with our French pairs, it suggested two possible tags to the following sentence: *Ce train va à Paris.* This train goes to Paris.

The first was: Ce(DET) train (NOUN) va (VERB) à (TO) Paris (NOUN)

And the second: Ce(DET) train (NOUN) 'va - %20à' (VERB) Paris (NOUN). With two different dependency roles: in the first case, the VERB is considered intransitive, Paris being a locative complement (LC) and the preposition à playing the role of a locative function. In the second case, the verb pair is considered as transitive, Paris being its Object Complement (OC), as in *I love Paris*.

Table 8. Verb-Preposition pairs in French and their popular English Translation

French Verb -preposition	Translation
venir de	come from
aller de	go from
partir de	move from
commencer de	begin from
finir de	finish
manger de	eat
venir à	come to
aller à	go to
partir à	move to
commencer à	begin to
finir à	end at
manger à	eat at
venir en	come by
aller en	go by
partir en	go by
commencer en	begin
finir en	end
manger en	eat

Some Drawbacks When we tried the translation of a few examples using the verbs in Table 8, several problems occurred.

1) A syntactic issue: OCP (verb complements that are infinitive propositions) did not translate well at all. For example: *Il vient de manger.* is normally translated by 'He has just eaten.' This means that considering *vient de* as translatable by *come from* is quite an erroneous translation. *vient de* could be translated by the insertion of *just* between the auxiliary and the participle, after transforming the infinitive tense into a pass tense.

2) A usage issue: we relied on the trends given by our first corpus and then considered the semantics of *de* as mostly designating the **source**, that of *à* the **goal** or **position**, and *en* used to describe the **means**, unfortunately, these roles are not univocal. Traditionally, one may say *venir à bicyclette* (*to*

come by bicycle) making the preposition *à* used for the means, something quite far from the empirical first guesses.

3) The multiplicity of equivalences, due to the ambiguity in semantic roles: In fact, the bi-gram verb-preposition is not necessarily less ambiguous than the composition of two words. For instance *venir en* would translate into 'come by' in the sentence *venir en train*, 'come in' in the sentence *venir en été* (to come in summer), with the adjunction of an adverb such as 'late' or 'early' respectively in the sentences: *venir en retard* (to come late) or *venir en avance* (to come early). In these last cases, we had a competition between bigrams which ended up into proposing two alternative parsing trees. One considered: Venir (VERB) "en - %20 avance" (ADVERB), and the other: "Venir -%20 en" (VERB) avance (NOUN).

The semantic roles are quite different. In the first case, *en avance* is a **temporal complement** (TC) and has the same semantics as the English preposition *in* as the introducer of a temporal noun phrase. In the second case, we have a noun phrase OC, and there is a rare, but existing occurrence of such a combination in a particular context, as in the sentence *Ce paiement est venu en avance sur le montant dû.* (This payment came as an advance to the amount due to ...)

A Few Rules. Although our bi-gram in vitro experiment did not seem conclusive, it was nevertheless worthy. It helped us building a set of translation heuristics requirements. It clearly showed that, in a verbal phrase containing a preposition, the nature of the following fragment, wether chunk of POS unit, is quite important. This led us to create separate translation rules for the sentences in which the complement is a noun phrase from those where the complement is an infinitive proposition, something that was already detected in the preceding experiment (see the OC and the OCP lines in Table 7), but was here clearly assessed.

Second, if prepositions act as introducers of complements, noun or verb, their semantics can be that of relations between the governing phrase and its complement. Therefore, if one can determine the nature of this relation, then it may lead to an appropriate choice of an equivalence in the target language. This is a symmetrical approach to the usual one which relies on some lexical clues to detect semantic relations. The determination of the relation greatly relies on lexical semantic networks (Here, WordNet has helped, as well as the semantic model used by the parser, derived from a Roget-based ontology.) Table 9 describes the relations and their instantiation with our prepositions (and some others we found while parsing sentences). The symbol "-" means that the preposition might be absent. To account for the corpus study distribution of the locative complement (LC) we specialized the spatial semantic relations into **source**, **goal** and **position**. Let us say that even the non spatial meaning of these relations (respectively the origin, the aim, the situation) translate quite similarly in English. In the experiment we run, we did not encounter all the possible equivalencies, but here are the most regular ones.

Table 9. Experimented Equivalencies according to the Semantic Relation (SR)

Prep	Trans- SR	Trans - SR	Trans- SR
à	at, in **location (position)** with, by **means**	to **goal** at **time moment**	- **hyponym**
de	of, - **part-of** from, out of **source** to **goal**	out of **material's ownership**	- **hypernym - attribute**
en	in **location (position)** in **manner** by **means**	in, during, at **time period** into state	while **nested actions**

4 Conclusion

The first semi-automatic study on a corpus seemed to show the emergence of a few regular patterns, strongly related to the dependency role in case of noun and locative complements, and better discriminated when the appropriate POS tag is included in the case of noun phrases object complements to verbal phrases. The case of a verbal complement seems to indicate a preferred translation (cf OCP line in Table 7). This study tended to mean that preposition translation obey to rules, and are far from being pure custom based (a case which would have favored a pure statistical approach). It also asserted that semantics have an important position [9]: Locative complements hint a spatial semantics, the behavior of *en* as a temporal complement appears in both noun phrases and OCP role and is translated by *in* or *during*, whereas its behavior as an instrument case indicates the use of *by*. There is a very clear frame that appears if a precious resource such as a semantic parser, able to assign casual roles to chunks in SL, is available as a pre-processor for machine aided translation. In our case, semantic roles were assigned in almost 90% of the correctly parsed OC and OCP, and all LC roles were correctly assigned. This gave us a valuable setting to produce transformation rules for our prototype. As an extension to our study, we followed a track that was hinted at us by a reviewer of the earlier paper. Can we use pairs made of verbs and prepositions in French the way they appear in English, i.e., as fixed bigrams with a given meaning? We tried to do so on a few examples, and we came up with the impression that French did not comply as well as English to such an approach. We don't know the reason why, but maybe the importance of function words in French seems to be greater than in English, and thus they tend to behave as semantic relation introducers, a thing that we tried to capture in summarizing the most used semantic relations for our three prepositions. However, following this track have led us improve OCP rules and question the ambiguity of the position of some prepositions. Do they belong to a locution or are they related to the verb? This question was raised when we tested temporal multiword expressions such as 'en avance'. This sheds an interesting light on the 'idiom' status of some colocations.

As a sequel to this study, we plan now to parse plain and random chosen texts in French, and to test our translation rules whenever we encounter our three prepositions, making the prototype 'learn' from its failures and improve its rules. With the (distant) goal that these rules may play the role of a filter for discriminating the best candidates.

References

1. Baldwin, T.: Distributional similarity and preposition semantics. In: Saint-Dizier, P. (ed.) Computational Linguistics Dimensions of Syntax and Semantics of Prepositions, pp. 197–210. Kluwer Academic Press (2006)
2. Chauché, J.: Un outil multidimensionnel d'analyse du discours. In: Proceedings of COLING-ACL, pp. 491–495 (1984). https://doi.org/10.3115/980491.980495
3. Chauché, J.: Un analyseur du Français en constituants et dépendances. Internal report LIRMM 40 p. Montpellier, France (2005)
4. Hadmeier, C., Guillou, L.: Pronoun Translation in English-French Machine Translation: An Analysis of Error Types. CoRR, volume abs/1808.10196 (2018)
5. Japkowicz, N., Wiebe, J.: A system for translating locative prepositions from English into French. In: Proceedings of the 29th Annual Meeting of the Association for Computational Linguistics, Berkeley, June, pp. 153–160 (1991)
6. Litkowski, K.C.: The Preposition Project, Proceedings of the Second ACL-SIGSEM Workshop on The Linguistic Dimensions of Prepositions and their Use in Computational Linguistics Formalisms and Applications, April 19–21, Colchester. University of Essex, England (2005)
7. Seretan, V., Nerima, L., Wehrli, E.: Multi-word collocation extraction by syntactic composition of collocation bigrams. In: Proceedings of RANLP, pp. 91–100 (2003)
8. Scherrer, Y., Russo, L., Goldman, J.P., Sanchez, S., Nerima, L., Wehrli, E.: La traduction automatique des pronoms: problémes et perspectives, pp. 185–190 (2011)
9. Taylor, J.: Prepositions: patterns of polysemization and strategies of disambiguation. In: The Semantics of Prepositions Cornelia Zelinsky-Wibbelt ed. Walter de Gruyter, pp. 151–178 (1993)
10. Trawinski, B., Sailer, M., Soehn, J.-P.: Combinatorial aspects of collocational prepositional phrases. In: Patrick, S.-D. (ed.), Computational Linguistics Dimensions of Syntax and Semantics of Prepositions. Kluwer Academic Press (2006)
11. Yeh, A.S., Vilain, M.B.: Some properties of preposition and subordinate conjunction attachments. In: Proceedings of COLING-ACL, vol. 2, pp. 1436–1442 (1998)
12. Villavicencio, A., Baldwin, T., Waldron, B.: A multilingual database of idioms. In: Proceedings of the 4th International Conference On Language Resources and Evaluation, LREC-2004, Lisboa (2004)
13. Wehrli, E.: Translating idioms. In: Proceedings of COLING-ACL, pp. 1388–1392 (1998)

Information Retrieval and Information Extraction

Sentence Answer Selection for Open Domain Question Answering via Deep Word Matching

Fabrizio Ghigi[(✉)], Diana Turcsany, Thomas Kaltenbrunner,
and Maurizio Cibelli

Hu:toma Artificial Intelligence SL, Consell de Cent 341, Barcelona, Spain
{fabrizio,diana,thomas,maurizio}@hutoma.ai
http://www.hutoma.ai

Abstract. Question answering is a challenging task due to the diversity of expression in human language. This work proposes a novel unsupervised approach for sentence answer selection, called Deep Word Matching (DWM), that uses both the string form and distributed representations of words, thereby capturing their latent semantic relatedness. Our method takes advantage of publicly available linguistic resources and word embeddings in order to explore various levels of word similarity and identify matching concepts between a question and the sentence that contains the answer. By evaluating on three large corpora (SQuAD, NewsQA and WikiQA), we show that the proposed method outperforms previously published baselines and is also task independent, eliminating the need for retraining and tuning on a new task. We obtain improvements between 5% and 11% on target baseline for two tasks, and best result on the third task.

Keywords: Question answering · Open domain · Word embeddings

1 Introduction

The challenge of answering questions in an open domain is a difficult task due to the richness and complexity of human communication. A single semantic concept can be expressed in various ways, using different word ordering, inflections, synonyms or paraphrase. This can result in sequences of text, with no observable common structure, having the same meaning. Question answering can be divided into three steps: document retrieval, sentence answer selection and exact match.

In this work, we focus on the second step, sentence answer selection, where the task is to determine which sentence from a paragraph(s) contains the answer to a question. To this end, we propose a novel algorithm that finds the answer by determining which sentence shares the highest score of semantic similarity with the question.

In a question answering system (e.g. a chatbot), presenting a sentence-level answer to an end user can be preferable to the exact match, as the correctness

© Springer Nature Switzerland AG 2020
Z. Vetulani et al. (Eds.): LTC 2017, LNAI 12598, pp. 291–303, 2020.
https://doi.org/10.1007/978-3-030-66527-2_21

of the answer is immediately verifiable by the user based on the context in the sentence (Fig. 1).

Question: When did Tesla enroll in Austrian Polytechnic?
Sentence Answer: In 1875, Tesla enrolled at Austrian Polytechnic in Graz, Austria, on a Military Frontier scholarship.
Exact Match: 1875

Fig. 1. Question with its sentence answer and exact match.

Also, the context can offer some additional useful information to the user, making the conversation more natural and stimulating further interaction.

Traditionally, sentence answer selection tasks have been carried out mainly using syntactic analysis with dependency tree matching, exploiting the structure in question and answer, performing a semantic analysis [9,13] or measuring the edit distance between parsing trees [6,11]. Nowadays, lexical and semantic resources are freely available for the most spoken languages. Furthermore, new resources in the form of word embeddings [8,10] have been introduced recently, adding the possibility to represent words as vectors, where the embeddings of semantically similar words are close within the vector space. In this work we show how finding word matches between a question and a candidate sentence, using different levels of similarity to score the word pairs, is an effective method to find the right answer to an input question. Furthermore we claim that this technique can be applied to any language where these resources are available, and has the advantage of requiring no previous training steps, in contrast with other approaches [24]. Through experiments, we show that our method significantly improves upon baselines for sentence answer selection on two different corpora (SQuAD [12], NewsQA [15]), and it also ranks among the best performing models on a third corpus (WikiQA [21]).

2   Related Work

First studies on questions answering during the '70s focused on restricted domains and used small datasets [19,20]. Early systems tackling open domain question answering can be traced back to 2000, with the appearance of TREC challenges [16]. While sentence answer selection needs both syntactic and semantic information to select the best answer in an effective way, early work on this matter focused mainly on syntactic similarity [6,11,17,18]. The importance of selecting the sentence including the correct answer was first studied by [18], where the correspondence between the question and the answer was determined by analysing their dependency trees using a generative probabilistic model [14]. This approach was then improved using Conditional Random Fields (CRF) [17]. Other approaches to sentence answer selection which adopted dependency trees applied a tree matching method measuring the tree edit distance [11], or used a tree kernel function to find the minimal edit sequences between parse trees;

extracted features from those sequences are then used to train a logistic regression classifier that selects the best answer. Although the main focus was still on syntactic structure, the addition of semantic value from *WordNet* [5] base knowledge (e.g. synonym, hypernym, antonym) to improve the performance in question answering was introduced in [22]. Deep Learning methods [2,3,23,24] have also been applied for question answering. [23] propose decomposing questions in entity mentions and relation patterns, then measure their similarity using a convolutional neural network. [2,3] use a Siamese Network to score the similarity of question and answer pairs, whereby each candidate pair is projected onto a joint space. [24] train a supervised model to learn matching between questions and answers using their semantic encoding via distributed representations of pre-built semantic word embeddings. In this work we present Deep Word Matching (DWM), an unsupervised method that combines words in string form employing lexical and syntactical features from base knowledge, with distributed representations of words in the form of word vectors [1,4]. Word vectors take into account the frequencies of the words appearing in the context around each target token. This allows to capture words similarities exploiting latent semantic information.

3   Deep Word Matching

Deep Word Matching (DWM) aims to estimate the similarity between two sentences by creating an alignment among the words of the input and the target sentence, and scoring each pair of matched words according to five different levels of similarity (Fig. 2).

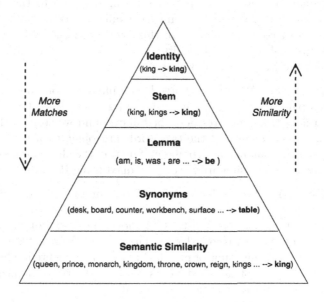

Fig. 2. The Deep Pyramid of Word Matching.

These layers are applied sequentially, until a condition is verified (i.e. two words are equals, or have the same stem, lemma, are synonyms etc.). We performed a coordinate-wise search to find the best score for each level of similarity, i.e. we varied the score for one level, and evaluated the change against the corpus. In this way we selected the scores for each level that obtained the best results in the evaluation.

Identity. The first feature we check is if words are equal. We assign to two identical words the maximum score 1.0.

Stemming. Reduce the inflected words to their word stem (root form). When the two words we are analysing are not equals, we check their stem and if matching, we assign a similarity score of 0.95.

Lemmatisation. Determines the lemma considering the intended meaning, which depends on correctly identifying its part-of-speech. As the third level of similarity, if both words have the same lemma, they get a score of 0.9.

Synonyms. The fourth similarity level takes into account the synonyms. If the words have different lexical forms, the algorithm checks if they have a synonym in common, and scores them accordingly if they do. In this work we use *WordNet* to check if two words are synonyms, and we assign the pair a similarity score of 0.85.

Semantic Similarity. Finally, we use word vectors [8] to compute the semantic similarity between two words. Word vectors (or embeddings) is the collective name for a set of language modelling and feature learning techniques in natural language processing where words or phrases from the vocabulary are mapped to a vector space. If none of the previous features applies to the current pair of words, we assign to the pair the semantic similarity score obtained from the word vectors using cosine similarity. In this work pretrained GloVe [10] vector embeddings are used in the final results.

The algorithm is divided into 3 phases:

Phase 1. Build Deep Word Matches. In this phase we compute the similarity score for each pair of words in the input and target sentences. For each pair we check sequentially the 5 features described above, and we stop at the first one that applies. Thus, we check if the two words are *equal*, then if they have the same *stem*, *lemma*, a *synonym* in common; if none of the above applies, we assign to the pair the *semantic* similarity score obtained using the word embeddings.

Phase 2. Best Matches Selection. Once the complete word matches list is built, we sort it by similarity score. We select the highest scored word match at each iteration from the sorted list, and remove the words composing the pair from the list of tokens we are considering. The score of the selected pair is added to the total score and the count of matched pairs is incremented. The algorithm thereby produces a list of matched words between the two sentences. Some of the words in the sentences may not have a match.

Phase 3. Apply Discount and Bonus. We compute the overall score for the entire sentence, normalising the total score by the number of matches. Then we apply a discount for unmatched words, computing the percentage of words in the sentence for which we could find a match; e.g. if we matched 8 out of 10 words, we apply a discount factor of 0.8 to the overall score. Finally a bonus for long words that have a high similarity score is applied (multiplying by 3 their weight in the final results), as they are a strong indication that the two sentences are related. Table 1 details the steps used in the DWM to compute the similarity between two sentences.

Table 1. Deep Word Matching (DWM) algorithm for sentence answer selection.

Deep Word Matching
Phase 0: Initialization

Initialize $S1,S2$ sentences, $T = \varnothing$
    $Words1 \leftarrow removeStopWords(tokenize(S1))$
    $Words2 \leftarrow removeStopWords(tokenize(S2))$

Phase 1: Build Deep Word Matches

For each $w1,w2$ **in** $Words1, Words2$
    if $w1$ equals $w2$: $similarityScore = 1.0$
    else if $stem(w1) == stem(w2)$: $similarityScore = 0.95$
    else if $lemma(w1) == lemma(w2)$: $similarityScore = 0.90$
    else if $syns(w1) \cap syns(w2) \neq \varnothing$: $similarityScore = 0.85$
    else: $similarityScore = gloveSimilarity(w1,w2)$
    $T \leftarrow T \cup [w1,w2,similarityScore]$

Phase 2: Best Matches Selection

$T \leftarrow sortBySimilarityScore(T)$
$totScore = 0$, $totMatches = 0$
Repeat Until ($T=\varnothing$) or ($Words1=\varnothing$) or ($Words2=\varnothing$)
    $BestMatch \leftarrow T.pop()$
    If $BestMatch.similarityScore > threshold$:
        $totScore = totScore + BestMatch.similarityScore()$
        $totMatches = totMatches +1$
        $Words1 \leftarrow Words1\text{-}BestMatch.w1$
        $Words2 \leftarrow Words2\text{-}BestMatch.w2$
$sentenceScore = totScore/totMatches$

Phase 3: Apply Discount and Bonus and Return
$sentenceScore\ -= discForUnmatchWords(Words1, Words2)$ $sentenceScore\ += bonusForMatchLongWords(thresh,\ len)$ **Return** $sentenceScore$

In Table 1 *discForUnmatchWords(Words1, Words2)* is a function that discounts value from the final score based on the percentage of unmatched words over the length of the sentence, and *bonusForMatchLongWords(thresh, len)* is a function that adds a bonus to the final score by assigning more weight to the word matches that have a length higher than *len* and a similarity score above *thresh*. The most similar sentence to the input sentence Q, will be the sentence s^* that gets the highest score among all the sentences in the candidates set S:

$$s^* = \arg\max_{S} \mathrm{DWM}(Q, s) \tag{1}$$

3.1 DWM Example

In this section we show an example using DWM to select the correct answer to the input question Q from the two candidate sentences $A1$ and $A2$ (Fig. 3).

Q: What was the colour of the apples on the table?
A1: On the desk in the kitchen there is a red apple.
A2: On the floor there are fruits of different colour.

Fig. 3. In this example we want to choose the most similar sentence to **Q** from **A1** and **A2**.

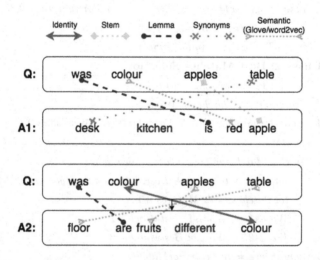

Fig. 4. Search for the best match for each word.

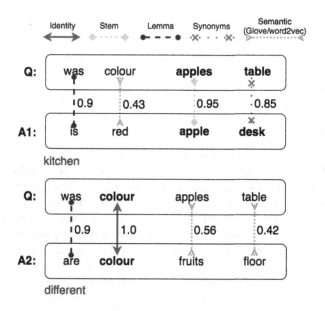

Fig. 5. Assign a score to the best matching word pairs based on the type of match. Grayed out words are words for which there is no correspondent match in the target sentence ('*kitchen*' for A1, '*different*' for A2).

$$\mathrm{DWM}(Q, S1) = \frac{\displaystyle\sum_{W_q, W_1} \mathrm{Score}(w_q, w_1)}{N} = \frac{0.9 + 0.43 + 0.95 + 0.85}{4} \quad (2)$$

$$= 0.78 + \mathrm{Bonus}('apples', 'table') - \mathrm{Disc}('kitchen')$$

$$\mathrm{DWM}(Q, S2) = \frac{\displaystyle\sum_{W_q, W_2} \mathrm{Score}(w_q, w_2)}{N} = \frac{0.9 + 1.0 + 0.56 + 0.42}{4} \quad (3)$$

$$= 0.72 + \mathrm{Bonus}('colour') - \mathrm{Disc}('different')$$

We search a match for each word $w1$ in the input sentence to a word $w2$ in the target sentence (Fig. 4), then sum up the scores of each match (Fig. 5), and average the results by the number of matches N to get an overall score for the sentence. Finally we apply a bonus for the long word matches that have a high similarity score, and a discount for the words that have no match in the corresponding sentence (Eq. 2, 3). The most similar sentence will be the one that maximizes this score (Eq. 1).

4   Evaluation

In this section we measure the performance of our approach, by comparing it to *"sentence level"* metrics measured in other works on three different QA corpora, *SQuAD* [12], *NewsQA* [15] and *WikiQA* [21].

4.1   Corpora

These are the 3 corpora used for the evaluation:

SQuAD (Stanford Question Answering Dataset) is a crowdsourced corpus which consist of 107 785 question-answer pairs constructed using the top 556 ranked articles from Wikipedia, divided into 23 215 paragraphs. The questions are formulated by users, who were presented with a paragraph from one of the Wikipedia articles. The answer is a handpicked text span, selected after a validation process which involves comparing more than one user-generated answer. Although crowd-workers were encouraged to generate the questions in their own words, the ability to read the entire paragraph influenced them (probably unconsciously) to reuse words they have just read. This significant amount of shared vocabulary between questions and answers resulted in high accuracy for sentence answer selection on this corpus even for simpler baseline methods such as word counting.

NewsQA is a set of question-answer pairs generated by crowd-sourced workers from 12 744 CNN articles. It contains 119 633 natural language questions. Like in SQuAD, the answers are spans of text within the article. There are two main differences between NewsQA and SQuAD:

- In NewsQA the annotators were not able to read the entire article. They were just provided with the article highlights. This generated a more varied vocabulary for the questions, increasing significantly the difficulty of the task.
- As the users could not read the entire text, they had to guess what information was contained in the article based only on the provided highlights. This resulted in a variety of questions without a correct answer.

WikiQA corpus was built by collecting factoid questions from Microsoft Bing query logs posed by at least 5 different users. Candidate answer sentences were generated by the summary section of the Wikipedia article listed in the Bing search. Crowdsourced workers were then responsible for labelling those sentences as correct or incorrect. Each question was labeled by at least 3 crowdsourcers. The final set is composed of 3 047 questions and 29 258 sentences, with 1 473 sentences labeled as correct answers.

4.2   Metrics

Three different metrics, Sentence Level Accuracy (SLA), Mean Average Precision (MAP) and Mean Reciprocal Rank (MRR), are used to compare DWM performance against systems from previous works.

SLA takes into account only the first sentence retrieved, while MAP and MRR evaluate the relative ranks of the correct answers in the candidate answer sentences of a question.

Sentence Level Accuracy. The SLA metric checks if the sentence retrieved by the system contains the expected answer.

$$SLA = \frac{|Sentence\ Containing\ Answer|}{|Total\ Questions|} \tag{4}$$

It is obtained by counting the number of correct sentence answers, divided by the total number of questions analysed.

Mean Average Precision. It takes into account the relative rank of the sentence containing the correct answer.

$$MAP = \frac{\sum_{q=1}^{Q} AveP(q)}{Q} \tag{5}$$

$$AveP = \frac{\sum_{k=1}^{n}(P(k) \times rel(k))}{number\ of\ relevant\ documents} \tag{6}$$

where $P(k)$ is the precision at k and $rel(k)$ is an indicator function equaling 1 if the item at rank k is a relevant document, zero otherwise.

Mean Reciprocal Rank. The mean reciprocal rank is the average of the reciprocal ranks of results for a sample of queries Q:

$$MRR = \frac{1}{|Q|} \sum_{i=1}^{|Q|} \frac{1}{rank_i}. \tag{7}$$

where $rank_i$ refers to the rank position of the first relevant document for the i-th query.

4.3 Results

SQuAD Sentence Level Accuracy. The first experiment we carried out was to compare the Sentence Level Accuracy (SLA) reported in two different papers [12,15] for the development set of the SQuAD corpus, to the SLA obtained by DWM. In [12] a Logistic Regression (LR) algorithm is used to get a baseline sentence level accuracy, while in [15] they use a technique similar to Inverse Document Frequency (IDF), denominated Inverse Sentence Frequency (ISF).

We can notice that in the SQuAD corpus the results for baselines are quite high, due to significant amount of shared vocabulary between questions and answers as stated in Sect. 4.1, that lower the task difficulty.

DWM approach achieves an accuracy at sentence level of 83.8% (Table 2), improving the previous results by almost 5%.

Table 2. Evaluation on the SQuAD corpus using SLA Metric.

$SQuAD$	SLA
ISF	79.4%
LR	79.3%
DWM	**83.8%**

NewsQA Sentence Level Accuracy. NewsQA corpus is considered much harder than SQuAD, due to the significant difference between the vocabularies of the questions and the answers [15]. Annotators were not provided the entire paragraph, only highlights, which resulted in questions using a different vocabulary from related text. This explains the significant difference in performances measured on the two corpora using the same metric. In this evaluation we take into account only questions that had at least one correct answer, any question with no correct answer was discarded.

Table 3. Evaluation on NewsQA corpus using SLA Metric.

$NewsQA$	SLA
ISF	35.4%
DWM	**46.4%**

The ISF algorithm used in [15] achieves just 35.4% on sentence level accuracy, leaving space for significant improvement on this harder corpus. We can observe a huge improvement of ~11% using DWM (Table 3), due to the ability of creating matches between words that have different form but similar semantic meaning introduced by the deepest levels of similarity described in Sec. 3.

WikiQA MAP and MRR. Finally we evaluate DWM on WikiQA corpus, using MAP and MRR metrics. In this task we have the opportunity to compare the results obtained by DWM against several other approaches reported in [21]. Those vary from simple word co-occurences (Word Count, Weighted Word Count), to LCLR [22], Paragraph Vectors (PV) [7] and Convolutional Neural Networks (CNN) [24]. PV and CNN are also improved by using word co-occurrences features (PV-Cnt, CNN-Cnt) combined in a logistic regression classifier.

In Table 4 we see that DWM achieves best result for MAP and MRR metrics on the test set of WikiQA corpus. Furthermore DWM doesn't need any specific training, while CNN was specifically trained using the training set of WikiQA corpus.

Table 4. Evaluation on WikiQA corpus using MAP and MRR metrics.

WikiQA	MAP	MRR
Word Cnt	0.4891	0.4924
Wgt Word Cnt	0.5099	0.5132
LCLR	0.5993	0.6086
PV	0.5110	0.5160
CNN	0.6190	0.6281
PV-Cnt	0.5976	0.6058
CNN-Cnt	0.6520	0.6652
DWM	**0.6733**	**0.6856**

5 Conclusions

In conclusion we have proposed a new approach for sentence answer selection (the task of selecting the sentence containing the answer to a specific question) that combines the analysis of sentence words in string form, rich lexical resources and distributed representation of words in order to find the most semantically similar sentence answer to the input question. The approach was evaluated on 3 different corpora, improving significantly upon the baseline in the first two tasks, and achieving best result among several approaches on the third one. Our approach has the advantages that it does not require training time and can be applied to any task and language where common lexical resources are available. Furthermore our approach is unsupervised and does not require any labeled data. This allows to easily adapt this method to new corpora and domains.

Although the DWM algorithm is able to catch with high accuracy the semantic similarity between the question and the corresponding sentence answer, at this point it still cannot understand the syntactic relations between words. In order to address this issue, syntactic dependencies (e.g. subject → verb → object) could be considered in future improvements.

Future work will include the exploration of new features for distributed sentence representation and the extraction of the exact answer text from a sentence answer. We will also test our technique on corpora of different languages and compositional question answering in order to answer questions which have information distributed in different parts of a paragraph.

References

1. Bengio, Y., Ducharme, R., Vincent, P., Jauvin, C.: A neural probabilistic language model. J. Mach. Learn. Res. **3**(6), 1137–1155 (2003)
2. Bordes, A., Chopra, S., Weston, J.: Question answering with subgraph embeddings. CoRR, abs/1406.3676 (2014)

3. Bordes, A., Weston, J., Usunier, N.: Open question answering with weakly supervised embedding models. In: Calders, T., Esposito, F., Hüllermeier, E., Meo, R. (eds.) ECML PKDD 2014. LNCS (LNAI), vol. 8724, pp. 165–180. Springer, Heidelberg (2014b). https://doi.org/10.1007/978-3-662-44848-9_11

4. Collobert, R., Weston, J.: A unified architecture for natural language processing: deep neural networks with multitask learning. In: Proceedings of the 25th International Conference on Machine Learning, ICML 2008. ACM, New York (2008)

5. Fellbaum, C.: WordNet: An Electronic Lexical Database. Bradford Books (1998)

6. Heilman, M., Smith, N.A.: Tree edit models for recognizing textual entailments, paraphrases, and answers to questions. In: Human Language Technologies: The 2010 Annual Conference of the North American Chapter of the Association for Computational Linguistics, HLT 2010. Association for Computational Linguistics, Stroudsburg (2010)

7. Le, Q., Mikolov, T.: Distributed representations of sentences and documents. In: Proceedings of the 31st International Conference on Machine Learning, ICML 2014, vol. 32. JMLR.org (2014)

8. Tomas, M., Chen, K., Corrado, G., Dean, J.: Efficient estimation of word representations in vector space. CoRR, abs/1301.3781 (2013)

9. Moldovan, D., Clark, C., Harabagiu, S., Hodges, D.: Cogex: a semantically and contextually enriched logic prover for question answering. J. Appl. Logic 5(1), 49–69 (2007)

10. Jeffrey, P., Socher, R., Manning, C.D.: Glove: global vectors for word representation. In: Empirical Methods in Natural Language Processing (EMNLP) (2014)

11. Vasin, P., Roth, D., Yih, W.T.: Mapping dependencies trees: an application to question answering. In: Proceedings of the 8th International Symposium on Artificial Intelligence and Mathematics, Fort (2004)

12. Pranav, R., Zhang, J., Lopyrev, K., Liang, P.: Squad: 100, 000+ questions for machine comprehension of text. CoRR, abs/1606.05250 (2016)

13. Shen, D., Lapata, M.: Using semantic roles to improve question answering. In: Proceedings of the 2007 Joint Conference on Empirical Methods in Natural Language Processing and Computational Natural Language Learning (EMNLP-CoNLL). Prague, Czech Republic: Association for Computational Linguistics (2007)

14. Smith, D.A., Jason, E.: Quasi-synchronous grammars: alignment by soft projection of syntactic dependencies. In: Proceedings of the Workshop on Statistical Machine Translation, StatMT 2006. Association for Computational Linguistics, Stroudsburg (2006)

15. Adam, T.: Newsqa: A machine comprehension dataset. CoRR, abs/1611.09830 (2016)

16. Voorhees, E.M., Tice, D.M.: Building a question answering test collection. In: Proceedings of the 23rd Annual International ACM SIGIR Conference on Research and Development in Information Retrieval, SIGIR 2000. ACM, New York (2000)

17. Wang, M., Manning, C.D.: Probabilistic tree-edit models with structured latent variables for textual entailment and question answering. In: Proceedings of the 23rd International Conference on Computational Linguistics, COLING 2010. Association for Computational Linguistics, Stroudsburg (2010)

18. Wang, M., Smith, N.A., Mitamura, T.: What is the jeopardy model? A quasi-synchronous grammar for QA. In: EMNLP-CoNLL, vol. 7 (2007)

19. Winograd, T.: Five lectures on artificial intelligence. Technical report, AI Lab, Stanford University (1974)

20. Woods, W.A.: Progress in natural language understanding: an application to lunar geology. In: Proceedings of the June 4–8, 1973, National Computer Conference and Exposition, AFIPS 1973. ACM, New York (1973)
21. Yi, Y., Yih, S.W., Meek, C.: Wikiqa: a challenge dataset for open-domain question answering. ACL "Association for Computational Linguistics" (2015)
22. Yih, W., Chang, M.-W., Meek, C., Pastusiak, A.: Question answering using enhanced lexical semantic models. In: Proceedings of the 51st Annual Meeting of the Association for Computational Linguistics (Volume 1: Long Papers). Association for Computational Linguistics, Sofia, Bulgaria (2013)
23. Yih, W., He, X., Meek, C.: Semantic parsing for single-relation question answering. In: Proceedings of the 52nd Annual Meeting of the Association for Computational Linguistics (Volume 2: Short Papers). Association for Computational Linguistics, Baltimore (2014)
24. Lei, Y., Hermann, K.M., Blunsom, P., Pulman, S.: Deep learning for answer sentence selection. CoRR, abs/1412.1632 (2014)

On the Contribution of Specific Entity Detection in Comparative Constructions to Automatic Spin Detection in Biomedical Scientific Publications

Anna Koroleva[1,2]([✉]) [iD] and Patrick Paroubek[3]

[1] Institute of Applied Simulation, School of Life Sciences and Facility Management,
Zurich University of Applied Sciences (ZHAW), 8820 Waedenswil, Switzerland
[2] Swiss Institute of Bioinformatics (SIB), 1015 Lausanne, Switzerland
aakorolyova@gmail.com
[3] LIMSI, CNRS, Université Paris-Saclay, 91405 Orsay, France

Abstract. In this article, we address the problem of providing automated aid for the detection of misrepresentation ("spin") of research results in scientific publications from the biomedical domain. Our goal is to identify automatically inadequate claims in medical articles, i.e. claims that present the beneficial effect of the experimental treatment to be greater than it is actually proven by the research results. To this end, we propose a Natural Language Processing (NLP) approach. We first make a review of related work and an NLP analysis of the problem; then we present our first results obtained on the articles that report results of Randomized Controlled Trials (RCTs), i.e. clinical trials comparing two or more interventions by randomly assigning them to patients. Our first experiments concern the identification of entities specific to RCTs (outcomes and patient groups), obtained with basic methods (local grammars) on a corpus extracted from the PubMed open archive. We explore the possibility to extract outcomes from comparative constructions that are commonly used to report results of clinical trials. Our second set of experiments consists in extracting outcomes from a manually annotated corpus using deep learning methods.

Keywords: Entity extraction · Biomedical text processing · Comparative constructions · Outcome extraction

---

This project has received funding from the European Union's Horizon 2020 research and innovation programme under the Marie Sklodowska-Curie grant agreement No 676207.
At the time of the reported work, Anna Koroleva was a PhD student at LIMSI-CNRS in Orsay, France and at the Academic Medical Center, University of Amsterdam in Amsterdam, the Netherlands.

Z. Vetulani et al. (Eds.): LTC 2017, LNAI 12598, pp. 304–317, 2020.
https://doi.org/10.1007/978-3-030-66527-2_22

1   Introduction

Interpretation of results in research publications is often affected by the presence of spin[1], i.e. beautifying the observed results. Spin can be introduced intentionally or unintentionally, as a result of the desire to prove the importance of one's research or from the lack of knowledge of the proper reporting practices. We address the problem of spin without any consideration of the presence or absence of the author's intention to mislead readers, judging only from the text of an article.

We are targeting spin the biomedical domain, namely in articles reporting Randomized Controlled Trials (RCTs) - clinical trials comparing two or more interventions by randomly allocating them to patients. In RCTs, spin consists in claiming that the treatment under study had a positive effect greater than the trial showed. Table 1 presents some examples of spin in RCTs [3,4,31].

Table 1. Examples of abstracts containing inappropriate claims and their version without spin rewritten by domain experts so that conclusions correspond to the actual research results (provided by I. Boutron from [3])

Abstract with inappropriate claims	Abstract rewritten without spin
Treatment A **may be useful** in controlling cancer-related fatigue **in patients who present with severe fatigue**	Treatment A **was not more effective than placebo** in controlling cancer-related fatigue.
This study demonstrated **improved PFS and response** for the treatment A compared with comparator B alone, although this did not result in improved survival	The treatment A was **not more effective** than comparator B on overall survival in patients with metastatic breast cancer previously treated with anthracycline and taxanes

Spin is more frequent in abstracts than in bodies of scientific articles [4]. Boutron et al. [3] studied articles reporting RCTs in oncology and showed that spin in abstracts influences the way doctors interpret the effects of the treatment studied, making them overestimate its efficacy. Spin can thus influence clinical decisions, causing a serious threat for the healthcare system. Furthermore, it is a common situation that only the abstract of an article is freely available, which aggravates the impact of spin in abstracts on the dissemination of research results.

This considerations lead to the conclusion that it is time to start investigating how to help scientific authors, reviewers and editors to identify probable occurrences of spin. We think that Natural Language Processing (NLP) and Machine Learning (ML) can provide helpful solutions for spin identification, both when writing or reading an article. The elaboration of such a solution will require, as a first step, building a corpus with appropriate annotations to model spin.

This article contains two main contributions.

---

[1] The term find its origins in the term "spin doctors", communication agents of public personalities particularly deft at improving the image of their clients.

First, we present a preliminary feasibility study for some basic NLP functionalities required by automatic spin detection: entity extraction of two main spin-related concepts (patient groups and outcomes) from comparative constructions, that are frequently used to present results of RCTs. The aim in this first phase is to gauge which types of phenomena are amenable to automatic processing with sufficient reliability and to assess their relative frequency in scientific publications. This step was aimed to support corpus building and annotation in the following second phase, before addressing the realization of spin detection algorithm in the final third phase.

Second, we introduce an corpus of sentences from abstracts of articles reporting RCTs, manually annotated with reported outcomes. We present the evaluation of the rule-based methods on this corpus and we report our experiments on extracting the outcomes using deep learning methods. We show that extracting reported outcomes from comparative sentences only, as we planned at the first step, is not sufficient, as outcomes occur in highly varying contexts. Our results suggest that rule-based methods are not efficient for extracting this type of entity, while deep learning methods achieve better results.

The structure of this paper is as follows. In Sect. 2 we present our current model of spin in the biomedical literature and address linguistic characteristics of spin with the corresponding kinds of language processing required. In Sect. 3, we present previous works related to our task. In Sect. 4 we present our entity extraction experiments (rule-based and deep learning) and corpus analysis which precedes our conclusion.

2 A Model of Spin

Previous studies proved that several kinds of medical trials are subject to spin (such as randomized control trials (RCTs) or diagnostic accuracy studies) and identified three main categories of spin, frequency of which varies in function of the trial type [3,4,31]:

1. Inappropriate **presentation** of research results, which declines into:
   - Negative effects of a treatment are not presented.
   - Some of the results are not evaluated, e.g. the primary outcome is not presented while the focus is put on significant secondary outcomes.
   - The presentation of the type of trial and its characteristics is incomplete or incorrect.
   - The description of the population studied is fuzzy, the focus is put on particular subgroups for which the results are statistically significant.
   - Linguistic spin (excessive use of positive comparisons or superlatives).
   - The limitations of the trial are not presented
   - Previous studies are partially cited (important articles are missing)
2. Inappropriate **interpretation** of research results, which may take the following forms:
   - The studied treatment is claimed to have a positive effect or an effect equivalent to the standard treatment in spite of non-significant results.

- The treatment is presented as safe while the results for safety are not significant.
- The treatment is presented as having positive effects without any comparative trial performed.
- Only the statistical significance is considered instead of the clinical pertinence.

3. Inappropriate **extrapolation**, which include:
   - Instead of the population, treatment or result evaluated, the author presents a different population, treatment or result.
   - The conclusions are inappropriate for clinical practice, for instance an advice to use the treatment not substantiated by sufficient evidences

We focus on spin in RCTs because they are the main source of information for Evidence-Based Medicine (EBM).

Spin is a complex phenomenon with heterogeneous aspects concerning syntax, semantics and pragmatics knowledge as well as inference. We focus here on the types of spin identifiable using only the text of the article, without using extra sources of information such as research protocol. From the listed types of spin we can deduce the following NLP functionalities required for automatic spin identification:

1. classification of biomedical articles according to the type of trial (up to now we have addressed the distinction between RCTs vs. other types);
2. extracting the treatment evaluation (positive/neutral/negative);
3. analysis of document structure (title/abstract/body);
4. entity extraction: studied outcomes (primary and secondary), population (with patients groups), statistical significance of results, trial restrictions, negative effects, treatments compared;
5. extracting the relations between the entities (e.g. between the results and their statistical significance);
6. paraphrase identification for comparing entity mentions from the abstract with those from the body;
7. syntactic analysis: identifying spin specific constructions, for instance concessive propositions often associated with a focus change:
   "This study demonstrates improved PFS and response for the treatment A compared with comparator B, **although** this did not result in improved survival" (*focus on secondary results*).

3 Previous Works

In this section we report on previous works addressing: (1) bias assessment (a task linked to spin detection), (2) entity and relation extraction and (3) comparatives.

3.1 Bias Assessment

In biomedical domain, the task closest to spin detection is bias assessment. Bias is defined as a systematic error or deviation with respect to truth in results or conclusions which may lead to under- or overestimation of the effect of the treatment evaluated [13]. Errors can concern study design, research implementation or analysis/presentation of results. The types of bias include: selection bias (generation of the random sequence, masking treatment assignment); performance bias (blind allocation of treatment to patient); detection bias (anonymizing results evaluation); attrition bias (incomplete data about outcomes); reporting bias (selective presentation of outcomes). We have a special interest in reporting bias because it falls under the definition of spin.

According to Higgins et al. [14], the evaluation of bias is more often done by experts who use scales or checklists than with an NLP approach. Examples of the latter are found in Marshall et al. [20]. The authors describe a corpus of systematic reviews archived by the Cochrane network[2]. The bias evaluation from the systematic reviews is used as gold standard annotation. Bias evaluation is subjective; the authors report that discrepancies are the largest for assessing reporting bias. An SVM based on words is used to classify articles according to their bias level.

The difference between bias assessment and spin detection is that the former matches the research protocol with the article, while for spin detection the abstract is compared to the body text.

3.2 Entities and Relations Extraction

In the biomedical domain the research on entity extraction deals mainly with gene, protein and medication names extraction, less attention was given to the entities which interest us [27]. The entities we address encompass specific Named Entities (e.g. medication brands etc.) and all nominal phrases associated to particular semantic roles in the description of an RCT, such as outcomes, patients groups, statistical indicators (p-value), cf. [15,21].

Entities representing characteristics of a trial or clinical situation have been addressed because they are od interest for many NLP tasks: summarization [26], systematic review elaboration[3], decision making aids, question answering systems, databases creation and querying. Different approaches have been used depending on the task addressed. For systematic reviews, at least four basic elements of a clinical study (known as "PICO framework") need to be identified: (1) Population/Problem, (2) "Intervention" (treatment), (3) Comparator treatment and (4) Outcome. This type of analysis does not always rely on entity extraction per se, since it is often enough to identify the sentences which contain

---

[2] Cochrane is an independent international network of researchers, health professionals and patients whose aim is to improve decision making in health care (http://www.cochrane.org).

[3] A systematic review is a type of scientific articles aimed at an exhaustive summary of the literature about a particular problem with statistical evaluation of the results.

the needed information [28]. With a focus on automatic summarization and with the limitation of using a small corpus of 20 abstracts, Dawes et al. [7] looked at a larger set of elements: patient-population-problem, exposure-intervention, comparison, outcome, duration and results and their associated co-texts. De Bruijn et al. [5] further extended the set of addressed elements to all the elements from the CONSORT statement (http://www.consort-statement.org) which contains among others: patients eligibility criteria, the treatment studied and the comparator treatment, the intervention parameters (dosing, frequency, etc.), financing information, publication metadata, etc. This system differs from the majority of other similar systems because it works on the entirety of an article, not only on its abstract. The approach has been further developed in Kiritchenko et al. [15], resulting in the ExaCT system.

Summerscales et al. [27] try to compute automatically summary statistics (reduction of absolute risk, number of patients to treat) from the articles on RCTs using Conditional Random Fields. They show that outcomes are among the most difficult elements to extract. Chung [6] addresses the extraction of the "branches" of an intervention (i.e. the application of the studied treatment and of the comparator treatment respectively) from coordinated constructions (using a maximum entropy classifier, a parser and the UMLS[4] accessed through the MetaMap application[5]) present in the methods section of the abstracts and the objectives, results and conclusion sections where the information is often explicitly mentioned. Other research looked at information about patient population: [29] extract the description of the population, the number of patients examined and the description of symptoms/diseases from RCTs using a Hidden Markov approach and parsing.

From this state of the art, we conclude that most of the research is focused on RCTs and work mainly with abstracts. A two-step approach is commonly used: first classifying the sentences then extracting the entities from the selected sentences, using a combination of Machine Learning and symbolic rule based algorithms. Systems like MetaMap are often used to match the article contents with the UMLS. The definition of the exact limits of an entity remains a difficult task; among various entities, the outcome is the most difficult to identify.

3.3 Comparatives

In NLP, comparative constructions have been studied early and mainly for English, e.g. [1,9,25] or [22]. Li et al. [19] extracted comparable entities from a corpus of questions using a minimal bootstrap approach. Hatzivassiloglou and Wiebe [12] focuses on gradable adjectives as subjectivity markers (defining a measure of gradability). Ganapathibhotla and Liu [10] mines opinions expressed about entities from comparative sentences, trying to determine which entity in a comparison is preferred by its author. More recently [30] addressed comparatives in Korean and extracted comparatives and comparisons between entities

---

[4] UMLS (Unified Medical Language System) is a compendium of several medical controlled vocabularies, https://www.nlm.nih.gov/research/um.

[5] https://metamap.nlm.nih.gov/.

from a corpus of questions. [11] addressed extracting compared entities and compared features (the feature with respect to which the entities are compared) from biomedical texts.

Comparative constructions can be divided into three types: morpho-lexical (e.g. with adjectives in comparative or superlative form), syntactic (with patterns like "as ADJ1 as ADJ2") or semantic (e.g. verbs or nouns indicating a change of state like "improved" or "improvement" comparing implicitly the current state of affairs with what it was in the past).

4   First Set of Experiments: A Rule-Based Approach

4.1   Entity Extraction

In this first set of experiments we used an approach combining manual exploration of the corpus and finite state automata (Unitex environment [23]) filtering in a bootstrapping approach alternatively relying on the phenomena targeted and on their co-text. The most important information for spin detection is the outcome. We provide here two examples of the outcome identification results performed with 9 Unitex graphs and using the following markup: PROL for the outcome marker and OUT for the mention of the OUTCOME:

The <PROL>primary outcome was < /PROL> <OUT type=PRIM> the remission of depressive symptoms at the 2-month follow up visit < /OUT>, defined as a HDRS score of 7 or less. <PROL> Secondary outcome parameters are < /PROL> <OUT type=SEC> overall mortality, severity of BPD, number of days on the ventilator, number of treatment failures, ventilation-induced lung injury and pulmonary hypertension < /OUT> according to clinical parameters.

From a subcorpus of 3,938 articles on RCTs from PubMed Central[6], we have identified 6,292 outcome occurrences.

4.2   Comparatives

For spin detection in RCTs, comparative constructions are of high interest because the main goal of RCTs is comparing two or more treatments with respect to a number of outcomes, and thus the results are often presented in the form of a comparative sentence.

Our first goal is to identify comparative sentences that state superiority of the experimental treatment over the control treatment, similarity between the two treatments, or some positive changes occurred under the experimental treatment. These sentences are considered as containing positive evaluation of the experimental treatment.

Our second goal is to extract the components of a comparison including a comparative word (such as "better"), compared entities and compared features [10]. In RCT reports, compared entities belong most often to one of three

---

[6] https://www.ncbi.nlm.nih.gov/pmc/.

types: compared treatments (example 1); patient groups that received the treatments (example 2); value of an outcome before and after a treatment (example 3). In first and second cases, the compared feature is typically an outcome, as "efficacy" in example 1 and "response rate" in example 2.

1. Treatment A was better that treatment B in terms of efficacy.
2. The group receiving treatment A showed better response rate than the group receiving treatment B.
3. PANSS score improved with treatment A.

The novelty of our work is that our aim is not only to extract the compared entities and features, but also to detect their type (treatment, patient group, outcome).

We performed our experiments on a corpus of 3934 abstracts of articles from Pubmed and 5005 abstracts of articles from Cochrane Schizophrenia group database.

We proceeded in two steps: 1) we collected concordances for a set of words (verbs and nouns with the semantics of change of a parameter such as "improve"/"improvement") and constructions (such as comparative adjective/adverb + "than"/"versus", etc.) that are likely to convey a comparative meaning; 2) we studied the concordances to identify typical ways of expressing different types of components of a comparison. Our analysis shows that each type of components is associated with a set of morphological, lexical, morphosyntactic and syntactic features of a phrase. Each feature may be associated with several types of components, but a whole set of features can determine the type with sufficient accuracy. For example, a group of preposition "in" may represent patient groups ("... improved **in aged subjects**") or outcomes ("improvement **in PANSS total score**"). Treatments may be subjects of the active transitive verbs (with an outcome as the direct object) or occur within groups of prepositions "by"/"with"/"on"/"over". Outcomes may be active or passive subjects, direct objects. Patient groups mentions have additional lexical feature: words "group", "population" or words denoting humans such as "patients" etc. Words "medication", "agent", etc. and suffixes "-one", "-ine" etc. are associated with treatments. Outcomes in general have fewer lexical and morphological features compared to patients and treatments.

We created a set of finite-state automata to extract the comparison components (noun and prepositional phrases) and detect their type using the above-mentioned features.

Our data show that verbs and nouns denoting a change occur more often that constructions with comparative adjectives/adverbs (21721 vs. 3660 occurrences), so the first set of experiments concerns these verbs and nouns. Extraction of components from constructions with adjectives/adverbs, as well as from statements of similarity, is our future work.

In the Table 2 we provide a preliminary statistics for constructions with outcome as a component of a comparison (where Out is outcome, Int is intervention). More precise evaluation is still work-in-progress.

5   Second Set of Experiments: A Deep Learning Approach

Our main focus in the second set of experiments in extracting reported outcomes. Outcomes are any variables measured during a clinical trial to assess the impact of the studied treatment on the patient population. We define reported outcomes as the mentions of outcomes that occur in contexts that report the results for the outcomes, e.g. (outcomes are in bold):

The HRQoL was higher in the experimental group.

The mean incremental QALY of intervention was 0.132 (95% CI: 0.104–0.286).

Table 2. Comparative constructions with outcomes

Type of components	Pattern	N of occurrences/percentage	Example
Outcome	noun.change + PP(in/on/of + OUT)	1991/29,9%	Reductions <Out> in plasma estrogen levels < /Out> and increases <Out> in bone-resorption < /Out> markers were comparable in both groups.
Intervention + Outcome	INT.subj + verb.change.active + OUT.obj	711/10,7%	<Int> Adjunctive treatment with galantamine < /Int> improves <Out> memory and attention < /Out> in patients with schizophrenia.
Outcome	OUT.subj + verb.change.active/ passive	1336/20%	<Out>HRQL< /Out> improves after successful treatment.
Outcome	Verb.change.prtcp + OUT	2306/34,6%	Three sessions of education led to significantly increased <Out> insight < /Out>.
Outcome	OUT = NP(noun + gain/elevation)	315/4,7%	Clozapine treatment is associated with <Out>weight gain < /Out>

Our annotated corpus and the experiments on extracting outcomes are presented in detail elsewhere [16]. Here we briefly summarize the corpus description and the experiments.

5.1   Manually Annotated Corpus

At the time of the beginning of our work, we were not able to identify a publicly available corpus with annotation of reported outcomes. Hence, we created our own corpus with the required annotation.

Our corpus consists of sentences from the abstracts of a dataset of 3,938 PMC[7] articles that have the PubMed publication type "Randomized controlled trial". It was infeasible to run a large-scale annotation project with several expert annotators, hence the annotation was performed by AK with guidance by domain experts. We developed an annotation tool [17] for the sake of simplicity, ease of format conversion and customizing. The final annotation uses a CoNLL-like representation scheme with B (begin) - I (inside) - O (outside) elements.

The final corpus contains 1,940 sentences from 402 articles. A total of 2,251 reported outcomes was annotated.

The analysis of the corpus shows that the ways of reporting outcomes are highly varied, with the same outcome being reported in different ways. Consider the following sentences:

Mean total nutrition knowledge score increased by 1.1 in intervention (baseline to follow-up : 28.3 to 29.2) and 0.3 in control schools (27.3 to 27.6).

*The increase in **mean total nutrition knowledge score** was 1.1 in intervention (baseline to follow-up : 28.3 to 29.2) and 0.3 in control schools (27.3 to 27.6).*

Both sentences report the same outcome but differ in their structure. In the second sentence, all the variants *"The increase in mean total nutrition knowledge score"*, *"mean total nutrition knowledge score"* and *"total nutrition knowledge score"* can be considered as the reported outcome. To preserve uniformity throughout the annotation of our corpus, we decided to annotate the smallest possible text fragment referring to an outcome (*"total nutrition knowledge score"* for the given example), as in this case the same outcome is annotated in all the variants of sentences.

Reported outcomes are highly diverse in terms of their syntactic characteristics. They can be represented by a noun phrase, a verb phrase or an adjective, cf. the examples:

1. *Overall **response rate** was 39.1% and 33.3% in 3-weekly and weekly arms.*
2. *No patients were **reintubated**.*
3. *The CSOM and MA appeared less **responsive** following a GLM-diet.*

These examples show that our initial hypothesis that outcome extraction can be limited to comparative sentences is wrong: outcomes are frequently reported in other types of statements (cf. examples 1 and 2 that do not contain any explicit comparison).

5.2 Deep Learning Algorithms

Pre-trained language models such as ELMO [24] and Google's BERT [8] have become very successful in several downstream NLP tasks, including named entity recognition. We adopted the fine-tuning approach, consisting in adjusting the pre-trained model parameters on a downstream task (outcome extraction).

---

[7] https://www.ncbi.nlm.nih.gov/pmc/.

We compared several pre-trained language models:

1. BERT (Bidirectional Encoder Representations from Transformers) models[8]. BERT provides cased and uncased models, pre-trained on general-domain texts.
2. BioBERT [18], a domain-specific analogue of BERT, pre-trained on a large (18B words) biomedical corpus in addition to BERT training data. BioBERT is based on the cased BERT model.
3. SciBERT [2], a domain-specific analogue of BERT, pre-trained on a corpus of scientific texts (3.1B) added to BERT training data. SciBERT provides both cased and uncased models, with two versions of vocabulary: BaseVocab (the initial BERT general-domain vocabulary) and SciVocab (the vocabulary built on the scientific corpus).

We evaluated the following models: BERT-base cased and uncased; BioBERT model; SciBERT cased and uncased models with the SciVocab vocabulary.

We explored two approached of using the BERT-based models. The first approach [8,18] consists in a simple fine-tuning of the models on an annotated dataset. The second approach [2] employs a minimal task-specific architecture on top of BERT-based embeddings. A representation of each token is a concatenation of its BERT embedding and a CNN-based character embedding. A multilayer bi-LSTM is applied to token embeddings, and a CRF is used on top of the bi-LSTM.

We report here the evaluation of the performance of all the models, along with the baseline rule-based approach (described in the previous sections). The algorithms were evaluated on the token level. Machine-learning algorithms were

Table 3. Reported outcome extraction: results

Algorithm	Precision	Recall	F1
SciBERT-uncased	81.17	78.09	79.42
BioBERT	79.61	77.98	78.6
SciBERT-cased	79.6	77.65	78.38
BERT-uncased	78.98	74.96	76.7
BERT-cased	76.63	74.25	75.18
SciBERT-uncased_biLSTM-CRF	68.44	73.47	70.77
BioBERT_biLSTM-CRF	70.18	71.43	70.63
SciBERT-cased_biLSTM-CRF	67.98	72.52	70.11
BERT-cased_biLSTM-CRF	65.98	65.54	65.64
BERT-uncased_biLSTM-CRF	64.6	66.73	65.4
Rule-based	26.69	55.73	36.09

---

[8] https://github.com/google-research/bert.

assessed using 10-fold cross-validation (train-dev-test split was done in proportion 8:1:1). We report the averaged results. We used Tensorflow for our experiments. Table 3 presents the results of outcome extraction.

6   Conclusion

In this paper, we presented our first approach to extracting reported trial outcomes from comparative constructions using a system of rules. We introduced an annotated corpus of reported outcomes and showed that the outcome extraction cannot be limited to comparative sentences only, as outcomes occur in highly varied contexts. We presented a deep learning approach to outcome extraction and compared several deep learning methods. All the deep learning models outperformed the rule-based approach, the best performance was shown by the fine-tuned SciBERT uncased model.

The developed outcome extraction model is a part of our spin detection pipeline.

References

1. Ballard, B.W.: A general computational treatment of comparatives for natural language question answering. In: Proceedings of the 26th Annual Meeting of the Association for Computational Linguistics, pp. 41–48. Association for Computational Linguistics, Buffalo (1988). https://doi.org/10.3115/982023.982029, http://www.aclweb.org/anthology/P88-1006
2. Beltagy, I., Cohan, A., Lo, K.: Scibert: Pretrained contextualized embeddings for scientific text. arXiv preprint arXiv:1903.10676 (2019)
3. Boutron, I., Altman, D., Hopewell, S., Vera-Badillo, F., Tannock, I., Ravaud, P.: Impact of spin in the abstracts of articles reporting results of randomized controlled trials in the field of cancer: the SPIIN randomized controlled trial. Journal of Clinical Oncology (2014)
4. Boutron, I., Dutton, S., Ravaud, P., Altman, D.: Reporting and interpretation of randomized controlled trials with statistically nonsignificant results for primary outcomes. JAMA **303**, 2058–2064 (2010)
5. Bruijn, B.D., Carini, S., Kiritchenko, S., Martin, J., Sim, I.: Automated information extraction of key trial design elements from clinical trial publications. In: Proceedings of the AMIA Annual Symposium (2008)
6. Chung, G.Y.C.: Towards identifying intervention arms in randomized controlled trials: Extracting coordinating constructions. J. Biomed. Inf. **42**(5), 790–800 (2009). https://doi.org/10.1016/j.jbi.2008.12.011. http://www.sciencedirect.com/science/article/pii/S1532046408001573
7. Dawes, M., Pluye, P., Shea, L., Grad, R., Greenberg, A., Nie, J.Y.: The identification of clinically important elements within medical journal abstracts: Patient-population-problem, exposure-intervention, comparison, outcome, duration and results (PECODR). J. Innov. Health Inf. **15**(1), 9–16 (2007). https://doi.org/10.14236/jhi.v15i1.640. https://hijournal.bcs.org/index.php/jhi/article/view/640
8. Devlin, J., Chang, M., Lee, K., Toutanova, K.: BERT: pre-training of deep bidirectional transformers for language understanding. CoRR abs/1810.04805 (2018). http://arxiv.org/abs/1810.04805

9. Friedman, C.: A general computational treatment of the comparative. In: 27th Annual Meeting of the Association for Computational Linguistics (1989). http://aclanthology.coli.uni-saarland.de/pdf/P/P89/P89-1020.pdf
10. Ganapathibhotla, M., Liu, B.: Mining opinions in comparative sentences. In: Proceedings of the 22nd International Conference on Computational Linguistics (Coling 2008), pp. 241–248. Coling 2008 Organizing Committee (2008). http://aclanthology.coli.uni-saarland.de/pdf/C/C08/C08-1031.pdf
11. Gupta, S., Mahmood, A.S.M.A., Ross, K.E., Wu, C.H., Vijay-Shanker, K.: Identifying comparative structures in biomedical text. In: Proceedings of the BioNLP 2017 Workshop, pp. 206–215 (2017)
12. Hatzivassiloglou, V., Wiebe, J.M.: Effects of adjective orientation and gradability on sentence subjectivity. In: COLING 2000 Volume 1: The 18th International Conference on Computational Linguistics (2000). http://www.aclweb.org/anthology/C00-1044
13. Higgins, J.P., Green, S. (eds.): Cochrane Handbook for Systematic Reviews of Interventions. Wiley, West Sussex (2008)
14. Higgins, J.P.T., et al.: The Cochrane collaboration's tool for assessing risk of bias in randomised trials. BMJ **343**, d5928 (2011). https://doi.org/10.1136/bmj.d5928. https://www.bmj.com/content/343/bmj.d5928
15. Kiritchenko, S., Bruijn, B.D., Carini, S., Martin, J., Sim, I.: Exact: automatic extraction of clinical trial characteristics from journal publications. BMC Med. Inf. Decis. Mak. **10**, 56 (2010). https://doi.org/10.1186/1472-6947-10-56
16. Koroleva, A., Kamath, S., Paroubek, P.: Extracting outcomes from articles reporting randomized controlled trials using pre-trained deep language representations. EasyChair Preprint no. 2940 (EasyChair, 2020)
17. Koroleva, A., Paroubek, P.: Demonstrating construkt, a text annotation toolkit for generalized linguistic contructions applied to communication spin. In: The 9th Language and Technology Conference (LTC 2019) Demo Session (2019)
18. Lee, J., Yoon, W., Kim, S., Kim, D., Kim, S., So, C.H., Kang, J.: Biobert: a pre-trained biomedical language representation model for biomedical text mining. arXiv preprint arXiv:1901.08746 (2019)
19. Li, S., Lin, C.Y., Song, Y.I., Li, Z.: Comparable entity mining from comparative questions. In: Proceedings of the 48th Annual Meeting of the Association for Computational Linguistics, pp. 650–658. Association for Computational Linguistics, Uppsala, Sweden, July 2010. http://www.aclweb.org/anthology/P10-1067
20. Marshall, I.J., Kuiper, J., Wallace, B.C.: Robotreviewer: evaluation of a system for automatically assessing bias in clinical trials. J. Am. Med. Inf. Assoc. JAMIA **23**, 193–201 (2015). https://doi.org/10.1093/jamia/ocv044
21. Nguyen, N., Miwa, M., Tsuruoka, Y., Tojo, S.: Open information extraction from biomedical literature using predicate-argument structure patterns. In: Proceedings of the 5th International Symposium on Languages in Biology and Medicine, pp. 51–55, December 2013
22. Olawsky, D.E.: The lexical semantics of comparative expressions in a multi-level semantic processor. In: 27th Annual Meeting of the Association for Computational Linguistics (1989). http://aclanthology.coli.uni-saarland.de/pdf/P/P89/P89-1021.pdf
23. Paumier, S.: Unitex 3.1 user manual (2016). http://unitexgramlab.org/releases/3.1/man/Unitex-GramLab-3.1-usermanual-en.pdf

24. Peters, M., et al.: Deep contextualized word representations. In: Proceedings of the 2018 Conference of the North American Chapter of the Association for Computational Linguistics: Human Language Technologies, vol. 1 (Long Papers) (2018). https://doi.org/10.18653/v1/n18-1202

25. Ryan, K.: Corepresentational grammar and parsing English comparatives. In: Proceedings of the 19th Annual Meeting of the Association for Computational Linguistics, pp. 13–18. Association for Computational Linguistics, Stanford, California, USA, June 1981. https://doi.org/10.3115/981923.981927, http://www.aclweb.org/anthology/P81-1003

26. Summerscales, R., Argamon, S., Hupert, J., Schwartz, A.: Identifying treatments, groups, and outcomes in medical abstracts. In: Proceedings of the Sixth Midwest Computational Linguistics Colloquium (MCLC) (2009)

27. Summerscales, R.L., Argamon, S.E., Bai, S., Hupert, J., Schwartz, A.: Automatic summarization of results from clinical trials. In: 2011 IEEE International Conference on Bioinformatics and Biomedicine, pp. 372–377 (2011)

28. Wallace, B.C., Kuiper, J., Sharma, A., Zhu, M., Marshall, I.J.: Extracting PICO sentences from clinical trial reports using supervised distant supervision. J. Mach. Learn. Res. **17**(1), 4572–4596 (2016). http://dl.acm.org/citation.cfm?id=2946645.3007085

29. Xu, R., Garten, Y., Supekar, K., Das, A., Altman, R., Garber, A.: Extracting subject demographic information from abstracts of randomized clinical trial reports. Stud. Health Technol. Inf. **129**, 550–4 (2007). https://doi.org/10.3233/978-1-58603-774-1-550

30. Yang, S., Ko, Y.: Extracting comparative entities and predicates from texts using comparative type classification. In: Proceedings of the 49th Annual Meeting of the Association for Computational Linguistics: Human Language Technologies, pp. 1636–1644. Association for Computational Linguistics (2011). http://aclanthology.coli.uni-saarland.de/pdf/P/P11/P11-1164.pdf

31. Yavchitz, A., et al.: A new classification of spin in systematic reviews and meta-analyses was developed and ranked according to the severity. J. Clin. Epidemiol **75**, 56–65 (2016)

Automatic Taxonomy Generation: A Use-Case in the Legal Domain

Cécile Robin[1](✉)(iD), James O'Neill[2](✉)(iD), and Paul Buitelaar[1](✉)(iD)

[1] Insight SFI Research Centre for Data Analytics, Data Science Institute,
National University of Ireland Galway, Galway, Ireland
{cecile.robin,paul.buitelaar}@insight-centre.org
[2] University of Liverpool, Liverpool, UK
james.o-neill@liverpool.ac.uk

Abstract. A key challenge in the legal domain is the adaptation and representation of the legal knowledge expressed through texts, in order for legal practitioners and researchers to access this information more easily and faster to help with compliance related issues. One way to approach this goal is in the form of a taxonomy of legal concepts. While this task usually requires a manual construction of terms and their relations by domain experts, this paper describes a methodology to automatically generate a taxonomy of legal noun concepts. We apply and compare two approaches on a corpus consisting of statutory instruments for UK, Wales, Scotland and Northern Ireland laws.

Keywords: Taxonomy generation · Legal domain · Hierarchical embedding clustering · Topic hierarchy

1 Introduction

A quicker understanding and comprehension of legal documents is an imperative for practitioners in the legal sector, who have witnessed a steep increase in legislation since the financial crisis in 2008. This can result in law containing even more ambiguous and complex expressions, which can subsequently lead to damaging non-compliance problems for financial institutions. A fundamental way of arranging the knowledge in legal texts to mitigate such problems is by representing the domain in the form of a taxonomy of legal concepts.

Our proposed approach tackles this issue through the automatic construction of a legal taxonomy, directly extracted from the content of the corpus of legal texts analysed. The idea here is to be able to create a classification based on the field of application of any type of legal documents, and facilitating the maintenance of the versions. This approach would help to track changes in regulations and to keep up-to-date with new ones, making this information easily searchable and browsable.

We compare two systems for automatic taxonomy generation applied to a small corpus of legal documents. First, we provide related work on automatic

© Springer Nature Switzerland AG 2020
Z. Vetulani et al. (Eds.): LTC 2017, LNAI 12598, pp. 318–328, 2020.
https://doi.org/10.1007/978-3-030-66527-2_23

taxonomy generation in general, and in the legal domain in particular. We then describe the two approaches chosen for our study. Next, we examine the experiments performed with both systems on a subset corpus of the UK Statutory Instruments, providing a comparative analysis of the results, before providing suggestions for future work.

2   Related Work

Generic Domain Approaches. Taxonomy construction is a relatively unexplored area, however [5] organised a related task in SemEval-2016: TExEval, where the aim was to connect given domain-specific terms in a hyperonym-hyponym manner (relation discovery), and to construct a directed acyclic graph out of it (taxonomy construction). Only one out of the 6 teams produced a taxonomy, focusing thus more on the relation discovery step. Most systems relied on WordNet [7] and Wikipedia resources.

[12] did a structured review of all the main types of approaches involved in the task of automatic taxonomy construction. It includes the use of WordNet, Natural Language Processing (NLP) techniques, tags from Web resources, or large external corpora. However, WordNet is a generic lexical resource and is not fitted for the legal language whose definitions and semantic relations are very specific to the domain, as well as constantly evolving. As for external annotated data, these are often non available and also non dynamic resources, therefore not well suited for our task.

[1] use Dynamic Hierarchical Dirichlet Process to track topics over time, documents can be exchanged however the ordering is intact. They also applied this to longitudinal *Neural Information Processing Systems* (NIPS) papers to track emerging and decaying topics (worth noting for tracking changing topics around compliance issues).

[11] described a semi-supervised method for constructing an *is-a* type relationship (i.e hypernym-hyponym relation) that uses *Global Vectors for Word Representation* (GloVe) vectors trained on a Wikipedia corpus. The approach attempts to represent these relations by computing an average offset for a set of 200 hypernym-hyponym vector pairs (sampled from *WordNet*). This offset distance is then added to each term so that hypernym-hyponyms relations could be identified outside of the 200 pairs which are averaged.

In the Legal Domain. Most work on taxonomy generation in the legal domain has involved manual construction of concept hierarchies by legal experts [6]. This task, besides being both tedious and costly in terms of time and qualified human resources, is also not easily adaptable to changes. Systems for automatic legal-domain taxonomy creation have on the contrary received very low attention so far. Only [2] worked on a similar task, and developed a machine learning-based system for scalable document classification. They constructed a hierarchical topic schemes of areas of laws and used proprietary methods of scoring and ranking to classify documents. However, this work has been deposited as a patent and is not freely available.

We will now introduce our two chosen methodologies, based on NLP and clustering techniques.

3   Automatic Taxonomy Construction

This section describes the two presented bottom-up approaches to taxonomy generation. We begin with an overview of *Hierarchical Embedding Clustering.*

3.1   Hierarchical Embedding Clustering

Hierarchical Embedding Clustering (HEC) is an agglomerative clustering method that we have used for encoding noun phrase predict vectors (i.e Skipgram trained vectors). We first identify noun phrases in the text by extracting n-grams and retaining only the pairs that contain nouns, determined by the NLTK Maximum Entropy PoS tagger[1]. This is followed by a filtering stage, whereby the top 5377 noun phrases are chosen, based on the highest Pointwise Mutual Information (PMI) scores within a range chosen through a distributional analysis as shown in Fig. 1. In this figure we present the scaled probability distribution (10^2) between noun phrase counts in the sample range [10–100]. The dashed line indicates the density, showing that most probability density is lying within the range [10–60]. This is a well established trend known as *Luhn's law* [10]. Thus, we choose a filtering range between 10–150 to allow for good coverage with still some degree of specificity, resulting in 5377 filtered words and phrases.

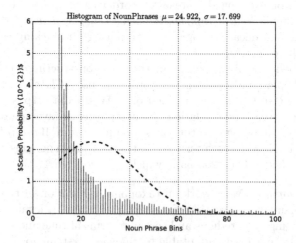

Fig. 1. Noun phrase distribution

Once the noun phrases are selected, we obtain embedded vectors. Each word within a noun phrase is averaged column-wise, therefore representing a whole

---

[1] https://textblob.readthedocs.io/en/dev/_modules/textblob/classifiers.html.

noun phrase as a single vector. This was carried out using both the corpus trained legal vector representations and the large scale pretrained vectors provided by GoogleNews[2]. However, we find that since the legal corpus was relatively small in comparison to Google vectors, it did not achieve the same coherency in grouping noun phrases in a hierarchical structure. Therefore, we focus only on results provided by Google's pretrained vectors. HEC is a bottom-up approach for creating taxonomies in the sense that each noun phrase is considered as its own cluster at the leaves, which are then incrementally merged until we arrive at the root.

3.2 Topical Hierarchy Generation

Saffron is a software tool[3] that aims to automatically construct a domain-specific topic hierarchy using domain modeling, term extraction and taxonomy construction.

Domain Modeling. In order to define the domain of expertise of the corpus, *Saffron* first builds a domain model, i.e. a vector of words representing the highest level of generality in this specific domain [3]. Candidate terms are first extracted using feature selection: giving more weight on part-of-speech carrying meaning, and selecting single words (for genericity) represented in at least a 1/4 of the corpus (for enough specificity to the domain). In order to filter the candidate words, [3] evaluates the coherence of a term within the domain based on [8]'s work on topic coherency, following the assumption that domain terms are more general when related to many specific ones. The domain model created is then used in the next phase for the extraction of topics which will make up the taxonomy.

Term Extraction. In the topic extraction phase, intermediate level terms of the domain are sought (as defined in [4]). It involves two approaches: one looking for domain model words in the context of the candidate terms (within a defined span size), and the second using the domain model as a base to measure the lexical coherence of terms by PMI calculation. At the end of this phase, all domain-specific topics have been extracted from the corpus, ready to be included in the taxonomy.

Taxonomy Construction. Building connections between the extracted topics is the next step toward the taxonomy construction. Edges are added in the graph for all pairs appearing together in at least three documents, and a generality measure allows to direct edges from generic concepts to more specific ones. A specific branching algorithm, successfully applied for the construction of domain taxonomies in [9] trims the noisy directed graph. This produces a tree-like structure where the root is the most generic topic, and the topic nodes are going from broader parent concepts to narrower children.

---

[2] https://code.google.com/archive/p/word2vec.
[3] http://saffron.insight-centre.org/.

3.3 Model Comparison

The two approaches show similarities and dissimilarities. While both systems use a basic term extraction approach for the selection of candidate noun phrases, and PMI for ranking and filtering them, their approach is different. *Saffron* applies PMI to calculate the semantic similarity of the terms to a domain model, while *HEC* uses the outcome of Luhn's cut analysis instead. As for taxonomy construction, both methods construct abstract and loosely related connections for the taxonomy hierarchy, instead of the traditional *is-a* relation type. However, *Saffron* defines a global generality measure using PMI to calculate how closely related a term is to other terms from the domain, following the assumption that generic terms are most often used along with a large number of specific terms. On the contrary, *HEC* relies on agglomerative clustering to detect these relations among embedded vector noun phrases, using cosine similarity as similarity measure. For this step, *Saffron* focuses rather on the hierarchy structure at the document level across the texts, whereas *HEC* works directly on all texts within the corpus. This results in abstract concepts at the intermediary levels of the clustering algorithm, and groupings of noun phrases at the leaves. In contrast, *Saffron* provides expressions from these groups at all levels, from the root to the tree.

4 Experimental Setup

This section gives a brief overview of the corpus used in our experiments. The experiments described here are a first step toward the larger objective of generating a taxonomy for legal corpora over a long time scale. We chose to test the two aforementioned approaches first on a subset of the available Statutory Instruments of Great Britain[4]. 41,518 documents have been produced between 2000 and 2016, each year being split in between UK, Scotland, Wales and Northern Ireland. For this experiment, we refine the analysis by selecting the most recent texts (i.e. 2016) of the UK Statutory Instruments (UKSI), that is 838 documents. We don't consider metadata (such as subject matters, directory codes) as they are not always available in legal texts. Furthermore, there is no agreed standard schema definition yet for describing legal documents across different jurisdictions. Our main goal is to compare the results provided by the two different techniques and determine which is the most suitable for the needs previously described, and focusing on the 2016 UK Statutory Instruments corpus eases the comparison towards that objective.

5 Results

In this section we analyse the noun concepts retrieved from each approach and the relations created in the automatically-built hierarchical taxonomy.

---

[4] http://www.legislation.gov.uk.

Hierarchical Clustering Approach. Figure 2 displays the overall results in the form of a greyscale heatmap where noun phrases (rows) and embedding dimension values (columns) are displayed. A filtering phase is performed on the corpus for the HEC approach to clean potential noisy legal domain syntax (such as the references to regulations e.g. *"Regulation EC No. 1370/2007 means Regulation 1370/2007 ... "* which is not meaningful in our case. Noun phrases which appear less than ten times and in less than five documents are also excluded from the analysis, as they are considered too specific and sparse.

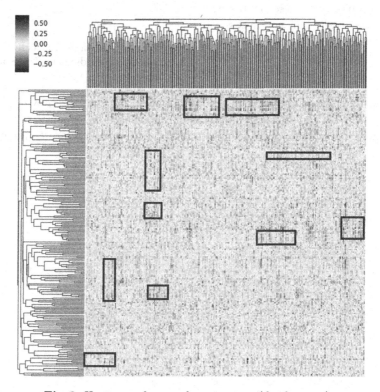

Fig. 2. Heatmap of noun phrase vectors (dendrogram)

The embedded dimensions are reduced representations of the words in an embedding space. Therefore, if the same dimensions of a noun phrase pair both have positively or negatively correlated values in particular dimensions, it means their context is similar in those elements of the vector, meaning that the two noun phrases are related within that given context.

From this figure, it can be identified that some noun phrases are merged due to a small number of dimensions being highly correlated in the embedding space, and not necessarily that all dimensions correlated consistently. This means that the noun phrases are very related only in certain contexts, based on the *Google-News* corpus which these vectors have been trained on, but not necessarily appearing together in other contexts. The rectangles within the heatmap aims at pointing out areas within the graph where this is particularly evident.

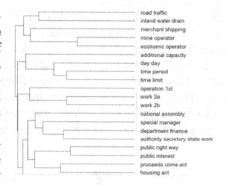

Fig. 3. Sample of hierarchical clustering of noun phrase embeddings for UK Statutory Instruments

Figure 3 shows a snapshot of the results obtained from the previous visualization. Here we can see some interesting groups based on semantic relatedness. The *crime act* and *housing act* have merged with *public interest* and *right of*

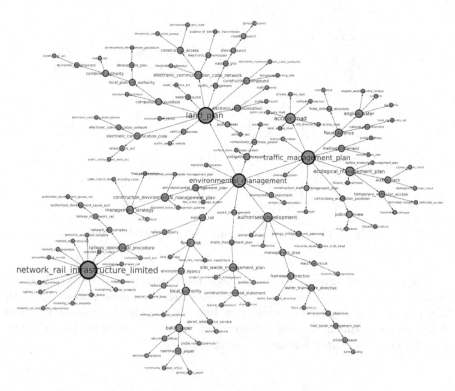

Fig. 4. Saffron taxonomy for the 2016 UK Statutory Instruments

way, which illustrates a topic within the corpus. Likewise, *mine operator* and *economic operator* have been combined with *merchant shipping*. This appears to show an organized relationship of these two noun concepts.

Saffron Approach. We visualize the representation of the taxonomy using an open source software platform, Cytoscape[5]. Nodes are topics, and the size of the nodes relates to the number of connections each topic shares with others. Figure 4 illustrates the whole taxonomy generated by *Saffron* for the corpus. Based on this representation, we detect the topics that are the most prominent in the 2016 UK Statutory Instruments, with four major themes shown in more detail in Fig. 5 (*network rail infrastructure limited, land plan, environmental management* and *traffic management plan*), included in their clusters of related topics. The proposed approach clearly shows the advantages of the hierarchical structure of the graph, which semantically merges topics from generic concepts to more specific ones, like in the *environmental management* node linking to *environmental management plan*, itself redirecting to *construction environmental management plan*, as we can see in Fig. 5.

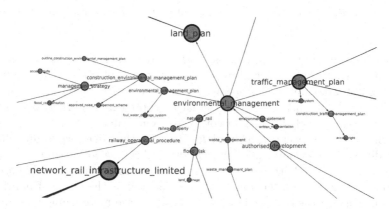

Fig. 5. Main topics from the 2016 UK Ireland Statutory Instruments

There is also a clear interest arising from connecting topics that appear together across the documents. This enables relations between concepts which might not be otherwise obvious to a legal practitioner. It would require carrying out an extensive amount of reading within a particular jurisdiction, while still being able to track links between various documents. For example, in Fig. 6, we observe that *traffic management plan* is connected within some regulation about *drainage system*, the accessibility of the road (*access road*), and is mentioned through the documents alongside with concepts of *ecological management plan*, and *flood defense*. This clearly shows the potential of such semantic processing as an assistance to legal practitioners to identify topics surrounding certain legal issues or for summarizing a whole jurisdiction.

---

[5] http://www.cytoscape.org/.

Fig. 6. *Traffic Management Plan* topic within the UK Statutory Instruments

Comparison. In the two approaches, both representations display connections between concepts in a different way. The approach of *Saffron*, involving the generation of a domain model, seems to retrieve concepts closer to a topic level than the HEC approach. However, the latter approach also brings up relations between terms worth being further considered. One can argue that both systems highlight different aspects of the legal domain from the same corpus, and allow to detect different behaviours, different relations and can be both useful to a domain expert. Furthermore, both methodologies show the importance of a hierarchical structure compared to a flat representation, as well as the usage of multi-word expressions as opposed to single word ones, which are more ambiguous and too broad for a practical use in such specific domain.

6   Conclusion and Future Work

This work has presented a comparison of two fully automated approaches for identifying and relating salient noun concepts in a taxonomy for the legal domain. The results show coherent groupings of words into legal concepts in both approaches, providing highlights on the emerging topics within the legal corpus. This motivates further research for automatic taxonomy construction to assist legal specialists in various applications. This kind of content management in the legal domain is essential for compliance, tracking change in law and terminology and can also assist legal practitioners in search.

Although both approaches seem to show interesting results in automatic taxonomy construction, there is a considerable difficulty in evaluating such systems in a quantitative way, due to the lack of benchmarks to evaluate taxonomies created for specific domains, and the low agreement between experts on fast changing areas. In [5], the authors evaluated expert agreement on the hierarchical relations between terms. The lowest was shown to be in the Science domain, highlighting the difficulty for experts to get a good overview of a domain which is subject to constant changes. Moreover, their approach to automatically evaluate the resulting hierarchies uses a gold standard taxonomy strictly extracted from *WordNet* [7]. This resource is too generic for the intermediate level of terms, on which we are focusing in this approach, specifically to the legal domain (eg. "notice of appeal", "housing allowance", "pension scheme"). However, we plan on carrying out further studies towards a formal representation of concepts within a domain, undertaken by domain experts. This kind of benchmark would establish an evaluation dataset for this domain, where the generated taxonomies are evaluated with taxonomy matching and alignment measures. We also consider establishing an expert user study to evaluate the generated results, with the idea to get legal domain practitioners' views on the practicability of such representation, and the pertinence of the relations established.

Acknowledgements. This publication has emanated from research supported in part by a research grant from Science Foundation Ireland (SFI) under Grant Number SFI/12/RC/2289_P2, co-funded by the European Regional Development Fund.

References

1. Ahmed, A., Xing, E.P.: Timeline: a dynamic hierarchical Dirichlet process model for recovering birth/death and evolution of topics in text stream. CoRR abs/1203.3463 (2012). http://arxiv.org/abs/1203.3463
2. Ahmed, S., Humphrey, T., Lu, X., Morelock, J., Peck, J., Wiltshire, J.: System and method for classifying legal concepts using legal topic scheme, 4 September (2002). https://encrypted.google.com/patents/EP1236175A1?cl=no, eP Patent App. EP20, 000, 952, 140
3. Bordea, G.: Domain adaptive extraction of topical hierarchies for Expertise Mining. Ph.D. thesis (2013)
4. Bordea, G., Buitelaar, P., Polajnar, T.: Domain-independent term extraction through domain modelling. In: Proceedings of the 10th International Conference on Terminology and Artificial Intelligence (2013)
5. Bordea, G., Lefever, E., Buitelaar, P.: Semeval-2016 task 13: taxonomy extraction evaluation (texeval-2). In: Proceedings of SemEval-2016, pp. 1081–1091. Association for Computational Linguistics (2016)
6. Buschettu, A., Concas, G., Pani, F.E., Sanna, D.: A kanban-based methodology to define taxonomies and folksonomies in KMS. In: Park, J.J.J.H., Stojmenovic, I., Jeong, H.Y., Yi, G. (eds.) Comput. Sci. Appl., pp. 539–544. Springer, Heidelberg (2015). https://doi.org/10.1007/978-3-662-45402-2_80
7. Fellbaum, C.: WordNet. Wiley Online Library, Hoboken (1998)

8. Mimno, D., Wallach, H.M., Talley, E., Leenders, M., McCallum, A.: Optimizing semantic coherence in topic models. In: Proceedings of the Conference on Empirical Methods in Natural Language Processing, pp. 262–272. EMNLP 2011. Association for Computational Linguistics, Stroudsburg (2011). http://dl.acm.org/citation.cfm?id=2145432.2145462

9. Navigli, R., Velardi, P., Faralli, S.: A graph-based algorithm for inducing lexical taxonomies from scratch. In: Proceedings of the Twenty-Second International Joint Conference on Artificial Intelligence, vol. 3, pp. 1872–1877. IJCAI 2011. AAAI Press (2011). https://doi.org/10.5591/978-1-57735-516-8/IJCAI11-313

10. Pao, M.L.: Automatic text analysis based on transition phenomena of word occurrences. J. Assoc. Inf. Sci. Technol. **29**(3), 121–124 (1978)

11. Pocostales, J.: Nuig-unlp at semeval-2016 task 13: a simple word embedding-based approach for taxonomy extraction. In: SemEval@ NAACL-HLT, pp. 1298–1302 (2016)

12. Sujatha, R., Bandaru, R., Rao, R.: Taxonomy construction techniques - issues and challenges. Civil Eng. 1–8 (2011)

Title Categorization Based on Category Granularity

Kazuya Shimura[1] and Fumiyo Fukumoto[2]

[1] Graduate School of Engineering, Kofu, Yamanashi, Japan
g17tk008@yamanashi.ac.jp
[2] Graduate Faculty of Interdisciplinary Research, University of Yamanashi,
Takeda, Kofu 4-3-11, Yamanashi, Japan
fukumoto@yamanashi.ac.jp

Abstract. We focus on a problem of short text categorization, *i.e.* categorization of newspaper titles, and present a method that maximizes the impact of informative words due to the sparseness of titles. We used the hierarchical structure of categories and a transfer learning technique based on pre-training and fine-tuning to incorporate the granularity of categories into categorization. According to the hierarchical structure of categories, we transferred trained parameters of Convolutional Neural Networks (CNNs) on upper layers to the related lower ones, and finely tuned parameters of CNNs. The method was tested on titles collected from the Reuters corpus, and the results showed the effectiveness of the method.

Keywords: Text categorization · Hierarchical structure of categories · Convolutional neural networks

1 Introduction

There has been a great deal of interest in short text categorization with the availability of online social networks. Automatic categorization of short text such as search snippets, reviews, and Web page titles supports many applications, *e.g.*, retrieving information on the Internet, and creating digital libraries. In this paper, we focus on news titles as short texts. A basic assumption in categorization with machine learning techniques is that the distribution of words between training and test data is identical. However, when the data consists of short texts, it is often the case that the word distribution between them is different from each other, and thus the performance of categorization is deteriorated. Moreover, similar to the categorization task with normal documents such as news articles and academic papers, titles are often assigned to several categories which make categorization task more problematic.

In this paper, we present a method for title categorization that maximizes the impact of informative words in titles. We used a transfer learning technique based on supervised pre-training and fine-tuning in Convolutional

© Springer Nature Switzerland AG 2020
Z. Vetulani et al. (Eds.): LTC 2017, LNAI 12598, pp. 329–340, 2020.
https://doi.org/10.1007/978-3-030-66527-2_25

Neural Networks (CNNs) to incorporate the granularity of title categories into categorization. The idea of supervised pre-training is to use a data-rich auxiliary dataset and task, to initialize the parameters of CNNs. The CNNs can then be used on the small dataset tagged by fine-grained categories, directly, as a feature extractor. The network can be updated by continued training on the small dataset. The process is so-called fine-tuning and it fits to learn the hierarchical structure of categories. Let us take a look at the RCV1 hierarchy. Suppose a hierarchy consisting of two top-level categories, "Corporate(Industrial)" and "Markets", and their second-level categories, "Corporate(Industrial)/Legal(Judicial)", "Corporate(Industrial)/Share Listings", "Markets/Equity Markets", and "Markets/Bond Markets". We can see that it is easy to make a distinction between "Corporate(Industrial)" and "Market" at the top-level, and they are more generic features. In contrast, "Equity Markets" and "Bond Markets" at the second level would be more specific and fine-grained features than "Market". Similarly, "Share Listings" and "Equity Markets" seem to be semantically close to each other, while they appear in the different top-level categories. They would be a useful feature for both the category "Corporate(Industrial)/Share Listings" and "Markets/Equity Markets". However, in a flat non-hierarchical model, it is difficult to make a clear distinction between "Corporate(Industrial)/Share Listings" and "Markets/Equity" because both of them are related to not only with each other but also to their top level categories.

The process for learning the hierarchical structure of categories is that we first learn to distinguish among categories at the top level of a hierarchy by using pre-training, then learns lower level distinctions by using fine-tuning only within the appropriate top level of the hierarchical structure. We repeat this process from the top to the bottom of a hierarchy. Each of the sub-problems can be solved more accurately, and efficiently as well.

2   Related Work

Short text categorization is widely studied since the recent explosive growth of e-commerce and online social network applications [18]. In contrast with documents such as news articles and academic papers, short texts are less topic-focused and informative words in texts as they consist of from a few words to a dozen words. Major attempts to tackle these problems are to expand short texts with knowledge extracted from the textual corpus, machine-readable dictionaries, e.g., CED and LDOCE, and thesauri such as WordNet. Chen et al. attempted to extract multiple granularity topics to generate features for short text by using auxiliary corpus, Wikipedia page collected by Phan [3,16]. Wang et al. presented a method of short text classification by using bag-of-concepts extracted from a large taxonomy knowledge base, Probase [21,25]. All of these approaches aimed at utilizing auxiliary data to leverage less informative words of short texts to improve the overall performance of short text classification. However, because of the domain-independent nature of machine-readable and thesauri, it is often the case that the data distribution of the external knowledge is different from the test data collected from some specific domain, which deteriorates the overall performance of classification. Moreover, manual collection and tuning the data is very expensive and time-consuming. The methodology which maximizes the impact of pre-defined domains/categories is needed to improve categorization performance.

Zhang *et al.* addressed this issue and proposed a classification method not utilizing auxiliary data but short texts only. They classified short text by detecting information path [30]. They assumed that instances of former classified subsets can assist classification of later subset followed by this path as ordered subsets of short text, *i.e.*, information path consists of sequential subsets in the test dataset. They reported that the method works well when the information path exists in the given dataset. However, in case of not existing the path, it needs to involve some existing techniques, including those leveraging auxiliary resources into their framework for further accuracy gains. Moreover, all the aforementioned methods treated words as independent features that ignore context.

More recently, deep learning techniques have been great successes for automatically extracting context-sensitive features from a textual corpus. Many authors have attempted to apply deep learning methods including CNNs [10,22,23,32–34], the attention-based CNNs [28], bag-of-words based CNN [6], and the combination of CNNs and recurrent neural network (RNN) [29] to text categorization. Most of these approaches based on deep learning techniques demonstrated that neural network models are powerful for learning effective features from a textual input, while most of them focused on single-label or a few labels problem.

Several efforts have been made to classify each test data into its most relevant class labels among a large number of labels, *i.e.*, multi-label text categorization [7,13,26,31]. One attempt is Johnson *et al.* method which makes use of a semi-supervised method with CNNs for categorization. Liu *et al.* explored a family of new CNNs models which are tailored for extreme multi-label classification [13]. They used a dynamic max pooling scheme, a binary cross-entropy loss, and a hidden bottleneck layer to improve the overall performance. The results by using six benchmark datasets where the label-set sizes are up 670 K showed that their method attained at the best or second-best in comparison with seven state-of-the-art methods including FastText [9] and bag-of-words based CNN [6]. All of these attempts aimed at utilizing a large volume of data set, *i.e.*, documents or reviews, to construct a high-quality classification model.

In contrast with the aforementioned works, here we propose a categorization method which addresses both of the multi-label and data sparsity in short texts.

3   Framework of the System

The method consists of three procedures, (1) Word embedding learning, (2) Transfer learning with CNN, and (3) multi-label categorization. Figure 1 illustrates our system.

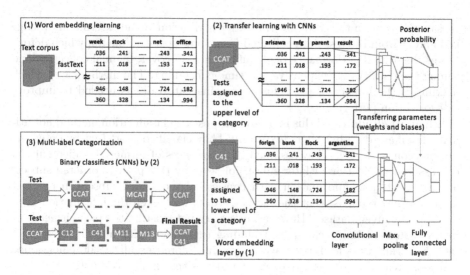

Fig. 1. Overview of the system

3.1 Word Embedding Learning

Word embedding models such as Word2vec by [14], GloVe [15], and fastText developed by [1,9] which are based on contexts of words in input data have become very popular as these models have demonstrated to improve the overall performance of many NLP tasks including short text classification [34], syntactic analysis [2], semantic similarity of documents [11], and machine translation [4,24].

We built word vectors by using fastText tools[1]. FastText developed by Joulin *et al.* is an approach to learn vector representations for each word using a neural network language model. The model consists of two layers; hidden layer and output layer and predicts nearby words. FastText uses (i) a hierarchical softmax [5] based on the Huffman coding tree [14], and (ii) a bag of n-grams as additional features to capture some partial information about the local word order, which makes the computational cost more efficiently. As illustrated in Fig. 1, we used all the texts included in the short texts and applied the skip-gram model provided in fastText to obtain distributed word representation. We denote the dimensionality of the word vectors by k. If the length of a given sentence is n, then the dimensionality of the sentence matrix is $k \times n$. As shown in (1) of Fig. 1, we used all the texts and applied fastText to obtain distributed word representation.

3.2 Transfer Learning with CNNs

Transfer learning is a learning technique that retains and applies the knowledge learned in one or more domains (henceforth we call it a source domain) to

---

[1] https://github.com/facebookresearch/fastText.

efficiently develop an effective hypothesis for a new domain (we call it a target domain). An essential requirement for successful knowledge transfer is that the source domain and the target domain should be closely related to [20]. We used titles consisting of source domain which is a coarse granularity compared to the target domain and learned a model by using CNNs. CNNs are a feedforward network equipped with convolution layers interleaved with pooling layers.

As shown in (2) of Fig. 1, we use word embeddings obtained by procedure (1) to initialize a lookup table, and lower levels extract more complex feature. Let $T = \{w_1, w_2, \cdots, w_N\}$ be a title. It's projected matrix $M \in R^{d*N}$ is obtained by a table of the word embedding layer, where d is the dimension of word embedding. The next layer, convolutional layer is to extract patterns of discriminative word sequences that occurred in the input titles across the training data. The output from the convolutional layer is passed to the pooling layer to reduce the representation. We used max pooling which returns the maximum value. Finally, pooling layers are passed to a fully connected softmax layer. It calculates the probability distribution over the labels as a classifier. For each category within the top-level hierarchy (CCAT) in Fig. 1, We applied this procedure. For the results, we repeatedly applied fine-tuning strategy and learned models for lower levels (C41) in Fig. 1.

Fine-tuning begins with transferring the parameters(weights and biases) from a pre-trained network to the network we wish to train [19]. After the parameters of the last few layers are initialized, the new network can be fine-tuned in a layer-wise manner. Fine-tuning is motivated by the observation that the earlier features of CNNs contain more generic features that should be effective for many tasks, but later layers of the CNNs becomes progressively more specific to the details of the classes contained in the original dataset. The motivation is identical to the hierarchical structure of categories, *i.e.* we first learn to distinguish among categories at the top level of a hierarchy, then learn lower level distinctions by using only within the appropriate top level of the hierarchical structure. Fine-tuning the last few layers is usually sufficient for transfer learning. However, if the distance between the upper and lower level of categories is significant, one may need to fine-tune the early layers as well. In this work, we transferred the convolutional layer, the last two layers within the fully connected layers.

3.3 Multi-label Categorization

The final procedure is multi-label categorization shown in Fig. 1. We compute the probabilities of the test title, being in each of the top-level categories, *e.g.*, CCAT and MCAT, and each of the second-level categories, *e.g.*, C12 and M13, respectively. We note that for the non-hierarchical case, models were learned to distinguish each category from all other categories. For the hierarchical case, models were learned to distinguish each category from only those categories within the same top-level probabilities from the first and second levels for the hierarchical approach. We compute a Boolean function. It shows that we first set a category at the top level, and only match second-level categories that pass this test.

4   Experiments

4.1   Data and CNNs Model Settings

We had experiments to evaluate our method. We used Reuters'96 corpus. The Reuters'96 corpus from 20th Aug. 1996 to 19th Aug. 1997 consists of 806,791 documents organized into coarse-grained categories, *i.e.*, 126 categories with a four-level hierarchy [12]. We selected 30 categories and used them in the experiments. The selection is made according to the number of documents belonging to the categories. We created three sets of categories, *i.e.*, a set of categories assigned to more than 15,000 documents, a set of categories assigned to less than 15,000 and more than 3,000 documents, and a set of categories with less than 3,000 documents. We call these, *large*, *medium*, and *small*, respectively. The data used in the experiments is shown in Table 1.

Table 1. The statistics of the data

Cat	# of doc	Level	Size	Cat	# of doc	Level	Size	Cat	# of doc	Level	Size
CCAT	315,946	Top	Large	C12	9,552	Second	Medium	C172	10,192	Third	Medium
MCAT	157,245	Top	Large	E13	5,747	Second	Medium	E131	4,943	Third	Medium
ECAT	77,557	Top	Large	C22	5,241	Second	Medium	C312	2,163	Third	Small
M11	41,138	Second	Large	C14	3,279	Second	Medium	E511	2,074	Third	Small
M13	37,639	Second	Large	C34	3,172	Second	Medium	E513	1,957	Third	Small
C17	31,215	Second	Large	E31	2,093	Second	Small	E311	1,514	Third	Small
C31	26,191	Second	Large	E61	218	Second	Small	C311	1,241	Third	Small
M12	16,811	Second	Large	M131	20,557	Third	Large	C313	959	Third	Small
E51	14,854	Second	Medium	M132	16,913	Third	Large	E132	768	Third	Small
C41	10,246	Second	Medium	C171	13,486	Third	Medium	E313	85	Third	Small

The total number of titles with Reuters corpus used in the experiment is 557,290, and the average number of categories per title is 1.496. We divided data into two: the training data consisting of 80% of the data, 445,832 titles, and the test data consisted of 20% of the data, 111,458 titles. We further divided the training data into two folds: we used 25% to tuning the parameters including hyper-parameters, and the remains to train the models. Our CNNs model is shown in Table 2. In Table 2, Feature maps refer to the number of feature maps for each filter region size and ReLU indicates to a rectified linear unit which is widely used as an activation function in CNN. Dropout rate1 shows dropout immediately after the embedding layer, and Dropout rate2 refers to dropout in a fully connected layer.

All the titles are tagged by using Tree Tagger [17]. We used nouns, verbs, and adjectives after performing inflectional stemming and conversion to lower case in the experiments. As shown in Table 1, the hierarchical level of the categories used in the experiments is three.

We examined two types of fine-tuning in the experiments. The first fine-tuning is initialize the parameters at each level. Namely, the model is learned

Table 2. CNN model settings

Description	Values
Input word vectors	fastText
Filter region size	(1, 2, 3)
Stride size	1
Feature maps	128
Filters	128×3
Activation function	ReLu
Pooling	1-max pooling
Dropout	Randomly selected
Dropout rate1	0.25
Dropout rate2	0.5
Hidden layers	1,024
Loss function	Binary cross-entropy loss over sigmoid activation
Batch sizes	100
Learning rate	Predicted by Adam
Epoch	40 with early stopping

at the top-level, then initializes the parameters obtained in the top-level and learned in the second level. The parameters obtained by the second-level are initialized, and models are learned in the third level of a hierarchy. We call it Gradual fine-tuning. The second type of fine-tuning is so-called Global fine-tuning, *i.e.*, models for the second and third levels are learned by initializing parameters which are obtained at the top level.

We compared our method with the results obtained by Support Vector Machines(SVMs) and fastText [8]. We used Bag-of-words for SVMs and distributed word representation for CNNs and fastText. We used all the Reuters corpus with titles and contents to learn word representation. We set word- embedding dimension to 300. The batch size of CNNs was set to 100, and the number of epoch was 40. The filter size of a convolutional layer is 300 $\times$3, and the number of filters is 128. These parameters are empirically determined. For all of three methods including our method, we evaluated results with non-hierarchical flat and hierarchical cases.

In the experiments, we used five cross-validation to evaluate the method. For evaluating the effectiveness of category assignments, we used the standard recall, precision, and $F1$ measures. Recall denotes the ratio of correct assignments by the system divided by the total number of correct assignments. Precision is the ratio of correct assignments by the system divided by the total number of the system's assignments. The $F1$ measure which combine recall(r) and precision(p) with an equal weight is $F1(r,p) = \frac{2rp}{r+p}$.

4.2   Results

The results are shown in Tables 3 and 4. Micro-$F1$ indicates the score computed globally over all the $n \times m$ binary decisions where n is the number of total test documents, and m is the number of categories in consideration. Macro-$F1$ refers to the score computed for the binary decisions on each category first, and then be averaged over categories [27]. "Flat" refers to the results of each method obtained by not applying category hierarchy. "Not fine-tuning" shows the results without fine-tuning.

Table 3. Categorization accuracy

Approach	Micro-$F1$	Macro-$F1$
SVMs (Flat)	0.625	0.085
SVMs (Hierarchy)	0.907	0.745
fastText (Flat)	0.691	0.297
fastText (Hierarchy)	0.909	0.747
CNNs (Flat)	0.754	0.238
CNNs (Not fine-tuning)	**0.930**	**0.795**
CNNs (Gradual fine-tuning)	**0.930**	0.794
CNNs (Global fine-tuning)	**0.932**	**0.798**

Table 4. Accuracy by category level (Average macro-F1)

Approach	Top	Second	Third
SVMs (Flat)	0.771	0.016	0.000
SVMs (Hierarchy)	0.933	0.744	0.703
fastText (Flat)	0.796	0.336	0.141
fastText (Hierarchy)	0.938	0.747	0.701
CNNs (Flat)	0.852	0.310	0.020
CNNs (Not fine-tuning)	**0.954**	**0.799**	0.753
CNNs (Gradual fine-tuning)	**0.954**	0.798	0.753
CNNs (Global fine-tuning)	**0.956**	**0.801**	**0.762**

We can see from Table 3 that the overall performance of CNNs was better than those of SVMs and fast Text. The best Micro-F1 was obtained by CNNs with a hierarchy, and that of Macro-$F1$ were CNN (Not fine-tuning) and CNNs (Global fine-tuning) with a hierarchical case. The results with hierarchy were better to those without hierarchy in all of the methods, as both of Micro-$F1$

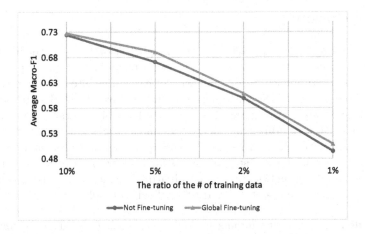

Fig. 2. Average macro-F1 against the ratio of the # of training data

and Macro-$F1$ with hierarchy outperformed than the flat non-hierarchical case. As shown in Table 4, there is a drop in performance in going from the top to the lower level of categories in all of the methods. However, the performance obtained by CNNs is still better than other methods. This demonstrates that CNNs are effective for categorization.

We recall that we used fine-tuning technique as it is effective for the small training dataset since it is motivated by the observation that the earlier features of a CNNs contain more generic features that should be effective for categorization, but later layers of the CNN is more specific to the details of the classes contained in the original dataset. Thus, it is important to compare the results with and without fine-tuning in the experiments. Table 4 shows that the results obtained by Not fine-tuning, Gradual fine-tuning and Global fine-tuning are not statistically significant with each other, especially Not fine-tuning and Global fine-tuning were the same accuracy, while they improved overall performance compared with other baseline methods. We examined how the number of training data affects the overall performance of these two methods. We decreased the number of training data ranging from 100% from 10% in steps 10%, and 10%, 5%, 2%, and 1%. There was no significant differences between Not-fine-tuning and Global fine-tuning methods when the ratio ranged from 100% to 10%. Figure 2 illustrates the average macro-$F1$ against the number of training data obtained by each method ranged from 10% to 1%. Each value of Macro-$F1$ indicates the results obtained by five cross-validation. The x-axis refers to the ratio of training data and y-axis shows the average Macro-$F1$. As shown in Figure 2, when the ratio is less than 10% the average macro-$F1$ obtained by Global fine-tuning was slightly better to those obtained by Not fine-tuning. From the results, We can conclude that when the training data consists of a small number of titles, fine-tuning works well.

5   Conclusions

We have developed an approach for titles categorization by using the hierarchical structure of categories and a transfer learning technique based on pre-training and fine-tuning to incorporate granularity of categories into categorization. The results showed that CNNs with hierarchical structure attained at 0.930 Micro-$F1$, and especially, it works well for the lower level of a hierarchy. Future work will include: (i) extending the method to make use of hierarchical structure of words, *e.g.*, WordNet structure, (ii) incorporating lexical semantics such as named entities and domain-specific senses for further improvement, (iii) exploring a novel scoring function, and (iv) evaluating the method by using other data such as the search snippets [16] and titles of scientific papers[2].

Acknowledgements. The authors would like to thank anonymous reviewers for their helpful comments. This work was supported by the Telecommunications Advancement Foundation, and Support Center for Advanced Telecommunications Technology Research, Foundation.

References

1. Bojanowski, P., Grave, E., Joulin, A., Mikolov, T.: Enriching Word Vectors with Subword Information. arXiv preprint arXiv:1607.04606 (2016)
2. Chen, D., Manning, C.D.: A Fast and accurate dependency parser using neural networks. In: Proceedings of the 2014 Conference on Empirical Methods in Natural Language Processing (EMNLP), pp. 740–750 (2014)
3. Chen, M., Jin, X., Shen, D.: Short text classification improved by learning multi-granularity topics. In: Proceedings of the 22nd International Joint Conference on Artificial Intelligence, pp. 1776–1781 (2011)
4. Devlin, J., Zbib, R., Huang, Z., Lamar, T., Schwartz, R., Makhoul, J.: Fast and robust neural network joint models for statistical machine translation. In: Proceedings of the 52nd Annual Meeting of the Association for Computational Linguistics, pp. 1370–1380 (2014)
5. Goodman, J.: Classes for fast maximum entropy training. In: IEEE International Conference on Acoustics, Speech, and Signal Processing, pp. 561–564 (2001)
6. Johnson, R., Zhang, T.: Effective use of word order for text categorization with convolutional neural networks. In: Proceedings of the 2015 Conference of the North American Chapter of the Association for Computational Linguistics: Human Language Technologies, pp. 103–112 (2015)
7. Johnson, R., Zhang, T.: Semi-supervised convolutional neural networks for text categorization vis region embedding. In: Proceedings of the Advances in Neural Information Processing Systems, vol. 28, pp. 919–927 (2015)
8. Joulin, A., Grave, E., Bojanowski, P., Mikolov, T.: Bag of Tricks for Efficient Text Classification. arXiv preprint arXiv:1607.01759 (2016)
9. Joulin, A., Grave, E., Bojanowski, P., Mikolov, T.: Bag of tricks for efficient text classification. In: Proceedings of the 15th Conference of the European Chapter of the Association for Conputational Linguistics, pp. 427–431 (2017)

---

[2] www.jst.go.jp/EN/index.html.

10. Kim, Y.: Convolutional neural networks for sentence classification. In: Proceedings of the 2014 Conference on Empirical Methods in Natural Language Processing, pp. 1746–1751 (2014)
11. Kusner, M.J., Sun, Y., Kolkin, N.L., Weinberger, K.Q.: From word embeddings to document distances. In: Proceedings of the 32nd International Conference on Machine Learning, pp. 957–966 (2015)
12. Lewis, D.D., Yang, Y., Rose, T.G., Li, F.: RCV1: a new benchmark collection for text categorization research. J. Mach. Learn. Res. **5**, 361–397 (2004)
13. Liu, J., Chang, W.C., Wu, Y., Yang, Y.: Deep learning for extreme multi-label text classification. In: Proceedings of the 40th International ACM SIGIR Conference on Research and Development in Information Retrieval, pp. 115–124 (2017)
14. Mikolov, T., Chen, K., Corrado, G., Dean, J.: Efficient estimation of word representations in vector space. In: Proceedings of the International Conference on Learning Representations Workshop (2013)
15. Pennington, J., Socher, R., Manning, C.D.: Glove: gloval vectors for word representation. In: Proceedings of the Empirical Methods in Natural Language Processing (EMNLP2014), pp. 1532–1543 (2014)
16. Phan, X.H., Nguyen, L.M., Horiguchi, S.: Learning to classify short and sparse text & web with hidden topics from large-scale data collections. In: Proceedings of the 17th International World Wide Web Conference, pp. 91–100 (2008)
17. Schmid, H.: Improvements in part-of-speech tagging with an application to German. In: Proceedings of the EACL SIGDAT Workshop, pp. 47–50 (1995)
18. Song, G., Ye, Y., Du, X., Huang, X., Bie, S.: Short text classification: a survey. Multimedia **9**(5), 635–643 (2014)
19. Tajbakhsh, N., et al.: Convolutional neural networks for medical image analysis: full training or fine tuning? IEEE Trans. Med. Imaging **35**(5), 1299–1312 (2016)
20. Tan, B., Zhang, Y., Pan, S.J., Yang, Q.: Distant domain transfer learning. In: Proceedings of the 31st AAAI Conference on Artificial Intelligence, pp. 2604–2610 (2017)
21. Wang, F., Wang, Z., Li, Z., Wen, J.R.: Concept-based short text classification and ranking. In: Proceedings of the 23rd ACM International Conference on Information and Knowledge Management, pp. 1069–1078 (2008)
22. Wang, J., Wang, Z., Zhang, D., Yan, J.: Combining knowledge with deep convolutional neural networks for short text classification. In: Proceedings of the 26th International Joint Conference on Artificial Intelligence, pp. 2915–2921 (2017)
23. Wang, P., et al.: Semantic clustering and convolutional neural network for short text categorization. In: Proceedings of the 53rd Annual Meeting of the Association for Computational Linguistics and the 7th International Joint Conference on Natural Language Processing, pp. 352–357 (2015)
24. Wang, Y., et al.: Dual transfer learning for neural machine translation with marginal distribution regularization. In: Proceedings of the 32nd AAAI Conference on Artificial Intelligence (2018)
25. Wu, W., Li, H., Wang, H., Zhu, K.Q.: A probabilistic taxonomy for text understanding. In: Proceedings of the 2012 ACM SIGMOD International Conference on Management of Data, pp. 481–492 (2012)
26. Xiao, L., Huang, X., Chen, B., Jing, L.: Label-specific document representation for multi-label text classification. In: Proceedings of the 2019 Conference on Empirical Methods in Natural Language Processing and the 9th International Joint Conference on Natural Language Processing, pp. 466–475 (2019)

27. Yang, Y., Lin, X.: A re-examination of text categorization. In: Proceedings of the 22nd International Conference on Research and Development in Information Retrieval, pp. 42–49 (1999)
28. Yang, Z., Yang, D., Dyer, C., He, X., Smola, A., Hovy, E.: Hierarchical attention networks for document classification. In: Proceedings of the 2016 Conference of the North American Chapter of the Association for Computational Linguistics Human Language Technologies, pp. 1480–1489 (2016)
29. Zhang, R., Lee, H., Radev, D.: Dependency sensitive convolutional neural networks for modeling sentences and documents. In: Proceedings of the 2016 Conference of the North American Chapter of the Association for Computational Linguistics Human Language Technologies, pp. 1512–1521 (2016)
30. Zhang, S., Jin, X., Shen, D., Cao, B., Ding, X., Zhang, X.: Short text classification by detecting information path. In: Proceedings of the 22rd ACM International Conference on Information and Knowledge Management, pp. 727–732 (2013)
31. Zhang, W., Yan, J., Wang, X., Zha, H.: Deep extreme multi-label learning. In: Proceedings of the ACM International Conference on Multimedia Retrieval, pp. 100–107 (2018)
32. Zhang, X., Zhao, J., LeCun, Y.: Character-level convolutional networks for text classification. In: Advances in Neural Information Processing systems, pp. 649–657 (2015)
33. Zhang, Y., Lease, M., Wallace, B.C.: Exploiting domain knowledge via grouped weight sharing with application to text categorization. In: Proceedings of the 55th Annual Meeting of the Association for Computational Linguistics, pp. 155–160 (2017)
34. Zhang, Y., Wallace, B.C.: A sensitivity analysis of (and practitioners' guide to) convolutional neural networks for sentence classification. In: Computing Research Repository (2015)

Identification of Domain-Specific Senses Based on Word Embedding Learning

Attaporn Wangpoonsarp⬚ and Fumiyo Fukumoto⁽⬚⁾⬚

Graduate Faculty of Interdisciplinary Research, University of Yamanashi, 4-3-11, Takeda, Kofu, Yamanashi 400-8510, Japan
{g16dhl01,fukumoto}@yamanashi.ac.jp

Abstract. This paper focuses domain-specific senses and presents a method for detecting predominant sense for each domain. The method is based on the similarity of senses which is obtained by word embedding learning. To detect domain-specific senses, we applied Markov Random Walk (MRW) model, which is a ranking algorithm that has been successfully used in the Web-link analysis. The approach decides the importance of a vertex within a graph based on global information drawn recursively from the entire graph. We applied the technique to rank senses for each domain. The method was tested on WordNet 3.1 and the Reuters corpus. The results obtained by comparing manual annotation of domain-specific senses showed the effectiveness of the method.

Keywords: Domain-specific sense · Markov Random Walk · Word sense disambiguation

1 Introduction

Identification of domain-specific senses has attracted the attention of NLP researchers as domain-specific sense of a word is crucial information for many NLP tasks, *e.g.*, Word Sense Disambiguation (WSD), Information Retrieval, and Machine Translation (MT). For example, the first sense heuristic applied to WordNet is often used as a baseline for supervised WSD systems, as the senses in WordNet are ordered according to the frequency data in the manually tagged resource SemCor [20]. In WSD, the heuristic of just choosing the most frequent sense of a word is very powerful, especially for words with highly skewed sense distributions [12,32]. Koeling *et al.* pointed out that only 5 out of the 26 systems in the SENSEVAL-3 English all-words task [30] outperformed the first sense heuristic as derived from SemCor [12]. The drawback in the first sense heuristic applied to WordNet is the small size of SemCor corpus. Therefore, we can not apply the first sense heuristic to the senses that do not appear in SemCor. Furthermore, identification of the first sense is not based on the domain but instead on the frequency of SemCor data. Consider the noun word, "ball". There are twelve noun senses of "ball" in the WordNet. The first sense of "ball" is "round object that is hit or thrown or kicked in games", and it is often used in

© Springer Nature Switzerland AG 2020
Z. Vetulani et al. (Eds.): LTC 2017, LNAI 12598, pp. 341–350, 2020.
https://doi.org/10.1007/978-3-030-66527-2_26

the "sports" domain rather than the "military" domain. In contrast, the second sense of "ball", *i.e.*, "a solid projectile that is shot by a musket" is more likely to be used in the "military" domain.

In this paper, we focus domain-specific noun senses and propose a method for detecting predominant sense in each domain/category. First, we extract noun words from Reuters news corpus [25], and collect their senses and gloss texts from the WordNet. Next, we calculated the similarity between senses by using Word Mover's Distance (WMD) [13]. WMD metric is based on Word2Vec embeddings. Gloss texts corresponding to the two senses are represented as a weighted point cloud of embedded words. The distance between two gloss texts T_i and T_j are the minimum cumulative distance that words from T_i need to travel to match exactly the point cloud of T_j. Finally, we rank scores for each sense with Markov Random Walk (MRW) model, and compared sense ranking among categories for automated domain labeling.

2   Related Work

In terms of domain-specific senses, Gale *et al.* first observed tendency to share sense in the same discourse [9]. To make the best use of the tendency, a method for automatically detecting the *one sense* given a document is required. Magnini *et al.* presented a lexical resource where WordNet 2.0 synsets were annotated with Subject Field Codes (SFC) by a procedure that exploits WordNet structure [2,15,16]. They annotated 96% of WordNet synsets of the noun hierarchy. However, mapping domain labels for word senses was semi-automated and required hand-labeling.

McCarthy *et al.* addressed the problem and proposed an automated method for assigning predominant noun senses [17]. They used a thesaurus acquired from raw textual corpora and WordNet similarity package that includes similarity measure such as lesk and jcn. They also used parsed data to find words with a similar distribution to the target word. Unlike [6] method, McCarthy *et al.* evaluated their method using publically available resources: the hand-tagged resources SemCor and the SENSEVAL-2 English all-words task. They reported that the method obtained precision of 64% on all-nouns task.

In terms of similarity metric, the Vector Space Model (VSM) has been widely used in NLP and IR research fields. It represents documents as vectors via a bag of words or by their term frequency feature. However, these features are often not suitable for document representation because of their frequent near-orthogonality [10]. There has been much attempts to avoid the problem by projecting document dimensional space into a lower dimensional space, *e.g.*, Latent Semantic Indexing (LSI) [7] and Latent Dirichlet Allocation (LDA) [3]. Mikolov *et al.* presented Word2Vec model that was a well known shallow model for training text and generate word embedding [19]. Learning the word embedding is unsupervised and it can be computed by using textual corpus. It can be used for finding a semantically related word from a given word. For example, vector ("Japan") − vector ("Yen") + vector ("Thailand") is close to vector ("Baht").

Word2Vec provides two architectures to generate word embedding: Continuous Bag-of-Words (CBOW) and skip-gram model. The CBOW predicts the center word given a representation of the surrounding words. The skip-gram model predicts contextual words given the center word.

Wan proposed Earth Mover's Distance (EMD) to measure document similarity [31]. The EMD is an improvement of optimal matching based similarity measure which takes into account the subtopic structures of documents [22]. It computes the minimum distance to move one distribution to another distribution. It allows many-to-many matching between subtopics. The results using TDT3 dataset[1] showed that the EMD measure outperformed all other existing similarity measures including Jaccard, BM26, and OM-based measure.

Kusner et al. presented WMD to compute the similarity between two sentences [13]. It is based on Word2Vec embeddings. It measures the dissimilarity between two sentences as the minimum amount of distance that the embedded words of one sentence reach the embedded words of another sentence. The results using eight real-world document classification data sets including Reuters and 20News in comparison with seven baselines including LDA show that the WMD attained at low k-nearest neighbor document classification error rates.

In terms of link analysis, the graph-based ranking method has been widely and successfully used in NLP and its applications, such as unsupervised WSD [21,29], text semantic similarity [23], query expansion in IR [14], and document summarization [18]. The basic idea is that of "voting" or "recommendations" between nodes. The model first constructs a directed or undirected graph to reflect the relationships between the nodes and then applies the graph-based ranking algorithm to compute the rank scores for the nodes. The nodes with large rank scores are chosen as important nodes. Reddy attempted to use the Personalized PageRank algorithm [1] over a graph representing WordNet to disambiguate ambiguous words [24]. They combine sense distribution scores and keyword ranking scores into the graph to personalize the graph for the given domain. The results showed that exploiting domain-specific information within the graph based methods produce better results than when this information is used individually. However, the sense distribution scores are based on the frequency of neighbors of the target word from the thesaurus.

There are three novel aspects in our method. Firstly, we propose a method for identifying domain-specific noun senses which make use of distributed representations of words and thus captures large semantic context. Secondly, the method does not require manual annotation of data, while Magnini et al. method required a considerable amount of hand-labeling. Finally, from the perspective of robustness, the method is automated and required only documents from the given domain/category, such as the Reuters corpus, and thesaurus with gloss texts such as WordNet. Therefore, it can be applied easily to a new domain or sense inventory, given sufficient documents.

---

[1] http://catalog.ldc.upenn.edu/LDC2001T58.

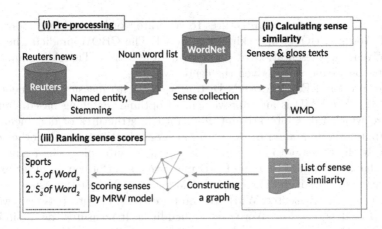

Fig. 1. Overview of the system

3   Framework of the System

The method consists of three procedures, (i) Pre-processing, (ii) Calculating sense similarity, and (iii) Ranking sense scores. Figure 1 illustrates an overview of the system.

3.1   Pre-processing

The goal of pre-processing is to make sense list for each category. Sense list consists of word senses with gloss texts in the WordNet. For each category in the Reuters, we collected the documents. They are tagged and lemmatized by using Tree Tagger [28]. We used noun words. We also used Named Entity Recognition (NER) to identify location, person, and organization by using Stanford Named Entity Recognizer [8] to tagging compound words such as "New York" and "Margaret Thatcher". Finally, for each word, we collected senses and their gloss texts to create sense list.

3.2   Calculating Sense Similarity

We used Markov Random Walk (MRW) model to identify domain-specific senses for each category. The input of MRW model is a graph consisting of vertices and edges with similarity value between vertices. As shown in (2) in Fig. 1, we calculated sense similarity by using WMD [13].

WMD measures the dissimilarity between two sentences as the minimum amount of distance that the embedded words of one sentence need to *travel* to reach the embedded words of another sentence. The word embedding is learned by using Word2Vec [19]. More precisely, Word2Vec learns vector representation of words from gloss text as the training documents. It is provided two models, CBOW and skip-gram.

We used CBOW model as it is reported to be faster and more suitable for larger datasets [19]. The CBOW model is based on the feedforward Neural Network Language Model (NNLM). The model predicts the center word w_t given a representation of the surrounding words. The hidden layer H is obtained by summing the embeddings of the context words:

$$H_{CBOWt} = \sum_{-R \leq j \leq R, j \neq 0} V(w_{t+j}) \tag{1}$$

R in Eq. (1) refers to the training window. $V(w_t) \in \mathbb{R}^d$ indicates d-dimensional word vector of w_t, and $H_{CBOW} \in \mathbb{R}^d$ shows d-dimensional vector. The model is learned by using negative-sampling approach. The objective of the model is to maximized L:

$$L = \sum_{w_t \in V} (\log(\tau(H_t W(w_t)))$$
$$+ \sum_{w_i \in N_t} \log(\tau(-H_t W(w_i)))) \tag{2}$$

t in Eq. (2) shows the tth sampling, and V refers to the number of all words in the training corpus. N_t indicates negative sampling of k when the center word is w_t. Here, w_t is a positive sample, and all words except for w_t are negative samples. The CBOW is trained using stochastic gradient descent. The gradient is computed using backpropagation rule [26].

Let $\mathbf{X} \in \mathbb{R}^{d \times n}$ be a Word2Vec embedding matrix for vocabulary size of n words. The ith column, $\mathbf{x_i} \in \mathbb{R}^d$ refers to the embedding of the ith word in d-dimensional space. We represent gloss text of each sense as normalized bag-of-words (nBOW) vector, $g \in \mathbb{R}^n$. The objective of the model is to minimize cumulative cost C of moving the gloss text g to g':

$$L = \sum_{i,j=1}^{n} \mathbf{T}_{i,j}\, c(i,j),$$
$$\text{subject to:} \sum_{j=1}^{n} \mathbf{T}_{ij} = g_i,\ \forall i \in \{1, \cdots, n\},$$
$$\sum_{i=1}^{n} \mathbf{T}_{ij} = g'_j.\ \forall j \in \{1, \cdots, n\}. \tag{3}$$

$\sum_{j=1}^{n} \mathbf{T}_{ij} = g_i$ indicates that outgoing flow from word i equals g_i. Similarly, $\sum_{i=1}^{n} \mathbf{T}_{ij} = g'_j$ shows that incoming flow to word j mush match g'_j. $c(i,j)$ in Eq. (3) refers to word travel cost which is defined by $c(i,j) = \| \mathbf{x}_i - \mathbf{x}_j \|_2$ [11].

3.3 Ranking Sense Score

The final procedure for detecting domain-specific senses which are shown in (3) in Fig. 1 is to score each sense for each domain/category. The MRW model we used decides the importance of a vertex within a graph based on global information drawn recursively from the entire graph [4]. The essential idea is that of "voting" between the vertices. An edge between two vertices is considered a vote cast from one vertex to the other. The score associated with a vertex is determined by the votes that are cast for it, and the score of the vertices casting these votes.

Given a set of senses S_d in the domain d, as shown in (iii) of Fig. 1, we construct a graph. $G_d = (V, E)$ is a graph reflecting the relationships between senses in the set. V is the set of vertices, and each vertex v_i in V is the gloss text assigned from WordNet. E is a set of edges, which is a subset of $V \times V$. Each edge e_{ij} in E is associated with an affinity weight $f(i \rightarrow j)$ between senses v_i and v_j ($i \neq j$). The weight is computed using the standard cosine measure between the two senses. Two vertices are connected if their affinity weight is larger than 0 and we let $f(i \rightarrow i) = 0$ to avoid self transition. The transition probability from v_i to v_j is then defined by $p(i \rightarrow j) = \frac{f(i \rightarrow j)}{\sum_{k=1}^{|V|} f(i \rightarrow k)}$, if $\Sigma f \neq 0$, otherwise, 0.

We used the row-normalized matrix $U_{ij} = (U_{ij})_{|V| \times |V|}$ to describe G with each entry corresponding to the transition probability, where $U_{ij} = p(i \rightarrow j)$. To make U a stochastic matrix, the rows with all zero elements are replaced by a smoothing vector with all elements set to $\frac{1}{|V|}$. The matrix form of the saliency score $Score(v_i)$ can be formulated in a recursive form as in the MRW model, $\boldsymbol{\lambda} = \mu U^T \boldsymbol{\lambda} + \frac{(1-\mu)}{|V|} \boldsymbol{e}$, where $\boldsymbol{\lambda} = [Score(v_i)]_{|V| \times 1}$ is the vector of saliency scores for the senses. $\boldsymbol{e}$ is a column vector with all elements equal to 1. μ is the damping factor. We set μ to 0.85, as in the PageRank [5]. The final transition matrix is given by $M = \mu U^T + \frac{(1-\mu)}{|V|} \boldsymbol{ee}^T$, and each score of the sense in a specific domain is obtained by the principal eigenvector of the matrix.

We applied the algorithm for each domain. We note that the matrix M is a high-dimensional space. Therefore, we used a ScaLAPACK, a library of high-performance linear algebra routines for distributed memory MIMD parallel computing [27], which includes routines for solving systems of linear equations, least squares, eigenvalue problems.

We selected the topmost $K\%$ words (senses) according to rank score for each domain and make a sense-domain list. For each word w in a document, find the sense s that has the highest score within the list. If a domain with the highest score of the sense s and a domain in a document appeared in the word w match, s is regarded as a domain-specific sense of the word w.

4   Experiments

We had experiments to evaluate our method. We used Reuters'96 corpus from 20th Aug. 1996 to 19th Aug. 1997, and WordNet 3.1. The corpus consists of 806,791 documents organized into 126 categories. There are no existing sense-tagged data for domains that could be used for evaluation. We thus used the Subject Field Codes (SFC) resource [15], which annotates WordNet 2.0 synsets with domain labels. The SFC consists of 115,424 words assigning 168 domain labels with a hierarchy. It contains some Reuters categories. We tested Reuters

Table 1. The Reuters and SFC category correspondences

SFC	Reuters	The # of doc
Tourism	Travel	680
Sports	Sports	35,225
Military	War	32,580
Law	Legal/Judicial	32,194
Economy	Economics	117,501
Politics	Politics	56,834

six categories corresponding to the SFC labels which are shown in Table 1. "The number of doc" in Table 1 shows the number of documents in each category. We set parameters used in the Word2Vec, *i.e.* the number of dimensions is 100, the window size is 5. We used CBOW learning model. For each category, we built individual models and we chose words whose frequencies in all of the documents are more than five. The results are shown in Table 2.

Table 2. The results of sense assignments (The top 20% words according to rank score)

SFC/Reuters	S	DSS	SFC	DSS (WMD)					DSS (Cos)					P_IRS
				Cor	Prec	Rec	F	IRS	Cor	Prec	Rec	F	IRS	
Tourism/Travel	78	16	18	15	.938	.833	**.882**	3.24	16	1	.941	**.970**	3.38	3.38
Sports/Sports	297	59	63	50	.847	.794	.820	4.40	45	.763	.763	.763	3.42	4.66
Military/War	548	110	115	85	.773	.739	.756	4.95	74	.673	.632	.652	4.58	5.28
Law/Law	740	148	153	101	.682	.663	.671	4.06	40	.270	.268	.269	1.21	5.58
Economy/Economics	577	115	120	77	.670	.642	.655	4.63	65	.565	.551	.558	3.74	5.33
Politics/Politics	815	163	174	107	.656	.615	.635	4.92	20	.123	.113	.118	0.89	5.67
Average	509	102	109	72	.761	.714	**.737**	**4.37**	43	.566	.545	**.555**	2.87	**4.98**

Table 2 shows the results obtained by the topmost 20% senses according to rank score. We also tested a baseline system which does not use WMD as a similarity measure but instead uses cosine measure, and compared it with our system. "S" shows the total number of senses which should be assigned to each category. "DSS (Domain-Specific Senses)" refers to the results obtained by our system. "SFC" indicates the number of senses appearing in the SFC resource. "Cor" denotes the number of senses appearing in both of the system (WMD/Cos) and SFC. "Prec" means the ratio of correct assignments by our system divided by the total number of the system's assignments. "Rec" is the ratio of correct assignments by the system divided by the total number of correct assignments. The F measure which combines recall (r) and precision (p) with an equal weight is $\mathrm{F}(r, p) = \frac{2rp}{r+p}$. "IRS" refers to Inverse Rank Score which is a measure of system performance by considering the rank of correct senses within the candidate collections. It is the sum of the inverse rank of each matching collections, and the higher the IRS value, the better the system performance. "P_IRS" indicates the perfect correct value of IRS.

We can see from Table 2 that the overall performance obtained by WMD was better to those by Cos except for "Tourism/Travel" domain as the average F attained at 0.737, while that of Cos was 0.555. Performance depends on the categories. In both methods, we obtained the best F score when we used category "Tourism/Travel". In contrast, the worst results obtained by WMD were categories "Economy/Economics" and "Politics/Politics", while these results were better to those of Cos. This is not surprising because they are semantically close with each other.

We examined how the topmost ratio of ranking affects the overall performance of the system. Figure 2 shows F-score against the ratio of topmost ranking. We can see from Fig. 2 that when the ratio increased, the F-score obtained by both methods dropped. However, the results obtained by WMD were still better to those obtained by Cosine measure. This shows that WMD works well to detect domain-specific senses.

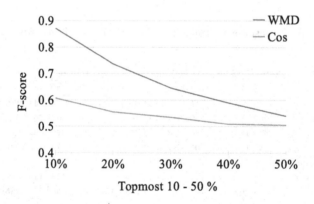

Fig. 2. F-score against the ratio of topmost ranking

5   Conclusion

We presented an approach for detecting domain-specific senses. We used the WMD to calculate the similarity between senses by using gloss text which is a short length of sentence. The results using Reuters 1996 corpus and WordNet 3.1 showed that embedding learning is effective for detecting domain-specific senses. Future work will include: (i) testing more than six domains and comparing with other algorithms, (ii) applying the method to other categories, and parts of speech for quantitative evaluation, (iii) investigating other types of a lexicon for further improvement, and (iv) conducting extrinsic evaluation *e.g.*, text classification and machine translation.

References

1. Agirre, E., Soroa, A.: Personalizing PageRank for word sense disambiguation. In: 12th Proceedings on Conference of the European Chapter of the Association for Computational Linguistics, Athens, pp. 33–41. ACL (2009)
2. Bentivogli, L., Forner, P., Magnini, B., Pianta, E.: Revising the WordNet domains hierarchy: semantics, coverage and balancing. In: Proceedings of the Workshop on Multilingual Linguistic Resources, Geneva, pp. 94–101. COLING (2004)
3. Blei, D.M., Ng, A.Y., Jordan, M.I.: Latent Dirichlet allocation. J. Mach. Learn. Res. **3**, 993–1022 (2003)
4. Bremaud, P.: Markov Chains: Gibbs Fields, Monte Carlo Simulation, and Queues. Springer, New York (1999). https://doi.org/10.1007/978-1-4757-3124-8
5. Brin, S., Page, L.: The anatomy of a large-scale hypertextual web search engine. J. Comput. Netw. ISDN Syst., 107–117 (1998)
6. Buitelaar, P., Sacaleanu, B.: Ranking and selecting synsets by domain relevance. In: Proceedings of the WordNet and Other Lexical Resources: Applications, Extensions and Customizations NAACL Workshop, pp. 119–124 (2001)
7. Deerwester, S., Dumais, S.T., Furnas, G.W., Landauer, T.K., Harshman, R.: Indexing by latent semantic analysis. J. Am. Soc. Inf. Sci. **41**, 391–407 (1990)
8. Finkel, J.R., Grenager, T., Manning, C.: Incorporating non-local information into information extraction systems by Gibbs sampling. In: Proceedings of the 43rd Annual Meeting on Association for Computational Linguistics, Michigan, pp. 363–370. ACL (2005)
9. Gale, W.A., Church, K.W., Yarowsky, D.: One sense per discourse. Proc. Speech Nat. Lang. Work. **1992**, 233–237 (1992)
10. Greene, D., Cunningham, P.: Practical solutions to the problem of diagonal dominance in kernel document clustering. In: Proceeding of the 23rd International Conference on Machine Learning, pp. 377–384. ACM Press (2006)
11. Hitchcock, F. L.: The Distribution of a Product from Several Sources to Numerous Localities, pp. 224–230. Wiley (1941)
12. Koeling, R., McCarth, D., Carroll, J.: Domain-specific sense distributions and predominant sense acquisition. In: Proceedings of Human Language Technology Conference and Conference on Empirical Methods in Natural Language Processing (HLT/EMNLP), Vancouver, pp. 419–426. ACL (2005)
13. Kusner, M., Sun Y., Kolkin, N., Weinberger, K.: From word embeddings to document distances. In: Proceedings of the 32nd International Conference on Machine Learning, pp. 957–966 (2015)
14. Lafferty, J., Zhai, C.: Document language modeling, query models, and risk minimization for information retrieval. In: Proceedings of 24th Annual International ACM SIGIR Conference on Research and Development in Information Retrieval, pp. 111–119. (2001)
15. Magnini, B., Cavaglia, G.: Integrating subject field codes into WordNet. In: Proceedings of International Conference on Language Resources and Evaluation 2000, Athens, pp. 1413–1418. ELRA (2000)
16. Magnini, B., Strapparava, C., Pezzulo, G., Gliozzo, A.: The role of domain information in word sense disambiguation. J. Nat. Lang. Eng. **8**, 359–373 (2002)
17. McCarthy, D., Koeling, R., Weeds, J., Carroll, J.: Finding predominant word senses in untagged text. In: Proceedings of the 42nd Annual Meeting on Association for Computational Linguistics, Barcelona, pp. 279–286 (2004)

18. Mihalcea R.: Language independent extractive summarization. In: Proceedings of the ACL Interactive Poster and Demonstration Sessions, pp. 49–52. (2005)

19. Mikolov, T., Chen, K., Corrado, G., Dean, J.: Efficient estimation of word representations in vector space. arXiv:1301.3781 (2013)

20. Miller, G.A.: WordNet: a lexical database for English. J. Commun. ACM **38**, 39–41 (1995)

21. Navigli, R., Lapata, M.: An experimental study of graph connectivity for unsupervised word sense disambiguation. J. IEEE Trans. Pattern Anal. Mach. Intell. **32**, 678–692 (2010)

22. Pele, O., Werman, M.: Fast and robust earth mover's distances. In: Proceedings of the 12th International Conference on Computer Vision, Kyoto, pp. 460–467. IEEE (2009)

23. Ramage, D., Rafferty, A.N., Manning, C.D.: Random walks for text semantic similarity. In: Proceedings of the 4th TextGraphs Workshop on Graph-based Algorithms in NLP, Suntec, pp. 23–31. ACL (2009)

24. Reddy, S., Inumella, A., McCarthy, D., Stevenson, M.: IIITH: domain specific word sense disambiguation. In: Proceedings of the 5th International Workshop on Semantic Evaluation, Uppsala, pp. 387–391. ACL (2010)

25. Rose, T., Stevenson, M., Whitehead, M.: The Reuters corpus volume 1 - from yesterday's news to tomorrow's language resources. In: Proceedings of Language Resources and Evaluation-2002, Las Palmas, ELRA (2002)

26. Rumelhart, D.E., Hinton, G.E., Williams, R.J.: Learning representation by back-propagating errors. J. Nat. **323**, 533–536 (1986)

27. Netlib: Netlib Repository at UTK and ORNL. http://www.netlib.org/scalapack/index.html. Accessed 19 Mar 2020

28. Schmid, H.: Probabilistic part-of-speech tagging using decision trees. In: Proceedings of the International Conference on New Methods in Language Processing, pp. 44–49 (1994)

29. Sinha, R., Mihalcea, R.: Unsupervised graph-based word sense disambiguation using measures of word semantic similarity. In: Proceedings of the International Conference on Semantic Computing, pp. 363–369 (2007)

30. Snyder, B., Palmer M.: The English all-words task. In: Proceedings of Senseval-3: Third International Workshop on the Evaluation of Systems for the Semantic Analysis of Text, Barcelona, pp. 41–43. ACL (2004)

31. Wan, X.: A novel document similarity measure based on earth mover's distance. J. Inf. Sci. **177**, 3718–3730 (2007)

32. Yarowsky, D., Florian, R.: Evaluating sense disambiguation performance across diverse parameter spaces. J. Nat. Lang. Eng., 293–310 (2002)

Author Index

Printed in the United States
By Bookmasters